Lecture Notes in Computer Science 9829

Commenced Publication in 1973
Founding and Former Series Editors:
Gerhard Goos, Juris Hartmanis, and Jan van Leeuwen

Sanjay Madria · Takahiro Hara (Eds.)

Big Data Analytics and Knowledge Discovery

18th International Conference, DaWaK 2016
Porto, Portugal, September 6–8, 2016
Proceedings

Springer

Editors
Sanjay Madria
University of Science and Technology
Rolla, MO
USA

Takahiro Hara
Osaka University
Osaka
Japan

ISSN 0302-9743 ISSN 1611-3349 (electronic)
Lecture Notes in Computer Science
ISBN 978-3-319-43945-7 ISBN 978-3-319-43946-4 (eBook)
DOI 10.1007/978-3-319-43946-4

Library of Congress Control Number: 2016946945

LNCS Sublibrary: SL3 – Information Systems and Applications, incl. Internet/Web, and HCI

Printed on acid-free paper

This Springer imprint is published by Springer Nature
The registered company is Springer International Publishing AG Switzerland

Preface

Big data are rapidly growing in all domains. Knowledge discovery using data analytics is important to several applications ranging from health care to manufacturing to smart city. The purpose of the International Conference on Data Warehousing and Knowledge Discovery (DAWAK) is to provide a forum for the exchange of ideas and experiences among theoreticians and practitioners who are involved in the design, management, and implementation of big data management, analytics, and knowledge discovery solutions.

We received 73 good-quality submissions, of which 25 were selected for presentation and inclusion in the proceedings after peer-review by at least three international experts in the area. The selected papers were included in the following sessions: Big Data Mining, Applications of Big Data Mining, Big Data Indexing and Searching, Graph Databases and Data Warehousing, and Data Intelligence and Technology.

Major credit for the quality of the track program goes to the authors who submitted quality papers and to the reviewers who, under relatively tight deadlines, completed the reviews. We thank all the authors who contributed papers and the reviewers who selected very high quality papers. We would like to thank all the members of the DEXA committee for their support and help, and particularly to Gabriela Wagner her endless support. Finally, we would like to thank the local Organizing Committee for the wonderful arrangements and all the participants for attending the DAWAK conference and for the stimulating discussions.

July 2016

Sanjay Madria
Takahiro Hara

Organization

Program Committee Co-chairs

Sanjay K. Madria   Missouri University of Science and Technology, USA
Takahiro Hara   Osaka University, Japan

Program Committee

Abelló, Alberto	Universitat Politecnica de Catalunya, Spain
Agrawal, Rajeev	North Carolina A&T State University, USA
Al-Kateb, Mohammed	Teradata Labs, USA
Amagasa, Toshiyuki	University of Tsukuba, Japan
Bach Pedersen, Torben	Aalborg University, Denmark
Baralis, Elena	Politecnico di Torino, Italy
Bellatreche, Ladjel	ENSMA, France
Ben Yahia, Sadok	Tunis University, Tunisia
Bernardino, Jorge	ISEC - Polytechnic Institute of Coimbra, Portugal
Bhatnagar, Vasudha	Delhi University, India
Boukhalfa, Kamel	USTHB, Algeria
Boussaid, Omar	University of Lyon, France
Bressan, Stephane	National University of Singapore, Singapore
Buchmann, Erik	Karlsruhe Institute of Technology, Germany
Chakravarthy, Sharma	The University of Texas at Arlington, USA
Cremilleux, Bruno	Université de Caen, France
Cuzzocrea, Alfredo	University of Trieste, Italy
Davis, Karen	University of Cincinnati, USA
Diamantini, Claudia	Università Politecnica delle Marche, Italy
Dobra, Alin	University of Florida, USA
Dou, Dejing	University of Oregon, USA
Dyreson, Curtis	Utah State University, USA
Endres, Markus	University of Augsburg, Germany
Estivill-Castro, Vladimir	Griffith University, Australia
Furfaro, Filippo	University of Calabria, Italy
Furtado, Pedro	Universidade de Coimbra, Portugal, Portugal
Goda, Kazuo	University of Tokyo, Japan
Golfarelli, Matteo	DISI - University of Bologna, Italy
Greco, Sergio	University of Calabria, Italy
Hara, Takahiro	Osaka University, Japan
Hoppner, Frank	Ostfalia University of Applied Sciences, Germany
Ishikawa, Yoshiharu	Nagoya University, Japan

Josep, Domingo-Ferrer	Rovira i Virgili University, Spain
Kalogeraki, Vana	Athens University of Economics and Business, Greece
Kim, Sang-Wook	Hanyang University, South Korea
Lechtenboerger, Jens	Westfälische Wilhelms - Universität Münster, Germany
Lehner, Wolfgang	Dresden University of Technology, Germany
Leung, Carson K.	University of Manitoba, Canada
Maabout, Sofian	University of Bordeaux, France
Madria, Sanjay Kumar	Missouri University of Science and Technology, USA
Marcel, Patrick	Université François Rabelais Tours, France
Mondal, Anirban	Shiv Nadar University, India
Morimoto, Yasuhiko	Hiroshima University, Japan
Onizuka, Makoto	Osaka University, Japan
Papadopoulos, Apostolos	Aristotle University, Greece
Patel, Dhaval	Indian Institute of Technology Roorkee, India
Rao, Praveen	University of Missouri-Kansas City, USA
Ristanoski, Goce	National ICT Australia, Australia
Rizzi, Stefano	University of Bologna, Italy
Sapino, Maria Luisa	Università degli Studi di Torino, Italy
Sattler, Kai-Uwe	Ilmenau University of Technology, Germany
Simitsis, Alkis	HP Labs, USA
Taniar, David	Monash University, Australia
Teste, Olivier	IRIT, University of Toulouse, France
Theodoratos, Dimitri	New Jersey Institute of Technology, USA
Vassiliadis, Panos	University of Ioannina, Greece
Wang, Guangtao	School of Computer Engineering, NTU, Singapore, Singapore
Weldemariam, Komminist	IBM Research Africa, Kenya
Wrembel, Robert	Poznan University of Technology, Poland
Zhou, Bin	University of Maryland, Baltimore County, USA

Additional Reviewers

Adam G.M. Pazdor	University of Manitoba, Canada
Aggeliki Dimitriou	National Technical University of Athens, Greece
Akihiro Okuno	The University of Tokyo, Japan
Albrecht Zimmermann	Université de Caen Normandie, France
Anas Adnan Katib	University of Missouri-Kansas City, USA
Arnaud Soulet	University of Tours, France
Besim Bilalli	Universitat Politecnica de Catalunya, Spain
Bettina Fazzinga	ICAR-CNR, Italy
Bruno Pinaud	University of Bordeaux, France
Bryan Martin	University of Cincinnati, USA
Carles Anglès	Universitat Rovira i Virgili, Spain
Christian Thomsen	Aalborg University, Denmark
Chuan Xiao	Nagoya University, Japan

Daniel Ernesto Lopez Barron	University of Missouri-Kansas City, USA
Dilshod Ibragimov	ULB Bruxelles, Belgium
Dippy Aggarwal	University of Cincinnati, USA
Djillali Boukhelef	USTHB, Algeria
Domenico Potena	Università Politecnica delle Marche, Italy
Emanuele Storti	Università Politecnica delle Marche, Italy
Enrico Gallinucci	University of Bologna, Italy
Evelina Di Corso	Politecnico di Torino, Italy
Fan Jiang	University of Manitoba, Canada
Francesco Parisi	DIMES - University of Calabria, Italy
Hao Wang	University of Oregon, USA
Hao Zhang	University of Manitoba, Canada
Hiroaki Shiokawa	University of Tsukuba, Japan
Hiroyuki Yamada	The University of Tokyo, Japan
Imen Megdiche	IRIT, France
João Costa	Polytechnic of Coimbra, ISEC, Portugal
Julián Salas	Universitat Rovira i Virgili, Spain
Khalissa Derbal	USTHB, Algeria
Lorenzo Baldacci	University of Bologna, Italy
Luca Cagliero	Politecnico di Torino, Italy
Luca Venturini	Politecnico di Torino, Italy
Luigi Pontieri	ICAR-CNR, Italy
Mahfoud Djedaini	University of Tours, France
Meriem Guessoum	USTHB, Algeria
Muhammad Aamir Saleem	Aalborg University, Denmark
Nicolas Labroche	University of Tours, France
Nisansa de Silva	University of Oregon, USA
Oluwafemi A. Sarumi	University of Manitoba, Canada
Oscar Romero	UPC Barcelona, Spain
Patrick Olekas	University of Cincinnati, USA
Peter Braun	University of Manitoba, Canada
Prajwol Sangat	Monash University, Australia
Rakhi Saxena	Desh Bandhu College, University of Delhi, India
Rodrigo Rocha Silva	University of Mogi das Cruzes, ADS - FATEC, Brazil
Rohit Kumar	Université libre de Bruxelles, Belgium
Romain Giot	University of Bordeaux, France
Sabin Kafle	University of Oregon, USA
Sergi Nadal	Universitat Politecnica de Catalunya, Spain
Sharanjit Kaur	AND College, University of Delhi, India
Souvik Shah	New Jersey Institute of Technology, USA
Swagata Duari	University of Delhi, India
Takahiro Komamizu	University of Tsukuba, Japan
Uday Kiran Rage	The University of Tokyo, Japan
Varunya Attasena	Kasetsart University, Thailand
Vasileios Theodorou	Universitat Politecnica de Catalunya, Spain

Victor Herrero   Universitat Politecnica de Catalunya, Spain
Xiaoying Wu   Wuhan University, China
Yuto Yamaguchi   National Institute of Advanced Industrial Science
                          and Technology (AIST), Japan
Yuya Sasaki   Osaka University, Japan
Zakia Challal   USTHB, Algeria
Ziouel Tahar   Tiaret University, Algeria

Contents

Graph Databases and Data Warehousing

Data Intelligence and Technology

Mining Big Data I

Mining Recent High-Utility Patterns from Temporal Databases with Time-Sensitive Constraint

Wensheng Gan[1], Jerry Chun-Wei Lin[1(✉)], Philippe Fournier-Viger[2],
and Han-Chieh Chao[1,3]

[1] School of Computer Science and Technology,
Harbin Institute of Technology Shenzhen Graduate School, Shenzhen, China
wsgan001@gmail.com, jerrylin@ieee.org
[2] School of Natural Sciences and Humanities,
Harbin Institute of Technology Shenzhen Graduate School, Shenzhen, China
philfv@hitsz.edu.cn
[3] Department of Computer Science and Information Engineering,
National Dong Hwa University, Hualien, Taiwan
hcc@ndhu.edu.tw

Abstract. Useful knowledge embedded in a database is likely to be changed over time. Identifying recent changes and up-to-date information in temporal databases can provide valuable information. In this paper, we address this issue by introducing a novel framework, named recent high-utility pattern mining from temporal databases with time-sensitive constraint (RHUPM) to mine the desired patterns based on user-specified minimum recency and minimum utility thresholds. An efficient tree-based algorithm called RUP, the global and conditional downward closure (GDC and CDC) properties in the recency-utility (RU)-tree are proposed. Moreover, the vertical compact recency-utility (RU)-list structure is adopted to store necessary information for later mining process. The developed RUP algorithm can recursively discover recent HUPs; the computational cost and memory usage can be greatly reduced without candidate generation. Several pruning strategies are also designed to speed up the computation and reduce the search space for mining the required information.

Keywords: Temporal database · High-utility patterns · Time-sensitive · RU-tree · Downward closure property

1 Introduction

Knowledge discovery in database (KDD) aims at finding meaningful and useful information from the amounts of mass data; frequent itemset mining (FIM) [7] and association rule mining (ARM) [2,3] are the fundamental issues in KDD. Instead of FIM or ARM, high-utility pattern mining (HUPM) [5,6,19] incorporates both quantity and profit values of an item/set to measure how "useful"

© Springer International Publishing Switzerland 2016
S. Madria and T. Hara (Eds.): DaWaK 2016, LNCS 9829, pp. 3–18, 2016.
DOI: 10.1007/978-3-319-43946-4_1

an item or itemset is. The goal of HUPM is to identify the rare items or itemsets in the transactions, and bring valuable profits for the retailers or managers. HUPM [5,15–17] serves as a critical role in data analysis and has been widely utilized to discover knowledge and mine valuable information in recent decades. Many approaches have been extensively studied. The previous studies suffer, however, from an important limitation, which is to utilize a minimum utility threshold as the measure to discover the complete set of HUIs without considering the time-sensitive characteristic of transactions. In general, knowledge found in a temporal database is likely to be changed as time goes by. Extracting up-to-date knowledge especially from temporal databases can provide more valuable information for decision making. Although HUPs can reveal more significant information than frequent ones, HUPM does not assess how recent the discovered patterns are. As a result, the discovered HUPs may be irrelevant or even misleading if they are out-of-date.

In order to enrich the efficiency and effectiveness of HUPM with time-sensitive constraint, an efficient tree-based algorithm named mining Recent high-Utility Patterns from temporal database with time-sensitive constraint (abbreviated as RUP) is developed in this paper. Major contributions are summarized as follows:

- A novel mining approach named mining Recent high-Utility Patterns from temporal databases (RUP) is proposed for revealing more useful and meaningful recent high-utility patterns (RHUPs) with time-sensitive constraint, which is more feasible and realistic in real-life environment.
- The RUP approach is developed by spanning the Set-enumeration tree named Recency-Utility tree (RU-tree). Based on this structure, it is unnecessary to scan databases for generating a huge number of candidate patterns.
- Two novel global and conditional sorted downward closure (GDC and CDC) properties guarantee the global and partial anti-monotonicity for mining RHUPs in the RU-tree. With the GDC and CDC properties, the RUP algorithm can easily discover RHUPs based on the pruning strategies to prune a huge number of unpromising itemsets and speed up computation.

2   Related Work

HUPM is different from FIM since the quantities and unit profits of items are considered to determine the importance of an itemset rather than only its occurrence. Chan et al. [6] presented a framework to mine the top-k closed utility patterns based on business objective. Yao et al. [19] defined utility mining as the problem of discovering profitable itemsets while considering both the purchase quantity of items in transactions (internal utility) and their unit profit (external utility). Liu et al. [16] then presented a two phases algorithm to efficiently discover HUPs by adopting a new transaction-weighted downward closure (TWDC) property and named this approach as transaction-weighted utilization (TWU) model. Tseng et al. then proposed UP-growth+ [17] algorithm to mine HUPs using an UP-tree structure. Liu et al. [15] proposed a novel list-based

Table 1. An example database

TID	Transaction time	Items with quantities
T_1	2016/1/2 09:30	a:2, c:1, d:2
T_2	2016/1/2 10:20	b:1, d:2
T_3	2016/1/3 19:35	b:2, c:1, e:3
T_4	2016/1/3 20:20	a:3, c:2
T_5	2016/1/5 10:00	a:1, b:3, d:4, e:1
T_6	2016/1/5 13:45	b:4, e:1
T_7	2016/1/6 09:10	a:3, c:3, d:2
T_8	2016/1/6 09:44	b:2, d:3
T_9	2016/1/6 16:10	c:1, d:2, e:2
T_{10}	2016/1/8 10:35	a:2, c:2, d:1

Table 2. Derived HUPs and RHUPs

Itemset	$r(X)$	$u(X)$	Itemset	$r(X)$	$u(X)$
(a)	2.9145	66	(ce)	1.2405	45
(c)	3.6235	100	(de)	1.3414	57
(d)	4.3626	112	(abd)	0.5314	37
(e)	2.3624	35	(acd)	1.9048	137
(ac)	2.3831	140	(ade)	0.5314	39
(ad)	2.4362	111	(bde)	0.5314	36
(bd)	1.6479	69	(cde)	0.81	34
(be)	1.5524	34	$(abde)$	0.5314	42
(cd)	2.7148	119			

algorithm named HUI-Miner to efficiently mine HUPs without generating candidates. Other algorithms were also extensively developed for various problems of HUPM [9,11,13,14,18], etc.

In real-world situations, knowledge embedded in a database is changed all the time. The discovered HUPs may be out of date or possibly invalid at present. Identifying recent changes and up-to-date information in temporal databases can provide valuable information. Recently, a new up-to-date high-utility pattern (UDHUP) [12] was proposed to reveal more useful and meaningful HUPs, while considering both the utility and the recency of patterns. The UDHUP reveals the patterns which are not HUPs in the entire databases but are HUPs in the recent intervals. Mining UDHUPs may, however, easily suffer from the "combination explosion problem" and return huge number of patterns which may not the interesting ones. The reason is that the patterns occurred in recent days are always considered as UDHUPs in this mining framework. Thus, it is a critical issue and a challenge to discover more reasonable recent HUPs with time-sensitive constraint.

3 Preliminaries and Problem Statement

Let $I = \{i_1, i_2, \ldots, i_m\}$ be a finite set of m distinct items in a temporal transactional database $D = \{T_1, T_2, \ldots, T_n\}$, where each transaction $T_q \in D$ is a subset of I, and has an unique identifier, TID and a timestamp. An unique profit $pr(i_j)$ is assigned to each item $i_j \in I$, and they are stored in a profit-table $ptable = \{pr(i_1), pr(i_2), \ldots, pr(i_m)\}$. An itemset $X \in I$ with k distinct items $\{i_1, i_2, \ldots, i_k\}$ is of length k and is referred to as a k-itemset. For an itemset X, let the notation $TIDs(X)$ denotes the $TIDs$ of all transactions in D containing X. As a running example, Table 1 shows a transactional database containing 10 transactions, which are sorted by purchase time. Assume that the $ptable$ is defined as $\{pr(a){:}6, pr(b){:}1, pr(c){:}10, pr(d){:}7, pr(e){:}5\}$.

Definition 1. The recency of each T_q is denoted as $r(T_q)$ and defined as:

$$r(T_q) = (1 - \delta)^{(T_{current} - T_q)}. \tag{1}$$

where δ is a user-specified time-decay factor ($\delta \in (0,1]$), $T_{current}$ is the current timestamp which is equal to the number of transactions in D, and T_q is the TID of the currently processed transaction which is associated with a timestamp.

Thus, a higher recency value is assigned to transactions having a time-stamp closer to the most recent time-stamp. When δ was set to 0.1, the recency values of T_1 and T_8 are respectively calculated as $r(T_1) = (1 - 0.1)^{(10-1)}(= 0.3874)$ and $r(T_8) = (1 - 0.1)^{(10-8)}(= 0.8100)$.

Definition 2. The recency of an itemset X in a transaction T_q is denoted as $r(X, T_q)$ and defined as:

$$r(X, T_q) = r(T_q) = (1 - \delta)^{(T_{current} - T_q)}. \tag{2}$$

Definition 3. The utility of an item i_j in a transaction T_q is denoted as $u(i_j, T_q)$, and is defined as:

$$u(i_j, T_q) = q(i_j, T_q) \times pr(i_j). \tag{3}$$

For example, the utility of item (c) in transaction T_1 is calculated as $u(c, T_1) = q(c, T_1) \times pr(c) = 1 \times 10 = 10$.

Definition 4. The utility of an itemset X in transaction T_q is denoted as $u(X, T_q)$, and defined as:

$$u(X, T_q) = \sum_{i_j \in X \wedge X \subseteq T_q} u(i_j, T_q). \tag{4}$$

For example, the utility of the itemset (ad) is calculated as $u(ad, T_1) = u(a, T_1) + u(d, T_1) = q(a, T_1) \times pr(a) + q(d, T_1) \times pr(d) = 2 \times 6 + 2 \times 7 = 26$.

Definition 5. The recency of an itemset X in a database D is denoted as $r(X)$, and defined as:

$$r(X) = \sum_{X \subseteq T_q \wedge T_q \in D} r(X, T_q). \tag{5}$$

Definition 6. The utility of an itemset X in a database D is denoted as $u(X)$, and defined as:

$$u(X) = \sum_{X \subseteq T_q \wedge T_q \in D} u(X, T_q). \tag{6}$$

For example, the utility of itemset (acd) is calculated as $u(acd) = u(acd, T_1) + u(acd, T_7) + u(acd, T_{10}) = 36 + 62 + 39 = 137$.

Definition 7. The transaction utility of a transaction T_q is denoted as $tu(T_q)$, and defined as:

$$tu(T_q) = \sum_{i_j \in T_q} u(i_j, T_q). \tag{7}$$

in which j is the number of items in T_q.

Definition 8. The total utility in D is the sum of all transaction utilities in D and denoted as TU, which can be defined as:

$$TU = \sum_{T_q \in D} tu(T_q). \tag{8}$$

For example, the transaction utilities for T_1 to T_{10} are respectively calculated as $tu(T_1) = 36$, $tu(T_2) = 15$, $tu(T_3) = 27$, $tu(T_4) = 38$, $tu(T_5) = 42$, $tu(T_6) = 9$, $tu(T_7) = 62$, $tu(T_8) = 23$, $tu(T_9) = 34$, and $tu(T_{10}) = 39$; the total utility in D is calculated as: $TU = 325$.

Definition 9. An itemset X in a database is a HUP iff its utility is no less than the minimum utility threshold $(minUtil)$ multiplied by the TU as:

$$HUP \leftarrow \{X | u(X) \geq minUtil \times TU\}. \tag{9}$$

Definition 10. An itemset X in a database D is defined as a RHUP if it satisfies two conditions: (1) $u(X) \geq minUtil \times TU$; (2) $r(X) \geq minRe$. The $minUtil$ is the minimum utility threshold and $minRe$ is the minimum recency threshold; both of them can be specified by users' preference.

For the given example, when $minRe$ and $minUtil$ are respectively set at 1.50 and 10%, the itemset (abd) is a HUP since its utility is $u(abd) = 57 >$ $(minUtil \times TU = 32.5)$, but not a RHUP since its recency is $r(abd)$ $(= 0.5314 < 1.5)$. Thus, the complete set of RHUPs is marked as red color and shown in Table 2.

Given a quantitative transactional database (D), a *ptable*, a user-specified time-decay factor $(\delta \in (0,1])$, a minimum recency threshold $(minRe)$ and a minimum utility threshold $(minUtil)$. The goal of RHUPM is to efficiently find out the complete set of RHUPs while considering both time-sensitive and utility constraints. Thus, the problem of RHUPM is to find the complete set of RHUPs, in which the utility of each itemset X is no less than $minUtil \times TU$ and its recency value is no less than $minRec$.

4   Proposed RUP Algorithm for Mining RHUPs

4.1   Proposed RU-tree

Definition 11 (Total order $\prec$ on items). Assume that the total order $\prec$ on items in the addressed RHUPM framework is the TWU-ascending order of 1-items.

Definition 12 (Recency-utility tree, RU-tree). A recency-utility tree (RU-tree) is presented as a sorted set-enumeration tree with the total order $\prec$ on items.

Definition 13 (Extension nodes in the RU-tree). The extensions of an itemset w.r.t. node X can be obtained by appending an item y to X such that y is greater than all items already in X according to the total order $\prec$. Thus, the all extensions of X is the all its descendant nodes.

The proposed RU-tree for the RUP algorithm can be represented as a set-enumeration tree [10] with the total order $\prec$ on items. For the running example, an illustrated RU-tree is shown in Fig. 1. As shown in Fig. 1, the all extension nodes w.r.t. descendant of node (ea) are (eac), (ead) and $(eacd)$. Note that all the supersets of node (ea) are (eba), (eac), (ead), $(ebac)$, $(ebad)$, $(eacd)$ and $(ebacd)$. Hence, the extension nodes of a node are a subset of the supersets of that node. Based on the designed RU-tree, the following lemmas can be obtained.

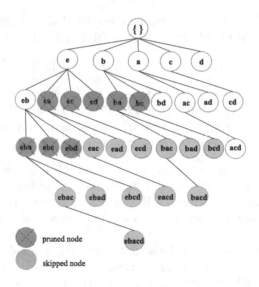

Fig. 1. The search space and pruned nodes in the RU-tree.

Lemma 1. *The complete search space of the proposed RUP algorithm for the addressed RHUPM framework can be represented by a RU-tree where items are sorted in TWU-ascending order of items.*

Lemma 2. *The recency of a node in the RU-tree is no less than the recency of any of its child nodes (extensions).*

Proof. Assume a node X^{k-1} in the RU-tree contains $(k-1)$ items, then any its child node can be denoted as X^k which containing k items and sharing with common $(k-1)$ items. Since $X^{k-1} \subseteq X^k$, it can be proven that:

$$r(X^k) = \sum_{X^k \subseteq T_q \wedge T_q \in D} r(X^k, T_q) \leq \sum_{X^{k-1} \subseteq T_q \wedge T_q \in D} r(X^{k-1}, T_q) \Longrightarrow r(X^k) \leq r(X^{k-1}).$$

Thus, the recency of a node in the proposed RU-tree is always no less than that of any of its extension nodes.

4.2   The RU-list Structure

The recency-utility list (RU-list) structure is a new vertical data structure, which incorporates the inherent recency and utility properties to keep necessary information. Let an itemset X and a transaction (or itemset) T such that $X \subseteq T$, the set of all items from T that are not in X is denoted as $T \backslash X$, and the set of all the items appearing after X in T is denoted as T/X. Thus, $T/X \subseteq T \backslash X$. For example, consider $X = \{bd\}$ and transaction T_5 in Table 1, $T_5 \backslash X = \{ae\}$, and $T_5/X = \{e\}$.

Definition 14 (Recency-Utility list, RU-list). The RU-list of an itemset X in a database is denoted as $X.RUL$. It contains an entry (element) for each transaction T_q where X appears ($X \subseteq T_q \wedge T_q \in D$). An element consists of four fields: (1) the **tid** of X in T_q ($X \subseteq T_q \wedge T_q \in D$); (2) the recency of X in T_q (**rec**); (3) the utilities of X in T_q (**iu**); and (4) the remaining utilities of X in T_q (**ru**), in which **ru** is defined as $X.ru(T_q) = \sum_{i_j \in (T_q/X)} u(i_j, T_q)$.

Thanks to the property of RU-list, the recency and utility information of the longer k-itemset can be built by join operation of $(k-1)$-itemset without rescanning the database. Details of the construction can be referred to Algorithm 3. The RU-list of the running example is constructed in TWU-ascending order as $(e \prec b \prec a \prec c \prec e)$ and shown in Fig. 2.

(e)				(b)				(a)				(c)				(d)			
tid	rec	iu	ru	tid	rec	iu	ru	tid	rec	iu	ru	tid	rec	iu	ru	tid	rec	iu	ru
3	0.4783	15	12	2	0.4305	1	14	1	0.3874	12	24	1	0.3874	10	14	1	0.3874	14	0
5	0.5905	5	37	3	0.4783	2	10	4	0.5314	18	20	3	0.4783	10	0	2	0.4305	14	0
6	0.6561	5	4	5	0.5905	3	34	5	0.5905	6	28	4	0.5314	20	0	5	0.5905	28	0
9	0.9000	10	24	6	0.6561	4	0	7	0.7290	18	44	7	0.7290	30	14	7	0.7290	14	0
				8	0.8100	2	21	10	1.0000	12	27	9	0.9000	10	14	8	0.8100	21	0
												10	1.0000	20	7	9	0.9000	14	0
																10	1.0000	7	0

$TWU(e) < TWU(b) < TWU(a) < TWU(c) < TWU(d)$

Fig. 2. Constructed RU-list of 1-items.

Definition 15. Based on the RU-list, the total recency of an itemset X in D is denoted as $X.RE$ (it equals to the r(X)), and defined as:

$$X.RE = \sum_{X \subseteq T_q \wedge T_q \in D} (X.rec). \tag{10}$$

Definition 16. Let the sum of the utilities of an itemset X in D denoted as $X.IU$. Based on the RU-list, it can be defined as:

$$X.IU = \sum_{X \subseteq T_q \wedge T_q \in D} (X.iu). \tag{11}$$

Definition 17. Let the sum of the remaining utilities of an itemset X in D denoted as $X.RU$. Based on the RU-list, it can be defined as:

$$X.RU = \sum_{X \subseteq T_q \wedge T_q \in D} (X.ru). \tag{12}$$

4.3 Proposed GDC and CDC Properties

Lemma 3. *The actual utility of a node/pattern in the RU-tree is (1) less than, (2) equal to, or (3) greater than that of any of its extension nodes (descendant nodes).*

Thus, the downward closure property of ARM could not be used in HUPM to mine HUPs. The TWDC property [16] was proposed in traditional HUPM to reduce the search space. Based on the RU-list and the properties of recency and utility, some lemmas and theorems can be obtained from the built RU-tree.

Definition 18. The transaction-weighted utility (TWU) of an itemset X is the sum of all transaction utilities $tu(T_q)$ containing X, which is defined as:

$$TWU(X) = \sum_{X \subseteq T_q \wedge T_q \in D} tu(T_q). \tag{13}$$

Definition 19. An itemset X in a database D is defined as a recent high transaction-weighted utilization pattern (RHTWUP) if it satisfies two conditions: (1) $r(X) \geq minRe$; (2) $TWU(X) \geq minUtil \times TU$.

Theorem 1 (Global downward closure (GDC) property). *Let X^k be a k-itemset (node) in the RU-tree and a $(k$-$1)$-itemset (node) X^{k-1} has the common $(k$-$1)$-items with X^k. The GDC property guarantees that: $TWU(X^k) \leq TWU(X^{k-1})$ and $r(X^k) \leq r(X^{k-1})$.*

Proof. Let X^{k-1} be a $(k$-$1)$-itemset and its superset k-itemset is denoted as X^k.

$$TWU(X^k) = \sum_{X^k \subseteq T_q \wedge T_q \in D} tu(T_q) \leq \sum_{X^{k-1} \subseteq T_q \wedge T_q \in D} tu(T_q) \Longrightarrow TWU(X^k) \leq TWU(X^{k-1}).$$

From Lemma 2, it can be found that $r(X^{k-1}) \geq r(X^k)$. Therefore, if X^k is a RHTWUP, any its subset X^{k-1} is also a RHTWUP.

Theorem 2 (RHUPs $\subseteq$ RHTWUPs). *Assume that the total order $\prec$ is applied in the RU-tree, we have that $RHUPs \subseteq RHTWUPs$, which indicates that if a pattern is not a RHTWUP, then none of its supersets will be RHUP.*

Proof. Let X^k be an itemset such that X^{k-1} is a subset of X^k. We have that $u(X) = \sum_{X \subseteq T_q \land T_q \in D} u(X, T_q) \leq \sum_{X \subseteq T_q \land T_q \in D} tu(T_q) = TWU(X)$; $u(X) \leq TWU(X)$. Besides, Theorem 2 shows that $r(X^k) \leq r(X^{k-1})$ and $TWU(X^k) \leq TWU(X^{k-1})$. Thus, if X^k is not a RHTWUP, none of its supersets are RHUPs.

Lemma 4. *The TWU of any node in the Set-enumeration RU-tree is greater than or equal to the sum of all the actual utility of any one of its descendant nodes, as well as the other supersets (which are not the descendant nodes in RU-tree).*

Proof. Let X^{k-1} be a node in the RU-tree, and X^k be a children (extension) of X^{k-1}. According to Theorem 1, we can get the relationship $TWU(X^{k-1}) \geq TWU(X^k)$. Thus, the lemma holds.

Theorem 3. *In the RU-tree, if the TWU of a tree node X is less than the $minUtil \times TU$, X is not a RHUP, and all its supersets (not only the descendant nodes, but also the other nodes which containing X) are not considered as RHUP either.*

Proof. According to Theorem 2, this theorem holds.

Theorem 4 (Conditional downward closure property, (CDC) property). *For any node X in the RU-tree, the sum of $X.IU$ and $X.RU$ in the RU-list is larger than or equal to utility of any one of its descendant nodes (extensions). It shows the anti-monotonicity of unpromising itemsets in RU-tree.*

The above lemmas and theorems ensure that all RHUPs would not be missed. Thus, the designed GDC and CDC properties guarantee the **completeness** and **correctness** of the proposed RUP approach. By utilizing the GDC property, we only need to initially construct the RU-list for those promising itemsets w.r.t. the $RHTWUPs$[1] as the input for later recursive process. Furthermore, the following pruning strategies are proposed in the RUP algorithm to speed up computation.

4.4 Proposed Pruning Strategies

Based on the above lemmas and theorems, several efficient pruning strategies are designed in the developed RUP model to early prune unpromising item sets. Thus, a more compressed search space can be obtained to reduce the computation.

Strategy 1. *After the first database scan, we can obtain the recency and TWU value of each 1-item in database. If the TWU of a 1-item i (w.r.t. $TWU(i)$) and the sum of all the recencies of i (w.r.t. $r(i)$) do not satisfy the two conditions of RHTWUP, this item can be directly pruned, and none of its supersets is concerned as RHUP.*

Strategy 2. *When traversing the RU-tree based on a depth-first search strategy, if the sum of all the recencies of a tree node X w.r.t. $X.RE$ in its constructed RU-list is less than the minimum recency, then none of the child nodes of this node is concerned as RHUP.*

Strategy 3. *When traversing the RU-tree based on a depth-first search strategy, if the sum of X.IU and X.RU of any node X is less than the minimum utility count, any of its child node is not a RHUP, they can be regarded as irrelevant and be pruned directly.*

Theorem 5. *If the TWU of 2-itemset is less than the minUtil, any superset of this 2-itemset is not a HTWUP and would not be a HUP either* [8].

According to the definitions of RHTWUP and RUP, Theorem 5 can be applied in the proposed RUP algorithm to further filter unpromising patterns. To effectively apply the EUCP strategy, a structure named Estimated Utility Co-occurrence Structure (*EUCS*) [8] is built in the proposed algorithm. It is a matrix that stores the TWU values of the 2-itemsets and will be applied to the Strategy 4.

Strategy 4. *Let X be an itemset (node) encountered during the depth-first search of the Set-enumeration tree. If the TWU of a 2-itemset $Y \subseteq X$ according to the constructed EUCS is less than the minimum utility threshold, X is not a RHTWUP and would not be a RHUP; none of its child nodes is a RHUP. The construction of the RU-lists of X and its children is unnecessary to be performed.*

Strategy 5. *Let X be an itemset (node) encountered during the depth-first search of the Set-enumeration tree. After constructing the RU-list of an itemset, if X.RUL is empty or the X.RE value is less than the minimum recency threshold, X is not a RHUP, and none of X its child nodes is a RHUP. The construction of the RU-lists for the child nodes of X is unnecessary to be performed.*

Based on the above pruning strategies, the designed RUP algorithm can prune the itemsets with lower recency and utility count early, without constructing their RU-list structures of extensions. For example in Fig. 1, the itemset (*eba*) is not considered as a RHUP since $(eba).AU + (eba).RU (= 42 > 32.5)$, but $(eba).RE (= 0.5314 < 1.50)$. By applying the Strategy 2, all the child nodes of itemset (*eba*) are not considered as the RHUPs since their recency values are always no greater than those of (*eba*). Hence, the child nodes (*ebac*), (*ebad*) and (*ebacd*) (the shaded nodes in Fig. 1) are guaranteed to be uninteresting and can be directly skipped.

4.5   Procedure of the RUP Algorithm and the Enhanced Algorithm

Based on the above properties and pruning strategies, the pseudo-code of the proposed RUP algorithm is described in Algorithm 1. The RUP algorithm first lets $X.RUL$, $D.RUL$ and $EUCS$ are initially set as an empty set (Line 1), then scans the database to calculate the $TWU(i)$ and $r(i)$ values of each item $i \in I$ (Line 2), and then find the potential 1-itemsets which may be the desired RHUP (Line 3). After sorting I^* in the total order $\prec$ (the TWU-ascending order, Line 4), the algorithm scans D again to construct the RU-list of each 1-item $i \in I^*$ and build the $EUCS$ (Line 5). The RU-list for all 1-extensions of $i \in I^*$ is recursively

---

Input: D; $ptable$; δ; $minRe$, $minUtil$.
Output: The set of complete recent high-utility patterns (RHUPs).
1 let $X.RUL \leftarrow \emptyset, D.RUL \leftarrow \emptyset, EUCS \leftarrow \emptyset$;
2 scan D to calculate the $TWU(i)$ and $re(i)$ of each item $i \in I$;
3 find I^* $(TWU(i) \geq minUtil \times TU) \wedge (r(i) \geq minRe)$, w.r.t. $RHTWUP^1$;
4 sort I^* in the designed total order $\prec$ (ascending order in TWU value);
5 scan D to construct the $X.RUL$ of each $i \in I^*$ and build the $EUCS$;
6 call **RHUP-Search**$(\phi, I^*, minRe, minUtil, EUCS)$;
7 return $RHUPs$;

Algorithm 1. RUP algorithm

processed by using a depth-first search procedure RHUP-Search (Line 6) and
the desired RHUPs are returned (Line 7).

As shown in RHUP-Search (cf. Algorithm 2), each itemset X_a is determined to directly produce the RHUPs (Lines 2 to 4). Two constraints are then
applied to further determine whether its child nodes should be executed for
the later depth-first search (Lines 5 to 12). If one itemset is promising, the

---

Input: X, $extendOfX$, $minRe$, $minUtil$, $EUCS$.
Output: The set of complete RHUPs.
1 for *each itemset $X_a \in extendOfX$* do
2   obtain the $X_a.RE$, $X_a.IU$ and $X_a.RU$ values from the built $X_a.RUL$;
3   if $(X_a.IU \geq minUtil \times TU) \wedge (X_a.RE \geq minRe)$ then
4          $RHUPs \leftarrow RHUPs \cup X_a$;

5   if $(X_a.IU + X_a.RU \geq minUtil \times TU) \wedge (X_a.RE \geq minRe)$ then
6          $extendOfX_a \leftarrow \emptyset$;
7   for *each $X_b \in extendOfX$ such that b after a* do
8   if $\exists TWU(a, b) \in EUCS \wedge TWU(a, b) \geq minUtil \times TU$ then
9                  $X_{ab} \leftarrow X_a \cup X_b$;
10                 $X_{ab}.RUL \leftarrow construct(X, X_a, X_b)$;
11   if $X_{ab}.RUL \neq \emptyset \wedge (X_a.RE \geq minRe)$ then
12                     $extendOfX_a \leftarrow extendOfX_a \cup X_{ab}.RUL$;

13   call **RHUP-Search** $(X_a, extendOfX_a, minRe, minUtil, EUCS)$;

14 return $RHUPs$;

Algorithm 2. RHUP-Search procedure

Input: X, an itemset; X_a, the extension of X with an item a; X_b, the extension of X with
    an item b $(a \neq b)$.
Output: $X_{ab}.RUL$, the RU-list of an itemset X_{ab}.
1 set $X_{ab}.RUL \leftarrow \emptyset$;
2 for *each element $E_a \in X_a.RUL$* do
3   if $\exists E_a \in X_b.RUL \wedge E_a.tid == E_b.tid$ then
4   if $X.RUL \neq \emptyset$ then
5   find $E \in X.RUL, E.tid = E_a.tid$;
6              $E_{ab} \leftarrow\ < E_a.tid, E_a.re, E_a.iu + E_b.iu - E.iu, E_b.ru >$;
7   else
8          $E_{ab} \leftarrow\ < E_a.tid, E_a.re, E_a.iu + E_b.iu, E_b.ru >$;
9      $X_{ab}.RUL \leftarrow X_{ab}.RUL \cup E_{ab}$;

10 return $X_{ab}.RUL$;

Algorithm 3. RU-list construction

$Construct(X, X_a, X_b)$ process (cf. Algorithm 3) is executed continuously to construct a set of RU-lists of all 1-extensions of itemset X_a (w.r.t. $extendOfX_a$) (Lines 9 to 13). Note that each constructed X_{ab} is a 1-extension of itemset X_a; it should be put into the set of $extendOfX_a$ for executing the later depth-first search. The **RHUP-Search** procedure then is recursively performed to mine RHUPs (Line 13).

5   Experimental Results

Substantial experiments were conducted to verify the effectiveness and efficiency of the proposed RUP algorithm and its improved algorithms. Note that only one previous study [12] has addressed the up-to-date issue but it mines the seasonal or periodic HUPs than the entire ones, which is totally different than the discovered RHUPs. The state-of-the-art FHM [8] was executed to derive HUPs, which can provide a benchmark to verify the efficiency of the proposed RUP algorithm. Note that the baseline RUP$_{baseline}$ algorithm does not adopt the pruning Strategies 4 and 5, the RUP1 algorithm does not use the pruning Strategy 5, and the RUP2 algorithm adopts the all designed pruning strategies. Two real-life datasets, foodmart [4] and mushroom [1], were used in the experiments. A simulation model [16] was developed to generate the quantities and profit values of items in transactions for mushroom dataset since the foodmart dataset already has real utility values. A log-normal distribution was used to randomly assign quantities in the [1,5] interval, and item profit values in the [1,1000] interval.

5.1   Runtime Analysis

With a fixed time-decay threshold δ, the runtime comparison of the compared algorithms under various minimum utility thresholds ($minUtil$) with a fixed minimum recency threshold ($minRe$) and under various $minRe$ with a fixed $minUtil$ are shown in Fig. 3. It can be observed that the proposed three RUP-based algorithms outperform the FHM algorithm, and the enhanced algorithms using different pruning strategies outperform the baseline one. In general, RUP2 is about one to two times faster than the state-of-the-art FHM algorithm. This is reasonable since both the utility and recency constraints are considered in the RHUPM to discover RHUPs while only the utility constraint is used in the FHM algorithm to find HUPs. With more constraints, the search space can be further reduced and fewer patterns are discovered. Besides, the pruning strategies used in the two enhanced algorithms are more efficient than those used by the baseline RUP$_{baseline}$ algorithm. The GDC and CDC properties provide a powerful way to prune the search space than that of FHM. It can be further observed that the runtime of FHM is unchanged, while runtimes of the proposed algorithms sharply decrease when $minRe$ is increased. It indicates that the traditional FHM do not effect by the time-sensitive constraint. Based on the designed RU-list, RU-tree structure and pruning strategies, the runtime of the proposed RUP algorithm can be greatly reduced.

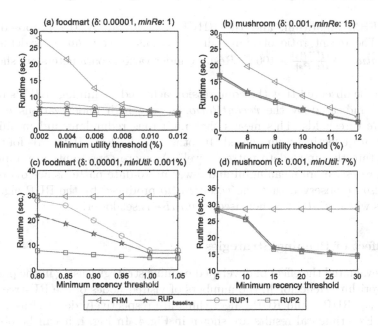

Fig. 3. Runtime under various parameters.

Table 3. Number of patterns under various parameters.

foodmart	0.002%	0.004%	0.006%	0.008%	0.010%	0.012%
HUPs	492041	267164	93467	26193	8365	3850
RHUPs	41539	33188	20515	11286	6222	3660
recentRatio	8.44%	12.42%	21.95%	43.09%	74.38%	95.06%
mushroom	7%	8%	9%	10%	11%	12%
HUPs	34331	22121	13953	7601	3420	1265
RHUPs	15321	9714	5745	2886	1162	388
recentRatio	44.63%	43.91%	41.17%	37.97%	33.98%	30.67%
foodmart	0.80	0.85	0.90	0.95	1.00	1.05
HUPs	547949	547949	547949	547949	547949	547949
RHUPs	547949	428809	296313	169378	42430	42430
recentRatio	100.00%	78.26%	54.08%	30.91%	7.74%	7.74%
mushroom	5	10	15	20	25	30
HUPs	34331	34331	34331	34331	34331	34331
RHUPs	33762	31862	15321	14618	13766	12892
recentRatio	98.34%	92.81%	44.63%	42.58%	40.10%	37.55%

5.2   Pattern Analysis

The derived patterns of HUPs and RHUPs are also evaluated to show the acceptable of the proposed RHUPM framework. Note that the HUPs are derived by the FHM algorithm, and the RHUPs are found by the proposed algorithms. The recent ratio of high-utility patterns ($recentRatio$) is defined as: $recentRatio = \frac{|RHUPs|}{|HUPs|} \times 100\%$. Results under various parameters are shown in Table 3.

It can be observed that the compression achieved by mining RHUPs instead of HUPs, indicated by the $recentRatio$, is very high under various $minUtil$ or $minRe$ thresholds. This means that numerous redundant and meaningless patterns are effectively eliminated. In other words, less HUPs are found but they capture the up-to-date patterns well. As more constraints are applied in mining process, more meaningful and fewer up-to-date patterns are discovered. It can also be observed that the $recentRatio$ produced by the RUP algorithm increases when $minUtil$ is increased or $minRe$ is set lower.

5.3   Effect of Pruning Strategies

We also evaluated the effect of developed pruning strategies using in the proposed RUP algorithm. Henceforth, the numbers of visited nodes of the RU-tree in the $RUP_{baseline}$, RUP1, and RUP2 algorithms are respectively denoted as N_1, N_2, and N_3. Experimental results are shown in Fig. 4. In Fig. 4, it can be observed that the various pruning strategies can reduce the search space from the RU-tree. It can also be concluded that the proposed extension of the pruning Strategy 5 in the RUP2 algorithm can efficiently prune a huge number of unpromising patterns, as shown for the foodmart dataset. The pruning Strategy 4 is no longer

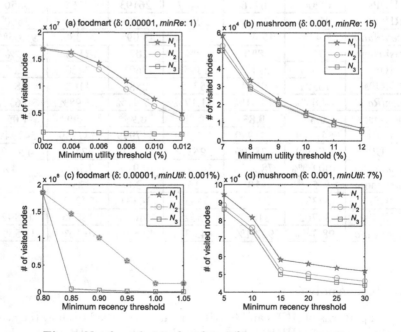

Fig. 4. Number of visited nodes under various parameters.

significantly effective for pruning unpromising patterns since those unpromising patterns can be efficiently filtered using Strategies 1 to 3. In addition, it can also be seen that pruning Strategy 1, which relies on TWU and recency values, can still prune some unpromising candidates early. This is useful since it allows avoiding the construction several RU-lists for items and their supersets. Although the number N_2 is only slightly less than N_1 on foodmart dataset, it can be seen that the number N_2 is quite less N_1 for the mushroom dataset, as shown in Fig. 4(b) and (d).

6 Conclusions

Since up-to-date knowledge is more interesting and helpful than outdated knowledge. In this paper, an enumeration tree-based algorithm nameed RUP is designed to discover RHUPs from temporal databases. A compact recency-utility tree (RU-tree) is proposed to make that the necessary information of itemsets from the databases can be easily obtained from a series of compact Recency-Utility (RU)-lists of their prefix itemsets. To guarantee the global and partial anti-monotonicity of RHUPs, two novel GDC and CDC properties are proposed for mining RHUP in the RU-tree. Several efficient pruning strategies are further developed to speed up the mining performance. Substantial experiments show that the proposed RUP algorithm can efficiently discover the RHUPs without candidate generation.

Acknowledgment. This research was partially supported by the Tencent Project under grant CCF-TencentRAGR20140114, and by the National Natural Science Foundation of China under grant No.61503092.

References

1. Frequent itemset mining dataset repository. http://fimi.ua.ac.be/data/
2. Agrawal, R., Imielinski, T., Swami, A.: Database mining: A performance perspective. IEEE Trans. Knowl. Data Eng. **5**(6), 914–925 (1993)
3. Agrawal, R., Srikant, R.: Fast algorithms for mining association rules in large databases. In: The International Conference on Very Large Data Bases, pp. 487–499 (1994)
4. Microsoft. Example database foodmart of microsoft analysis services. http://www.Almaden.ibm.com/cs/quest/syndata.html
5. Ahmed, C.F., Tanbeer, S.K., Jeong, B.S., Lee, Y.K.: Efficient tree structures for high utility pattern mining in incremental databases. IEEE Trans. Knowl. Data Eng. **21**(12), 1708–1721 (2009)
6. Chan, R., Yang, Q., Shen, Y.D.: Mining high utility itemsets. In: The International Conference on Data Mining, pp. 19–26 (2003)
7. Han, J., Pei, J., Yin, Y., Mao, R.: Mining frequent patterns without candidate generation: a frequent-pattern tree approach. Data Min. Knowl. Discov. **8**(1), 53–87 (2004)

8. Fournier-Viger, P., Wu, C.-W., Zida, S., Tseng, V.S.: FHM: faster high-utility itemset mining using estimated utility co-occurrence pruning. In: Andreasen, T., Christiansen, H., Cubero, J.-C., Raś, Z.W. (eds.) ISMIS 2014. LNCS, vol. 8502, pp. 83–92. Springer, Heidelberg (2014)

9. Fournier-Viger, P., Lin, J.C.W., Gueniche, T., Barhate, P.: Efficient incremental high utility itemset mining. In: ASE BigData & Social Informatics, p. 53 (2015)

10. Rymon, R.: Search through systematic set enumeration. Technical Reports (CIS), 297 (1992)

11. Lan, G.C., Hong, T.P., Tseng, V.S.: Discovery of high utility itemsets from on-shelf time periods of products. Expert Syst. Appl. **38**(5), 5851–5857 (2011)

12. Lin, J.C.W., Gan, W., Hong, T.P., Tseng, V.S.: Efficient algorithms for mining up-to-date high-utility patterns. Adv. Eng. Inf. **29**(3), 648–661 (2015)

13. Lin, J.C.W., Gan, W., Fournier-Viger, P., Hong, T.P.: Mining high-utility itemsets with multiple minimum utility thresholds. In: ACM International Conference on Computer Science & Software Engineering, pp. 9–17 (2015)

14. Lin, J.C.W., Gan, W., Fournier-Viger, P., Hong, T.P., Tseng, V.S.: Fast algorithms for mining high-utility itemsets with various discount strategies. Adv. Eng. Inf. **30**(2), 109–126 (2016)

15. Liu, M., Qu, J.: Mining high utility itemsets without candidate generation. In: ACM International Conference on Information and Knowledge Management, pp. 55–64 (2012)

16. Liu, Y., Liao, W., Choudhary, A.K.: A two-phase algorithm for fast discovery of high utility itemsets. In: Ho, T.-B., Cheung, D., Liu, H. (eds.) PAKDD 2005. LNCS (LNAI), vol. 3518, pp. 689–695. Springer, Heidelberg (2005)

17. Tseng, V.S., Shie, B.E., Wu, C.W., Yu, P.S.: Efficient algorithms for mining high utility itemsets from transactional databases. IEEE Trans. Knowl. Data Eng. **25**(8), 1772–1786 (2013)

18. Tseng, V.S., Wu, C.W., Fournier-Viger, P., Yu, P.S.: Efficient algorithms for mining top-K high utility itemsets. IEEE Trans. Knowl. Data Eng. **28**(1), 54–67 (2016)

19. Yao, H., Hamilton, J., Butz, C.J.: A foundational approach to mining itemset utilities from databases. In: SIAM International Conference on Data Mining, pp. 211–225 (2004)

TopPI: An Efficient Algorithm for Item-Centric Mining

Martin Kirchgessner[1]($\boxtimes$), Vincent Leroy[1], Alexandre Termier[2], Sihem Amer-Yahia[1], and Marie-Christine Rousset[1]

[1] Université Grenoble Alpes, LIG, CNRS, Grenoble, France
{martin.kirchgessner,vincent.leroy,sihem.amer-yahia,
marie-christine.rousset}@imag.fr
[2] Université Rennes 1, INRIA/IRISA, Rennes, France
alexandre.termier@irisa.fr

Abstract. We introduce TopPI, a new semantics and algorithm designed to mine long-tailed datasets. For each item, and regardless of its frequency, TopPI finds the k most frequent closed itemsets that item belongs to. For example, in our retail dataset, TopPI finds the itemset "nori seaweed, wasabi, sushi rice, soy sauce" that occurrs in only 133 store receipts out of 290 million. It also finds the itemset "milk, puff pastry", that appears 152,991 times. Thanks to a dynamic threshold adjustment and an adequate pruning strategy, TopPI efficiently traverses the relevant parts of the search space and can be parallelized on multi-cores. Our experiments on datasets with different characteristics show the high performance of TopPI and its superiority when compared to state-of-the-art mining algorithms. We show experimentally on real datasets that TopPI allows the analyst to explore and discover valuable itemsets.

Keywords: Frequent itemset mining · Top-K · Parallel data mining

1 Introduction

Over the past twenty years, pattern mining algorithms have been applied successfully to various datasets to extract frequent itemsets and uncover hidden associations [1,9]. As more data is made available, large-scale datasets have proven challenging for traditional itemset mining approaches. Indeed, the worst-case complexity of frequent itemset mining is exponential in the number of items in the dataset. To alleviate that, analysts use high threshold values and restrict the mining to the most frequent itemsets. But many large datasets exhibit a long tail distribution, characterized by the presence of a majority of infrequent items [5]. Mining at high thresholds eliminates low-frequency items, thus ignoring the majority of them. In this paper we propose TopPI, a new semantics that is more appropriate to mining long-tailed datasets, and the corresponding algorithm.

A common request in the retail industry is finding a product's associations with other products. This allows managers to obtain feedback on customer behavior and to propose relevant promotions. Instead of mining associations between

© Springer International Publishing Switzerland 2016
S. Madria and T. Hara (Eds.): DaWaK 2016, LNCS 9829, pp. 19–33, 2016.
DOI: 10.1007/978-3-319-43946-4_2

popular products only, TopPI extracts itemsets for all items. By providing the analyst with an overview of the dataset, it facilitates the exploration of the results.

We hence formalize the objective of TopPI as follows: extract, for each item, the k most frequent closed itemsets containing that item. This semantics raises a new challenge, namely finding a pruning strategy that guarantees correctness and completeness, while allowing an efficient parallelization, able to handle web-scale datasets in a reasonable amount of time. Our experiments show that TopPI can mine 290 million supermarket receipts on a single server. We design an algorithm that restricts the space of itemsets explored to keep the execution time within reasonable bounds. The parameter k controls the number of itemsets returned for each item, and may be tuned depending on the application. If the itemsets are directly presented to an analyst, $k = 10$ would be sufficient, while $k = 500$ may be used when those itemsets are post-processed.

The paper is organized as follows. Section 2 defines the new semantics and our problem statement. The TopPI algorithm is fully described in Sect. 3. In Sect. 4, we present experimental results and compare TopPI against a simpler solution based on TFP [6]. Related work is reviewed in Sect. 5, and we conclude in Sect. 6.

2   TopPI Semantics

The data contains *items* drawn from a set $\mathcal{I}$. Each item has an integer identifier, referred to as an index, which provides an order on $\mathcal{I}$. A *dataset* $\mathcal{D}$ is a collection of *transactions*, denoted $\langle t_1, ..., t_n \rangle$, where $t_j \subseteq \mathcal{I}$. An *itemset* P is a subset of $\mathcal{I}$. A transaction t_j is an *occurrence* of P if $P \subseteq t_j$. Given a dataset $\mathcal{D}$, the *projected dataset* for an itemset P is the dataset $\mathcal{D}$ restricted to the occurrences of P: $\mathcal{D}[P] = \langle t \mid t \in \mathcal{D} \wedge P \subseteq t \rangle$. To further reduce its size, all items of P can be removed, giving the *reduced dataset* of P: $\mathcal{D}_P = \langle t \setminus P \mid t \in \mathcal{D}[P] \rangle$.

The number of occurrences of an itemset in $\mathcal{D}$ is called its *support*, denoted $support_{\mathcal{D}}(P)$. Note that $support_{\mathcal{D}}(P) = support_{\mathcal{D}[P]}(P) = |\mathcal{D}_P|$. An itemset P is said to be closed if there exists no itemset $P' \supset P$ such that $support(P) = support(P')$. The greatest itemset $P' \supseteq P$ having the same support as P is called the *closure* of P, further denoted as $clo(P)$. For example, in the dataset shown in Table 1a, the itemset $\{1, 2\}$ has a support equal to 2 and $clo(\{1, 2\}) = \{0, 1, 2\}$.

Problem Statement: Given a dataset $\mathcal{D}$ and an integer k, TopPI returns, for each item in $\mathcal{D}$, the k most frequent closed itemsets (CIS) containing this item.

In this paper, we use TopPI to designate the new mining semantics, this problem statement, and our algorithm. Table 1b shows the solution to this problem applied to the dataset in Table 1a, with $k = 2$. Note that we purposely ignore itemsets that occur only once, as they do not show a behavioral pattern.

As the number of CIS is exponential in the number of items, we cannot firstly mine all CIS and their support, then sort the top-k frequent ones for each item. The challenge is instead to traverse the small portions of the solutions space which contains our CIS of interest.

Table 1. Sample dataset

TID	Transaction
t_0	$\{0, 1, 2\}$
t_1	$\{0, 1, 2\}$
t_2	$\{0, 1\}$
t_3	$\{2, 3\}$
t_4	$\{0, 3\}$

item	$top(i)$: P, $support(P)$	
i	1st	2nd
0	$\{0\}, 4$	$\{0, 1\}, 3$
1	$\{0, 1\}, 3$	$\{0, 1, 2\}, 2$
2	$\{2\}, 3$	$\{0, 1, 2\}, 2$
3	$\{3\}, 2$	

(a) Input $\mathcal{D}$ (b) TopPI results for $k = 2$

3 The TopPI Algorithm

After an general overview in Sect. 3.1, this section details TopPI's functions and their underlying principles. Section 3.2 shows how we shape the CIS (closed itemsets) space as a tree. Then Sect. 3.3 presents *expand*, TopPI's tree traversal function. Section 3.4 shows an example traversal, to highlight the challenges of finding pruning opportunities specific to item-centric mining. The *startBranch* function, which implements the dynamic threshold adjustment, is detailed in Sect. 3.5. Section 3.6 presents the *prune* function and the prefix short-cutting technique, which allows TopPI to evaluate quickly and precisely which parts of the CIS tree can be pruned. We conclude in Sect. 3.7 by showing how TopPI can leverage multi-core systems.

3.1 Overview

TopPI adapts two principles from LCM [14] to shape the CIS space as a tree and enumerate CIS of high support first. Similarly to traditional top-k processing approaches [4], TopPI relies on heap structures to progressively collect its top-k results, and outputs them once the execution is complete. More precisely, TopPI stores traversed itemsets in a *top-k collector* which maintains, for each item $i \in \mathcal{I}$, $top(i)$, a heap of size k containing the current version of the k most frequent CIS containing i. We mine all the k-lists simultaneously to maximize the amortization of each itemset's computation. Indeed, an itemset is a candidate for insertion in the heap of all items it contains.

TopPI introduces an adequate pruning of the solutions space. For example, we should be able to prune an itemset $\{a, b, c\}$ once we know it is not a top-k frequent for a, b nor c. However, as highlighted in the following example, we cannot prune $\{a, b, c\}$ if it precedes interesting CIS in the enumeration. TopPI's pruning function tightly cuts the CIS space, while ensuring results' completeness. When pruning we can query the top-k-collector through $min(top(i))$, which is the kth support value in $top(i)$, or 2 if $|top(i)| < k$.

The main program, presented in Algorithm 1, initializes the collector in lines 2 and 3. Then it invokes, for each item i, $startBranch(i, \mathcal{D}, k)$, which enumerates itemsets P such that $max(P) = i$. In our examples, as in TopPI, items are

---

Algorithm 1. TopPI's main function

Data: dataset $\mathcal{D}$, integer k
Result: Output top-k CIS for all items of $\mathcal{D}$
1 **begin**
2   **foreach** $i \in \mathcal{I}$ **do**   `// Collector instantiation`
3 | initialize $top(i)$, heap of max size k
4   **foreach** $i \in \mathcal{I}$ **do**   `// In increasing item order`
5 | $startBranch(i, \mathcal{D}, k)$

---

represented by integers. While loading $\mathcal{D}$, TopPI indexes items by decreasing frequency, hence 0 is the most frequent item. Items are enumerated in their natural order in line 4, thus items of greatest support are considered first.

TopPI does not require the user to define a minimum frequency, but we observe that the support range in each item's top-k CIS varies by orders of magnitude from an item to another. Because filtering out less frequent items can speed up the CIS enumeration in some branches, *startBranch* implements a *dynamic threshold adjustment*. The internal frequency threshold, denoted ε, defaults to 2 because we are not interested in itemsets occurring once.

3.2   Principles of the Closed Itemsets Enumeration

Several algorithms have been proposed to mine CIS in a dataset [6,12,13]. We borrow two principles from the LCM algorithm [14]: the *closure extension*, that generates new CIS from previously computed ones, and the *first parent* that avoids redundant computation.

Definition 1. *An itemset* $Q \subseteq \mathcal{I}$ *is a* closure extension *of a closed itemset* $P \subseteq \mathcal{I}$ *if* $\exists e \notin P$, *called an* extension item, *such that* $Q = clo(P \cup \{e\})$.

TopPI enumerates CIS by recursively performing closure extensions, starting from the empty set. In Table 1a, $\{0, 1, 2\}$ is a closure extension of both $\{0, 1\}$ and $\{2\}$. This example shows that an itemset can be generated by two different closure extensions. Uno et al. [14] introduced two principles which guarantee that each closed itemset is traversed only once in the exploration. We adapt their principles as follows. First, extensions are restricted to items smaller than the previous extension. Furthermore, we prune extensions that do not satisfy the *first-parent* criterion:

Definition 2. *Given a closed itemset* P *and an item* $e \notin P$, $\langle P, e \rangle$ *is the* first parent *of* $Q = clo(P \cup \{e\})$ *only if* $max(Q \setminus P) = e$.

These principles shape the CIS space as a tree and lead to the following property: by extending P with e, TopPI can only recursively generate itemsets Q such that $max(Q \setminus P) = e$. This property is extensively used in our algorithms, in order to predict which items can be impacted by recursions.

Both TopPI and LCM rely on the prefix extension and first parent test principles. However, in TopPI CIS are not outputted as they are traversed.

They are instead inserted in the top-k-collector. This allows TopPI to determine if deepening closure extensions may enhance results held in the top-k-collector, or if the corresponding sub-branch can be pruned. These two differences impact the execution of the CIS enumeration function.

3.3 CIS Enumeration for Item-Centric Mining

TopPI traverses the CIS space with the *expand* function, detailed in Algorithm 2. *expand* performs a depth-first exploration of the CIS tree, and backtracks when no frequent extensions remain in $\mathcal{D}_J$ (line 6). Additionally, in line 7 the *prune* function (presented in Sect. 3.6) determines if each recursive call may enhance results held in the top-k-collector, or if it can be avoided.

Algorithm 2. TopPI's CIS exploration function

1 **Function** $expand(P, e, \mathcal{D}_P, \varepsilon)$
 Data: CIS P, extension item e, reduced dataset $\mathcal{D}_P$, frequency threshold ε
 Result: If $\langle e, P \rangle$ is a relevant closure extension, collects CIS containing $\{e\} \cup P$ and items smaller than e
2 **begin**
3 $\quad Q \leftarrow closure(\{e\} \cup P)$ // Closure extension
4 $\quad$ **if** $max(Q \setminus P) = e$ **then**   // First-parent test
5 $\quad\quad collect(Q, support_\mathcal{D}(Q), true)$
6 $\quad\quad$ **foreach** $i < e \mid support_{\mathcal{D}_Q}[i] \geq \varepsilon$ **do**   // In increasing item order
7 $\quad\quad\quad$ **if** $\neg prune(Q, i, \mathcal{D}_Q, \varepsilon)$ **then**
8 $\quad\quad\quad\quad expand(Q, i, \mathcal{D}_Q, \varepsilon)$

Upon validating the closure extension Q, TopPI updates $top(i)$, $\forall\, i \in Q$, via the *collect* function (line 5). The support computation exploits the fact that $support_\mathcal{D}(Q) = support_{\mathcal{D}_P}(e)$, because $Q = closure(\{e\} \cup P)$. The last parameter of *collect* is set to **true** to point out that Q is a closed itemset (we show in Sect. 3.5 that it is not always the case).

When enumerating items in line 6, TopPI relies on the items' indexing by decreasing frequency. As extensions are only done with smaller items this ensures that, for any item $i \in \mathcal{I}$, the first CIS containing i enumerated by TopPI combine i with some of the most frequent items. This heuristic increases their probability of having a high support, and overall raises the support of itemsets in the top-k-collector.

In *expand*, as in all functions detailed in this paper, operations like computing $clo(P)$ or $\mathcal{D}_P$ rely on an item counting over the projected dataset $\mathcal{D}[P]$. Because it is resource-consuming, in our implementation item counting is done only once over each $\mathcal{D}[P]$, and kept in memory while relevant. The resulting structure and accesses to it are not explicited for clarity.

3.4 Finding Pruning Opportunities in an Example Enumeration

We now discuss how we can optimize item-centric mining in the example CIS enumeration of Fig. 1, when $k = 2$. Items are already indexed by decreasing frequency. Candidate extensions of steps ③ and ⑨ are not collected as they fail the first-parent test (their closure is $\{0, 1, 2, 3\}$).

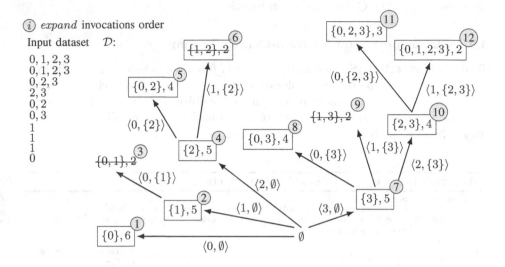

Fig. 1. An example dataset and its corresponding CIS enumeration tree with our *expand* function. Each node is an itemset and its support. $\langle i, P \rangle$ denotes the closure extension operation. Striked out itemsets are candidates failing the first-parent test (Algorithm 2, line 4).

In frequent CIS mining algorithms, the frequency threshold allows the program to lighten the intermediate datasets ($\mathcal{D}_Q$) involved in the enumeration. In TopPI our goal is to increase ε above 2, in some branches. In our example, before step ④ we can compute items' supports in $\mathcal{D}[2]$ — these supports are re-used in $expand(\emptyset, 2, \mathcal{D}, \varepsilon)$ — and observe that the two most frequent items in $\mathcal{D}[2]$ are 2 and 0, with respective supports of 5 and 4. These will yield two CIS of supports 5 and 4 in $top(2)$. The intuition of dynamic threshold adjustment is that 4 might therefore be used as a frequency threshold in this branch. It is not possible in this case because a future extension, 1, does not have its k itemsets at step ④. This is also the case at step ⑦. The dynamic threshold adjustment done by the *startBranch* function takes this into account.

After step ⑧, $top(0)$, $top(2)$ and $top(3)$ already contain two CIS, as required, all having a support of 4 or more. Hence it is tempting to prune the extension $\langle \{3\}, 2 \rangle$ (step ⑩), as it cannot enhance $top(2)$ nor $top(3)$. However, at this step, $top(1)$ only contains a single CIS and 1 is a future extension. Hence ⑩ cannot be pruned: although it yields an useless CIS, one of its extensions leads to a useful one (step ⑫). In this tree we can only prune the recursion towards step ⑪.

This example's distribution is unbalanced in order to show TopPI's corner cases with only 4 items; but in real datasets, with hundreds of thousands of items, such cases regularly occur. This shows that an item-centric mining algorithm requires rigorous strategies for both pruning the search space and filtering the datasets.

3.5 Dynamic Threshold Adjustment

If we initiate each CIS exploration branch by invoking $expand(\emptyset, i, \mathcal{D}, 2), \forall i \in \mathcal{I}$, then *prune* would be inefficient during the k first recursions — that is, until $top(i)$ contains k CIS. For frequent items, which yield the biggest projected datasets, letting the exploration deepen with a negligible frequency threshold is particularly expensive. Thus it is crucial to diminish the size of the dataset as often as possible, by filtering out less frequent items that do not contribute to our results. Hence we present the *startBranch* function, in Algorithm 3, which performs the dynamic threshold adjustment and avoids the cold start situation.

Algorithm 3. TopPI's CIS enumeration branch preparation

```
1  Function startBranch(i, D, k)
       Data: root item i, dataset D, integer k
       Result: Enumerates CIS P such that max(P) = i
2      begin
3          foreach j ∈ topDistinctSupports(D[i], k) do  // Pre-filling with partial itemsets
4          |   collect({i, j}, support_{D[i]}(j), false)
5          ε_i ← min_{j≤i}(min(top(j)))                  // Dynamic threshold adjustment
6          expand(∅, i, D, ε_i)
```

Given a CIS $\{i\}$ and an extension item $e < i$, computing $Q = clo(\{e\} \cup \{i\})$ is a costly operation that requires counting items in $\mathcal{D}_{\{i\}}[e]$. However we observe that $support(Q) = support_{\mathcal{D}}(\{e\} \cup \{i\}) = support_{\mathcal{D}[i]}(e)$, and the latter value is computed by the items counting, prior to the instantiation of $\mathcal{D}_{\{i\}}$. Therefore, when starting the branch of the enumeration tree rooted at i, we can already know the supports of some of the upcoming extensions.

The function *topDistinctSupports* counts items' frequencies in $\mathcal{D}[i]$ — resulting counts are re-used in *expand* for the instantiation of $\mathcal{D}_{\{i\}}$. Then, in lines 3–4, TopPI considers items j whose support in $\mathcal{D}[i]$ is one of the k greatest, and stores the partial itemset $\{i, j\}$ in the top-k collector (this usually includes $\{i\}$ alone). We call these itemsets partial because their closure has not been evaluated yet, so the top-k collector marks them with a dedicated flag: the third argument of *collect* is `false` (line 4). Later in the exploration, these partial itemsets are either ejected from $top(i)$ by more frequent CIS, or replaced by their closure upon its computation (Algorithm 2, line 5).

Thus $top(i)$ already contains k itemsets at the end of the loop of lines 3–4. The CIS recursively generated by the *expand* invocation (line 6) may only contain items lower than i. Therefore the lowest $min(top(j)), \forall j \leq i$, can be used as a frequency threshold in this branch. TopPI computes this value, ε_i, on line 5, This combines particularly well with the frequency-based iteration order, because $min(top(i))$ is relatively high for more frequent items. Thus TopPI can filter the biggest projected datasets as a frequent CIS miner would.

Algorithm 4. TopPI's pruning function

---

1 **Function** $prune(P, e, D_P, \varepsilon)$

    Data: itemset P, extension item e, reduced dataset D_P, minimum support
        threshold ε

    Result: **true** if $expand(P, e, D_P, \varepsilon)$ will not provide new results to the
        top-k-collector, **false** otherwise

2   **begin**

3   **if** $support_{D_P}(\{e\})) \geq min(top(e))$ **then**

4   **return** *false*

5   **foreach** $i \in P$ **do**

6   **if** $support_{D_P}(\{e\})) \geq min(top(i))$ **then**

7   **return** *false*

8   **foreach** $i < e \mid support_{D_P}(i) \geq \varepsilon$ **do**

9           $bound \leftarrow min(support_{D_P}(\{i\}), support_{D_P}(\{e\}))$

10   **if** $bound \geq min(top(i))$ **then**

11   **return** *false*

12   **return** *true*

Note that two partial itemsets $\{i, j\}$ and $\{i, l\}$ of equal support may in fact have the same closure $\{i, j, l\}$. Inserting both into $top(i)$ could lead to an overestimation of the frequency threshold and trigger pruning of legitimate top-k CIS of i. This is why TopPI only selects partial itemsets with *distinct* supports.

3.6   Pruning Function

As shown in the example of Sect. 3.4, TopPI cannot prune a sub-tree rooted at P by observing P alone. We also have to consider itemsets that could be enumerated from P through first-parent closure extensions. This is done by the *prune* function presented in Algorithm 4. It queries the collector to determine whether $expand(P, e, D_P, \varepsilon)$ and its recursions may impact the top-k results of an item. If it is not the case then *prune* returns **true**, thus pruning the sub-tree rooted at $clo(\{e\} \cup P)$.

The anti-monotony property [1] ensures that the support of all CIS enumerated from $\langle e, P \rangle$ is smaller than $support_{D_P}(\{e\})$. It also follows from the definition of *expand* that the only items potentially impacted by the closure extension $\langle e, P \rangle$ are in $\{e\} \cup P$, or are inferior to e. Hence we check $support_{D_P}(\{e\})$ against $top(i)$ for all concerned items i.

The first case, considered in lines 3 and 5, checks $top(e)$ and $top(i), \forall i \in P$. Smaller items, which may be included in future extensions of $\{e\} \cup P$, are considered in lines 8–11. It is not possible to know the exact support of these CIS, as they are not yet explored. However we can compute, as in line 9, an upper bound such that $bound \geq support(clo(\{i, e\} \cup P))$. If this bound is smaller than $min(top(i))$, then extending $\{e\} \cup P$ with i cannot provide a new CIS to $top(i)$. Otherwise, as tested in line 10, we should let the exploration deepen by returning `false`. If this test fails for all items i, then it is safe to prune because all $top(i)$ already contain k itemsets of greater support.

The inequalities of lines 3, 6 and 10 are not strict to ensure that no partial itemset (inserted by the *startbranch* function) remains at the end of the exploration. We can also note that the loop of lines 8–11 may iterate on up to $|\mathcal{I}|$ items, and thus may take a significant amount of time to complete. Hence our implementation of the *prune* function includes an important optimization.

Avoiding Loops with Prefix Short-Cutting: we can leverage the fact that TopPI enumerates extensions by increasing item order. Let e and f be two items successively enumerated as extensions of a CIS P (Algorithm 2 line 6). As $e < f$, in the execution of $prune(P, f, \mathcal{D}_P, \varepsilon)$ the loop of lines 8–11 can be divided into iterations on items $i < e \wedge i \notin P$, and the last iteration where $i = e$. We observe that the first iterations were also performed by $prune(P, e, \mathcal{D}_P, \varepsilon)$, which can therefore be considered as a prefix of the execution of $prune(P, f, \mathcal{D}_P, \varepsilon)$.

To take full advantage of this property, TopPI stores the smallest *bound* computed line 9 such that $prune(P, *, \mathcal{D}_P, \varepsilon)$ returned `true`, denoted $bound_{min}(P)$. This represents the lowest known bound on the support required to enter $top(i)$, for items $i \in \mathcal{D}_P$ ever enumerated by line 8. When evaluating a new extension f by invoking $prune(P, f, \mathcal{D}_P, \varepsilon)$, if $support_{\mathcal{D}_P}(f) \leq bound_{min}(P)$ then f cannot satisfy tests of lines 6 and 10. In this case it is safe to skip the loop of lines 5–7, and more importantly the prefix of the loop of lines 8–11, therefore reducing this latter loop to a single iteration. As items are sorted by decreasing frequency, this simplification happens very frequently.

Thanks to prefix short-cutting, most evaluations of the pruning function are reduced to a few invocations of $min(top(i))$. This allows TopPI to guide the itemsets exploration with a negligible overhead.

3.7 Parallelization

As shown by Négrevergne et al. [11], the CIS enumeration can be adapted to shared-memory parallel systems by dispatching *startBranch* invocations (Algorithm 1, line 5) to different threads. When multi-threaded, TopPI ensures that the dynamic threshold computation (Algorithm 3, line 5) can only be executed for an item i once all items lower than i are done with the top-k collector prefilling (Algorithm 3, line 3).

Sharing the collector between threads does not cause any congestion because most accesses are read operations from the *prune* function. Preliminary experiments, not included in this paper for brevity, show that TopPI shows an excellent speedup when allocated more CPU cores. Thanks to an efficient evaluation of *prune*, the CIS enumeration is the major time consuming operation in TopPI.

4   Experiments

We now evaluate TopPI's performance and the relevance of its results, with three real-life datasets on a multi-core machine. We start by comparing its performance to a simpler solution using a *global* top-k algorithm, in Sect. 4.1. Then we observe the impact of our optimizations on TopPI's run-time, in Sect. 4.2. Finally Sect. 4.3 provides a few itemsets examples, confirming that TopPI highlights patterns of interest about long tailed items. We use 3 real datasets:

- *Tickets* is a 24 GB retail basket dataset collected from 1884 supermarkets over a year. There are 290,734,163 transactions and 222,228 items.
- *Clients* is *Tickets* grouped by client, therefore transactions are two to ten times longer. It contains 9,267,961 transactions in 13.3 GB, each representing the set of products bought by a single customer over a year.
- *LastFM* is a music recommendation website, on which we crawled 1.2 million public profile pages. This results in a 277 MB file where each transaction contains the 50 favorite artists of a user, among 1.2 million different artists.

All measurements presented here are averages of 3 consecutive runs, on a single machine containing 128 GB of RAM and 2 Intel Xeon E5-2650 8-cores CPUs with Hyper Threading. We implemented TopPI in Java and will release its source upon the publication of this paper.

4.1   Baseline Comparison

We start by comparing TopPI to its baseline, which is the most straightforward solution to item-centric mining: it applies a *global* top-k CIS miner on the projected dataset $\mathcal{D}[i]$, for each item i in $\mathcal{D}$ occurring at least twice.

We implemented TFP [6], in Java, to serve as the top-k miner. It has an additional parameter l_{min}, which is the minimal itemset size. In our case l_{min} is always equal to 1 but this is not the normal use case of TFP. For a fair comparison, we added a major pre-filtering: for each item i, we only keep items having one of the k highest supports in $\mathcal{D}[i]$. In other words, the baseline also benefits from a dynamic threshold computation. This is essential to its ability to mine a dataset like *Tickets*. The baseline also benefits from the occurrence delivery provided by our input dataset implementation (*i.e.* instant access to $\mathcal{D}[i]$). Its parallelization is obvious, hence both solutions use all physical cores available on our machine.

Figure 2 shows the run-times on our datasets when varying k. Both solutions are equally fast for $k = 10$, but as k increases TopPI shows better performance. The baseline even fails to terminate in some cases, either taking over 8 h to complete or running out of memory. Instead TopPI can extract even 500 CIS per item out of the 290 million receipts of *Tickets* in less than 20 min, or 500 CIS per item out of *Clients* in 3 h.

For $k \geq 200$, as the number of items having less than k CIS increases, more and more CIS branches have to be traversed completely. This explains the exponential increase of run-time. However we usually need 10 to 50 CIS per item, in which case such complete traversals only happens in extremely small branches. During this experiment, TopPI's memory usage remains reasonable: below 50 GB for *Tickets*, 30 GB for *Clients* and 10 GB for *LastFM*.

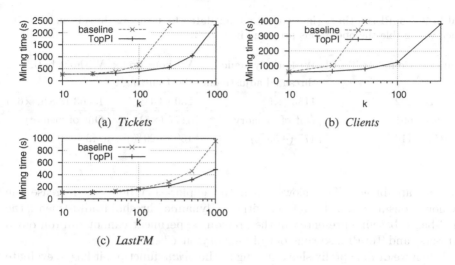

(a) *Tickets*   (b) *Clients*

(c) *LastFM*

Fig. 2. TopPI and baseline run-times using 16 threads

We also observe that the baseline enumerates many more intermediate solutions. Ideally, an algorithm would only enumerate outputted solutions. But, as shown in Sect. 3.4, item-centric mining requires the enumeration of a few additional itemsets to reach some solutions. On *LastFM*, when $k = 100$, the output contains approximately 12.8 million distinct itemsets. The baseline enumerated 50.6 million itemsets and TopPI only 17.4 millions. As each $\mathcal{D}[i]$ is mined independently for all items i, the baseline cannot amortize results from a branch to another, so this result would likely be also observed with another top-k CIS mining algorithm. This highlights the need for a specific item-centric CIS mining algorithm. Thanks to the use of appropriate heuristics to guide the exploration, TopPI only enumerates a small fraction of discarded itemsets.

These results show that TopPI is fast and scalable. Even on common hardware: TopPI is able to mine *LastFM* with $k = 50$ and $\varepsilon = 2$ on a laptop with 4 threads (Intel Core i7-3687U) and 6 GB of RAM in 16 min.

4.2 Contributions Impact

We now validate the individual impact of our contributions: the dynamic threshold adjustment in *startBranch*, described in Sect. 3.5, and the prefix short-cutting in *prune*, presented at the end of Sect. 3.6. To do so, we make these features optional in TopPI's implementation and evaluate their impact on the execution time. Finally we disable both features, to observe how TopPI would perform as a simple pruning by top-k-collector polling in LCM.

Table 2 compares the run-time measured for these variants against the fully optimized version of TopPI's, on all our datasets when using the full capacity of our server. We use $k = 50$, which is sufficient to provide interesting results.

Disabling dynamic threshold adjustment implies that all projected datasets created during the CIS exploration carry more items. Hence intermediate

Table 2. TopPI run-times (in seconds) on our datasets, using 32 threads and $k = 50$, when we disable the operations proposed in Sect. 3.

Dataset	Complete TopPI	Without dynamic threshold adjustment	Without prefix short-cutting	Without both
Tickets	222	1136 ($\times$5)	230 (+4%)	13863 (3.8 h, $\times$62)
Clients	661	Out of memory	4177 ($\times$6)	Out of memory
LastFM	116	177 (+53%)	150 (+29%)	243 ($\times$2)

datasets are bigger. This slows down the exploration but also increase the memory consumption. Therefore, without dynamic threshold adjustment, the TFP-based baseline presented in the previous experiment cannot run (on our 3 datasets) and TopPI also runs out of memory on *Clients*.

When we disable prefix short-cutting in the *prune* function, it has to evaluate more extensions in Algorithm 4, lines 8–11. Hence we observe greater slowdowns on datasets having longer transactions: on average, transactions contain 12 items in *Tickets*, 50 items in *LastFM*, and 213 in *Clients*.

The *prune* function is even more expensive if we disable both optimizations, as potential extensions are not only all evaluated but also more numerous. This experiment shows that it's the combined usage of dynamic threshold adjustment and prefix short-cutting that allows TopPI to mine large datasets efficiently.

4.3 Example Itemsets

From Tickets: Itemsets with high support can be found for very common products, such as milk: "milk, puff pastry" (152,991 occurrences), "milk, eggs" (122,181) and "milk, chocolate spread" (98,982). Although this particular milk product was bought 5,402,063 times (i.e. in 1.85% of the transactions), some of its top-50 associated patterns would already be out of reach of traditional CIS algorithms: "milk, chocolate spread" appears in 0.034% transactions.

Interesting itemsets can also be found for less frequent (tail) products. For example, "frangipane, puff pastry, sugar" (522), shows the basic ingredients for french king cake. We also found evidence of some sushi parties, with itemsets such as "nori seaweed, wasabi, sushi rice, soy sauce" (133). We observe similar patterns in *Clients*.

From LastFM: TopPI finds itemsets grouping artists of the same music genre. For example, the itemset "Tryo, La Rue Ketanou, Louise Attaque" (789 occurrences), represents 3 french alternative bands. Among the top-10 CIS that contain "Vardoger" (a black-metal band from Norway which only occurs 10 times), we get the itemset "Vardoger, Antestor, Slechtvalk, Pantokrator, Crimson Moonlight" (6 occurrences). TopPI often finds such itemsets, which, in the case of unknown artists, are particularly interesting to discover similar bands.

5 Related Work

The first algorithms in the field were *frequent* itemsets mining algorithms, whose parameter is a minimum support threshold. By definition, items that belong to the long tail are excluded from their results. Generate-and-test approaches like Apriori [1], if we want to keep a majority of items in the results, cannot be used on our datasets because the candidate generation would exhaust our machines' capacity. Pasquier et al. identified *closed* itemsets [12], which are less numerous while conveying the same information. CLOSET [13] is a CIS mining algorithm which uses prefix trees to store the initial and projected datasets. However prefix trees' instantiation is not amortized on our datasets, where between 90 % (*Tickets*) and 99 % (*Clients, LastFM*) of transactions are unique. Hence TopPI inherits its CIS enumeration from LCM [14], whose structures (transactions concatenated in an array) are straightforward and CPU cache-friendly.

Given a new dataset, it can be difficult to select an appropriate minimum frequency threshold. Therefore Han et al. therefore proposed TFP [6], an algorithm that returns the k most frequent closed itemsets in $\mathcal{D}$ containing at least l_{min} items. We show in Sect. 4.1 how it can be adapted to emulate TopPI, but is not as robust. In particular, this method runs out of memory if we do not let it benefit from our dynamic threshold adjustment.

TopPI relies on frequency to rank each k itemsets associated with an item. Another possibility could be to rank them by statistical correlation with the item, for example by p-value. But it implies the search of an adequate (usually low) frequency threshold, as in LAMP [10], which is not affordable at our scale. Le Bras et al. show how 13 other quality measures can be integrated in a generate-and-test mining algorithm [8]. But, as with Apriori, the candidate generation step is unfeasible at our scale. As TopPI is able to analyze a year of activity of 1884 supermarkets in minutes on a single server, ranking by p-value (or another quality measure) can instead be implemented as a post-processing of TopPI's results. We discuss in [7] which quality measures are adapted to retail data.

To the best of our knowledge, PFP [9] is the only other item-centric itemset mining algorithm: it returns a maximum of k itemsets per item. However PFP does not ensure that these itemsets are the most frequent, nor that they are closed, and is developed for the MapReduce platform [3]. Preliminary experiments, not shown here due to space restrictions, show that a single server running TopPI can outperform a 50 machines cluster running PFP.

6 Conclusion

To the best of our knowledge, TopPI is the first algorithm to formalize and solve at scale the problem of mining item-centric top-k closed itemsets, a semantic more appropriate to mining long-tailed datasets. TopPI is able to operate efficiently on long tail content, which is out of reach of standard mining and global top-k algorithms. Instead of generating millions of itemsets only containing the very few frequent items, TopPI spreads its exploration evenly to mine a fixed number of k itemsets for each item in the dataset, including rare ones. This is particularly important in the context of Web datasets, in which the long tail contains most of the information [5], and in the retail industry, where it can account for a large fraction of the revenue [2].

Acknowledgments. This work was partially funded by the Datalyse PIA project.

References

1. Agrawal, R., Srikant, R.: Fast algorithms for mining association rules in large databases. In: Proceedings of the 20th International Conference on Very Large Data Bases (VLDB), pp. 487–499 (1994)
2. Anderson, C.: The Long Tail: Why the Future of Business Is Selling Less of More. Hyperion, New York (2006)
3. Dean, J., Ghemawat, S.: Mapreduce: simplified data processing on large clusters. In: Proceedings of the 6th Symposium on Operating System Design and Implementation (OSDI) (2004)
4. Fagin, R., Lotem, A., Naor, M.: Optimal aggregation algorithms for middleware. In: Proceedings of the Symposium on Principles of Database Systems (PODS) (2001)
5. Goel, S., Broder, A., Gabrilovich, E., Pang, B.: Anatomy of the long tail: ordinary people with extraordinary tastes. In: Proceedings of the Third International Conference on Web Search and Data Mining (WSDM), pp. 201–210 (2010)
6. Han, J., Wang, J., Lu, Y., Tzvetkov, P.: Mining top-k frequent closed patterns without minimum support. In: Proceedings of the International Conference on Data Mining (ICDM), pp. 211–218. IEEE (2002)
7. Kirchgessner, M., Mishra, S., Leroy, V., Amer-Yahia, S.: Testing interestingness measures in practice: a large-scale analysis of buying patterns (2016). http://arxiv.org/abs/1603.04792
8. Le Bras, Y., Lenca, P., Lallich, S.: Mining interesting rules without support requirement: a general universal existential upward closure property. In: Stahlbock, R., Crone, S.F., Lessmann, S. (eds.) Data Mining. Annals of Information Systems, vol. 8, pp. 75–98. Springer, New York (2010)
9. Li, H., Wang, Y., Zhang, D., Zhang, M., Chang, E.Y.: PFP: parallel FP-growth for query recommendation. In: Proceedings of the Second Conference on Recommender Systems (RecSys), pp. 107–114 (2008)
10. Minato, S., Uno, T., Tsuda, K., Terada, A., Sese, J.: A fast method of statistical assessment for combinatorial hypotheses based on frequent itemset enumeration. In: Calders, T., Esposito, F., Hüllermeier, E., Meo, R. (eds.) ECML PKDD 2014, Part II. LNCS, vol. 8725, pp. 422–436. Springer, Heidelberg (2014)
11. Négrevergne, B., Termier, A., Méhaut, J.F., Uno, T.: Discovering closedfrequent itemsets on multicore: parallelizing computations and optimizing memory accesses. In: Proceedings of the International Conference on High Performance Computing and Simulation (HPCS). pp. 521–528 (2010)
12. Pasquier, N., Bastide, Y., Taouil, R., Lakhal, L.: Discovering frequent closed itemsets for association rules. In: Beeri, C., Bruneman, P. (eds.) ICDT 1999. LNCS, vol. 1540, pp. 398–416. Springer, Heidelberg (1998)

13. Pei, J., Han, J., Mao, R.: Closet: an efficient algorithm for mining frequent closed itemsets. In: ACM SIGMOD Workshop on Research Issues in Data Mining and Knowledge Discovery, vol. 4, pp. 21–30 (2000)
14. Uno, T., Asai, T., Uchida, Y., Arimura, H.: An efficient algorithm for enumerating closed patterns in transaction databases. In: Suzuki, E., Arikawa, S. (eds.) DS 2004. LNCS (LNAI), vol. 3245, pp. 16–31. Springer, Heidelberg (2004)

A Rough Connectedness Algorithm for Mining Communities in Complex Networks

Samrat Gupta[✉], Pradeep Kumar, and Bharat Bhasker

Indian Institute of Management, Lucknow, India
{samratgupta,pradeepkumar,bhasker}@iiml.ac.in

Abstract. Mining communities is essential for modern network analysis so as to understand the dynamic processes taking place in the complex real-world networks. Though community detection is a very active research area, most of the algorithms focus on detecting disjoint community structure. However, real-world complex networks do not necessarily have disjoint community structure. Concurrent overlapping and hierarchical communities are prevalent in real-world networked systems. In this paper, we propose a novel algorithm based on rough sets that is capable of detecting disjoint, overlapping and hierarchically nested communities in networks. The algorithm is initiated by constructing granules of neighborhood nodes and representing them as rough sets. Subsequently, utilizing the concept of constrained connectedness, upper approximation is computed in an iterative manner. We also introduce a new metric based on relative connectedness which is used as the merging criteria for sets during iterations. Experiments conducted on nine real-world networks, including a large word association network and protein-protein interaction network, demonstrate the effectiveness of the proposed algorithm. Moreover, it is observed that the proposed algorithm competes favorably with five relevant methods of community detection.

Keywords: Complex networks · Overlapping communities · Community detection · Rough set · Algorithms

1 Introduction

Community structure representing functional or organizational groups is an essential feature of network data in different areas like information science, social science, economics and biology. Amongst the modern data-driven methods for analyzing complex networked systems, community detection has gained a lot of attention. Most of the existing community detection algorithms aim at extracting discrete partitions, where each node belongs to a single community. However, a node in a network might have diverse roles leading to memberships in multiple communities. For example, the membership of individuals in a social network is not limited to a single community since they belong to multiple social groups like family, professional colleagues, college friends etc. simultaneously. These shared affiliations or overlaps are significant characteristics of complex real-world networks. Furthermore, the objective of conventional clustering schemes is to exhaustively assign each and every node to a community. This can lead to spurious results as the network data might not support assignment of each node to a community.

© Springer International Publishing Switzerland 2016
S. Madria and T. Hara (Eds.): DaWaK 2016, LNCS 9829, pp. 34–48, 2016.
DOI: 10.1007/978-3-319-43946-4_3

In this paper, we propose a novel algorithm based on a theoretical foundation that has not previously been applied to community detection. The proposed algorithm detects co-occurring overlapping and hierarchical communities and does not necessarily assign each node to a community. Experimental tests were performed on nine different datasets from diverse domains. To explain the working of the proposed algorithm we illustrate it on Risk game network [1]. The results of experiments followed by a comparative analysis with other relevant algorithms establish the usefulness of the proposed rough set based community detection algorithm.

This paper is organized into six sections. Section 2 discusses extant literature on the community detection problem. The proposed algorithm and its complexity are defined in Sect. 3. Section 4 explains the working of the proposed algorithm on Risk dataset. Section 5 presents experimental results and a comparative analysis of the proposed algorithm with five state-of-the-art algorithms. Finally, conclusion and potential future directions are discussed in Sect. 6.

2 Related Work

During the past years, a variety of methods have been proposed for identifying community structure in complex networks. These methods were mostly based on graph partitioning, partitional clustering, hierarchical clustering and spectral clustering [2]. The last decade saw a surge in methods especially meant for detecting disjoint community structure. In 2004, Newman and Girvan proposed and used "modularity" metric for disjoint community detection [3]. Since then numerous methods using modularity either as an evaluation metric for validation or as an objective function for optimization have been proposed [2]. However, modularity based algorithms suffer from several drawbacks and does not guarantee the correct division of a network [1]. Some algorithms operate on graph diffusion based seed expansion strategy [4, 5]. These algorithms mostly use conductance as the optimization objective because conductance is considered the best scoring function when the network contains disjoint communities [4]. An elaborate survey of the community detection problem with methods for solving it has been published by Fortunato [2].

Since the complex real-world networks typically have an overlapping community structure, identifying overlapping communities has been attracting increased attention of researchers recently. The clique percolation method (CPM) was the first attempt to detect overlapping community structure [6]. An algorithm named CONGA extended the Girvan & Newman disjoint community detection algorithm for overlapping community detection [7]. Another algorithm known as EAGLE forms initial communities by identifying maximal cliques and then performs pairwise merging, based on similarity [8]. Ball et al. proposed a method based on maximum-likelihood and expectation-maximization [9]. Some algorithms like OSLOM and LFM are based on the optimization of benefit function for local expansion [10, 11]. Algorithms based on Non-Negative Matrix Factorization (NMF) such as Symmetric NMF (SBMF) and Bayesian NMF (BNMF) have been proposed [12, 13]. Other algorithms such as, ones based on spectral clustering like MOSES and OSBM [14, 15], label propagation like COPRA and SLPA [16, 17] and density based shrinkage like DenShrink [18] have also been proposed to detect overlapping communities. Another class of algorithms use the concept of membership degree of node and extend fuzzy clustering for overlapping community detection [19, 20]. A review of overlapping community detection algorithms has been published by Xie et al. [21].

Rough set theory introduced by Pawlak [22] is a powerful mathematical approach to model vagueness and uncertainty. It uses concepts such as, equivalence relations, approximations, boundary region, belief function and reducts to find hidden patterns in data [23]. In the past, rough set theory has been used extensively for web mining [23, 24]. Recently, researchers have begun using it for community detection in social networks [25].

The proposed algorithm uses the notion of rough sets for initialization and constrained connectedness for iteratively merging community components. A network is first modeled in the framework of rough sets and a family of subsets (granules, in rough set terminology) of neighborhood nodes is generated. Then a modified equivalence relation called neighborhood relation and a topological operation called upper approximation are used to identify meaningful communities in a network. To the best of our knowledge, this work proposes one of the first algorithms based on the neighborhood rough set model to reveal communities in complex networks.

3 A Rough Connectedness Algorithm for Community Detection

In this section, we present the proposed approach in detail. Mathematically, the problem of community detection can be defined as follows: Let $G = (V, E)$ be a network with a vertex set $V(G) = \{v_1, v_2, v_3 \ldots v_n\}$ and a set of edges $E(G) = \{e_1, e_2, e_3 \ldots e_m\}$. The total number of vertices is $n = |V|$ and the total number of edges is $m = |E|$. Each edge is a connection of a vertex pair (v_i, v_j) such that $v_i, v_j \in V$. We have to find a set of communities $C(G)$ which correspond to the best set of communities in the network. So, if $p = |C|$ communities are detected in a network, we shall have $C_1, C_2 \ldots C_p \in G$ corresponding to the optimal community structure in that network. The community structure thus detected should coincide with the ground-truth community structure of a network which is known a priori.

In Rough Set theory, a granule is considered to be a chunk of objects in the universe of discourse U, drawn together by indiscernibility relation [26]. This indiscernibility can be due to proximity, similarity or functionality [26]. An equivalence relation R on the universe $R \subseteq U \times U$ can generally be regarded as an indiscernibility relation [22]. Thus, any two elements in the same equivalence class are indiscernible or indistinguishable.

From a network viewpoint, we can consider an edge of a particular vertex v_i as the indiscernibility relation such that all the vertices directly connected to this vertex through an edge represent a neighborhood class. Thus a basic granule of knowledge is formed by using neighborhood connectedness around each node. Some definitions which are used in the proposed algorithm are presented below.

Definition 1 (Neighborhood Connectedness Subset). For any network G, let R be a neighborhood relation defined on G. Then N represents a family of subsets of G induced by R such that $\cup N = V(G)$. A neighborhood connectedness subset (NCS) of vertex v_i is represented by $N(v_i)$ and it consists of all the vertices adjacent to vertex v_i including itself. There are n such subsets in N corresponding to the n vertices of G. The size of $|N(v_i)|$ is $(d(v_i) + 1)$, where $d(v_i)$ is the degree of vertex v_i.

So, $N = \{N(v_1), N(v_2), N(v_3)\ldots\ldots N(v_n)\}$ for all $v_i \in V(G)$ and $N(v_i) \neq \emptyset$. Each $N(v_i)$ is a neighborhood class consisting of vertex v_i and all the vertices connected to v_i with an edge.

Definition 2 (Connectedness Upper Approximation). Given a network G and a neighborhood relation R over G, for any subset $N(v_i) \subseteq G$, the connectedness upper approximation (CUA) of $N(v_i)$ is defined as:

$$\overline{N}(v_i) = \{v_i | N(v_i) \cap G \neq \emptyset, v_i \in V(G)\} \tag{1}$$

where $N(v_i)$ denotes the neighborhood connectedness of v_i. Therefore, CUA of vertex v_i consists of all its 2-hop neighbors (path length of 2 between v_i and any element of its CUA).

The subsets obtained through first CUA may share some vertices. These shared vertices (called boundary elements in rough set terminology) may belong to any of the subsets in second or higher CUAs. Therefore we calculate the strength of association of an element to a CUA by using a measure called relative connectedness as explained in Definition 3.

Definition 3 (Relative Connectedness). For any two subsets $N(v_i)$, $N(v_j) \in G$ the relative connectedness of v_i with respect to v_j is given by:

$$RelCon(v_i, v_j) = \frac{|N(v_i) \cap N(v_j)|}{\min(|N(v_i) - N(v_j)|, |N(v_j) - N(v_i)|)} \tag{2}$$

where $RelCon(v_i, v_j)$ lies in interval $[0, \infty)$.

The formula for relative connectedness is quite intuitive. In real-world, relative connectedness between two objects is measured by their common connections as well as different connections such that two objects are said to be highly connected if the number of their common connections is more than the number of their exclusive connections. Mathematically, the formula of relative connectedness measures the ratio of the shared region between two sets and the difference of two sets. Since $N(v_i) - N(v_j)$ is not same as $N(v_j) - N(v_i)$, we take the minimum of the two differences in the denominator of the formula of relative connectedness. This makes the formula of Relative Connectedness symmetric i.e., $RelCon(v_i, v_j) = RelCon(v_j, v_i)$. Relative connectedness is used after each iteration of the proposed algorithm to measure the extent of connectedness between any two vertices.

Definition 4 (Constrained Connectedness Upper Approximation). Given a network G and a neighborhood relation R over G, for any subset $N(v_i) \subseteq G$, the constrained connectedness upper approximation (CCUA) of $N(v_i)$ is defined as:

$$\overline{\overline{N}}(v_i) = \left\{ v_j \in \bigcup_{v_l \in \overline{N}(v_i)} \overline{N}(v_l) | RelCon(v_i, v_j) \geq 1 \right\} \tag{3}$$

In simple words, the elements of each subset in connectedness upper approximation are constrained using the relative connectedness measure as equal to one. The process is repeated and successive CCUAs are computed for each subset until two consecutive CCUAs are found to be the same. When the CCUAs of all subsets become stable, algorithm converges. As the number of iterations increases, the number of upper approximation computations decreases, thus accelerating convergence of the algorithm.

The CCUA technique renders several rough clusters wherein an element may have multiple cluster memberships. First of all, we remove the redundant clusters, thus retaining all the unique clusters. These unique clusters might consist of some distinct clusters with minor overlap among their elements and some non-distinct clusters with high overlap among their elements.

In order to resolve the ambiguity posed by non-distinct clusters, we employ fine-tuning method based on the concept of conductance. Conductance captures the cluster quality in a graph and is widely used to evaluate communities in a network. Since a community in a network represents a set of nodes with better internal connectivity than external connectivity, conductance can be quantitatively defined as the ratio of the number of edges leaving the community to the number of edges inside the community [27]. Therefore, community node sets have lower conductance. Mathematically, conductance is defined as:

$$\emptyset(C_i) = \frac{cut(C_i)}{\min\{\deg(C_i), \deg(\overline{C}_i)\}} \tag{4}$$

where $cut(C_i)$ denotes the size of cut induced by $C_i \in G$ [27]. The complement set of C_i is denoted by $\overline{C}_l$ and sum of degrees of vertices in C_i is denoted by $\deg(C_i)$.

Fine-tuning method merges a group of non-distinct clusters into a single cluster. The method first selects the cluster with minimum conductance in a particular group and then lists all the additional nodes from that group. Then it tests the conductance of the selected cluster after addition of each node from the list. If adding a node to the cluster decreases its conductance, that node is retained in the cluster, otherwise it is removed. The combination of clusters thus arising from non-distinct clusters (after the fine-tuning process) and distinct clusters gives the best set of communities in the input network. The resulting community structure might consist of disjoint, overlapping or hierarchical communities. For better understanding, a flow diagram of the proposed algorithm is shown in Fig. 1.

Let us consider the time complexity of the proposed algorithm. The key steps of the proposed algorithm are relative connectedness computation, CUA computation, merging operation and fine-tuning operation. Given a network G with n as the total number of nodes and m as the total number of edges, let the average degree of nodes in network G be $l = 2\ m/n$. The complexity of relative connectedness computation is $O(n^2 log_2 l)$ [23]. The complexity of CUA computation is $O(|G/R|)$, which is same as $O(n)$ because $|G/R|$ is bounded from above by $|G|$ [28]. Assuming an average of k clusters are merged in each iteration, the complexity of merging is $O(klogk)$ [29]. Since the maximum number of iterations may reach n/k, the complexity of merging

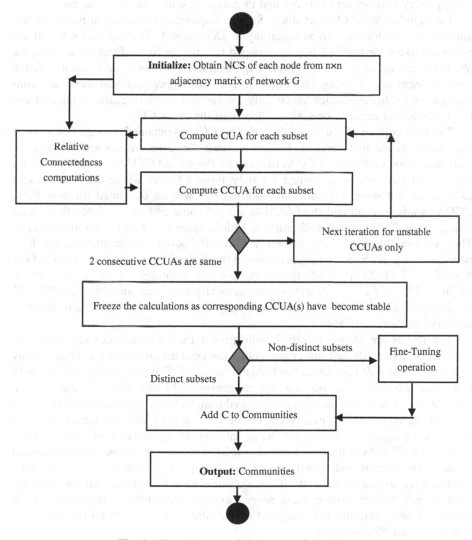

Fig. 1. Flow diagram of the proposed algorithm

procedure is in the order of $O(nlogk)$. Finally, the complexity of the fine tuning operation is $O(n + dn)$. After summing up and omitting dn, the total complexity of the proposed rough set based algorithm is in the order of $O(n^2 log_2 l) + O(nlogk) + O(n)$.

4 Pilot Experiment

We performed the first experiment on Risk game network [1]. Risk is a strategy game played on a board depicting a political map of the earth. It consists of 42 territories grouped into six continents. The objective of this game is global domination i.e., to occupy every territory on the board and in doing so, eliminate other players.

The Neighborhood Connectedness Subsets (equivalence classes, in rough set terminology) of nodes are formed according to Definition 1. Thus all nodes which are connected to a node within 1-hop are grouped to form the NCS of that node. Now, we obtain the first upper approximations as explained in Definition 2 and calculate relative connectedness matrix using Definition 3. The first upper approximations are constrained with value one which means only the nodes for which relative connectedness is greater than or equal to one are candidates of the first CCUA.

The third column of Table 1 shows the first CCUAs obtained through Definition 4 and completes the first iteration. Now, we repeat the same process and perform the second-iteration on the first CCUAs to compute the second CCUAs (second iteration). The third CUA will be computed for only those CCUAs whose first and second CCUAs are not same. Similarly, the fourth CUA will be computed for only those CCUAs whose second and third CCUAs are not same and so on. Since there is no change in the CCUA of any node after the fifth iteration we cease the iterative process. Then we use the fine-tuning method and identify seven communities in the Risk network. These seven communities consist of three disjoint communities (*1 2 3 4 5 6 7 8 9, 10 11 12 13, 39 40 41 42*), two overlapping communities (*14 15 16 17 18 19 20, 18 20 21 22 23 24 25 26 33*) and two hierarchically nested communities (*27 28 29 35 37, 28 30 31 32 34 36 38*). To save space, only the first and fifth CCUAs of the Risk network are shown in Table 1.

The results are shown in Fig. 2 with ground truth communities represented in different colors and dashed lines representing the detected communities. There is only one node i.e., Middle East which has been misclassified. This misclassification is due to confusion about the link between the territories of Middle East and East Africa. The link was removed in Risk II and a prior edition, however was later confirmed to be a manufacturing error. The experiment on the Risk network affirmed the capability of the proposed algorithm in detecting the actual community structure within a network.

The fifth CCUAs of the Risk network consist of redundant, distinct and non-distinct clusters. This can be understood through the example shown in Table 2. Redundant clusters have all similar elements, distinct clusters have only three common elements whereas non-distinct clusters have seven common elements as shown in bold in Table 2. Fine-Tuning method is applied on the fifth CCUAs to obtain the final communities in the Risk network.

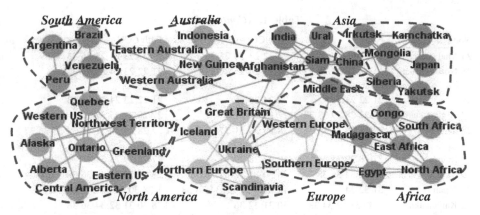

Fig. 2. Ground truth vs. detected communities in the risk network

Table 1. Experiment on the risk network

Nodes	NCS	First CCUA	Fifth CCUA
1 Alaska	1 2 6 32	1 7 31	1 2 3 4 5 6 7 8 9
2 Alberta	1 2 6 7 9	2 4 5 6 7	1 2 3 4 5 6 7 8 9
3 Central America	3 4 9 13	3 7	1 2 3 4 5 6 7 8 9
4 Eastern US	3 4 7 8 9	2 4 5 7 9	1 2 3 4 5 6 7 8 9
5 Greenland	5 6 7 8 15	2 4 5 7	1 2 3 4 5 6 7 8 9
6 Northwest Terr.	1 2 5 6 7	2 6 7 8 9	1 2 3 4 5 6 7 8 9
7 Ontario	2 4 5 6 7 8 9	1 2 3 4 5 6 7 8 9	1 2 3 4 5 6 7 8 9
8 Quebec	4 5 7 8	6 7 8 9	1 2 3 4 5 6 7 8 9
9 Western US	2 3 4 7 9	4 6 7 8 9	1 2 3 4 5 6 7 8 9
10 Argentina	10 11 12	10 11 12 13 25	10 11 12 13
11 Brazil	10 11 12 13 25	10 11 12	10 11 12 13
12 Peru	10 11 12 13	10 11 12	10 11 12 13
13 Venezuela	3 11 12 13	10 13	10 11 12 13
14 Great Britain	14 15 16 17 20	14 16 17 18 19	14 15 16 17 18 19 20
15 Iceland	5 14 15 17	15 16	14 15 16 17 18 19 20
16 Northern Europe	14 16 17 18 19 20	14 15 16 17 20	14 15 16 17 18 19 20
17 Scandinavia	14 15 16 17 19	14 16 17 18 20	14 15 16 17 18 19 20
18 Southern Europe	16 18 19 20 23 25 33	14 17 18 20 22 23	14 15 16 17 18 19 20 21 22 23
19 Ukraine	16 17 18 19 27 33 37	14 19 20 23 29	14 15 16 17 18 19 20 21 22 23
20 Western Europe	14 16 18 20 25	16 17 18 19 20 23	14 15 16 17 18 19 20 23
21 Congo	21 22 25 26	21 22 23 24	18 19 21 22 23 24 25 26 33
22 East Africa	21 22 23 24 25 26 33	18 21 22 23 24 26	18 19 21 22 23 24 25 26 33
23 Egypt	18 22 23 25 33	18 19 20 21 22 23 24 25 33	18 19 20 21 22 23 24 25 26 33

(Continued)

Table 1. (*Continued*)

Nodes	NCS	First CCUA	Fifth CCUA
24 Madagascar	22 24 26	21 22 23 24 25 26 33	21 22 23 24 25 26 33
25 North Africa	11 18 20 21 22 23 25	10 23 24 25 26 33	21 22 23 24 25 26 33
26 South Africa	21 22 24 26	22 24 25 26	21 22 23 24 25 26 33
27 Afghanistan	19 27 28 29 33 37	27 29 35 37	27 28 29 35 37
28 China	27 28 29 34 35 36 37	28 29 30 31 37	27 28 29 30 31 32 34 35 36 37 38
29 India	27 28 29 33 35	19 27 28 29 37	27 28 29 35 37
30 Irkutsk	30 32 34 36 38	28 30 31 32 34 36 38	28 30 31 32 34 36 38
31 Japan	31 32 34	1 28 30 31 32 34 36 38	1 28 30 31 32 34 36 38
32 Kamchatka	1 30 31 32 34 38	30 31 32 36	28 30 31 32 34 36 38
33 Middle East	18 19 22 23 27 29 33	23 24 25 33 37	21 22 23 24 25 26 33
34 Mongolia	28 30 31 32 34 36	30 31 34 37 38	28 30 31 32 34 36 38
35 Siam	28 29 35 40	27 35	27 28 29 35 37
36 Siberia	28 30 34 36 37 38	30 31 32 36	28 30 31 32 34 36 38
37 Ural	19 27 28 36 37	27 28 29 33 34 37	27 28 29 35 37
38 Yakutsk	30 32 36 38	30 31 34 38	28 30 31 32 34 36 38
39 East. Australia	39 41 42	39 40 41 42	39 40 41 42
40 Indonesia	35 40 41 42	39 40	39 40 41 42
41 New Guinea	39 40 41 42	39 41 42	39 40 41 42
42 West. Australia	39 40 41 42	39 41 42	39 40 41 42

Table 2. Example of redundant, distinct and non-distinct clusters in fifth CCUA of risk

Redundant clusters	Distinct clusters	Non-distinct clusters
14 15 16 17 18 19 20 21 22 23	14 15 16 17 **18 19** 20 **23**	**14 15 16 17 18 19 20**
14 15 16 17 18 19 20 21 22 23	**18 19** 21 22 **23** 24 25 26 33	**14 15 16 17 18 19 20** 21 22 23

5   Experimental Results

To evaluate the performance of proposed algorithm, experiments and comparative analysis were performed on networks with disjoint ground-truth communities: American political books [30], American football team [31] and Political blogs [32] as well as networks with hierarchical and overlapping ground-truth community structure: Karate club [33], Dolphin's association [34] and Les Miserables [35]. Further experiments were conducted on datasets with unknown ground truth viz. PPI network [36] and Roget's thesaurus [35]. The R version 3.2.3 running on a system with Intel Core i5 processor and 8.00 GB RAM was used for experiments. An overview of the experimental datasets, number of iterations required for convergence and execution time are shown in Table 3.

Table 3. Summary of experiments

Network dataset	# Nodes	# Edges	Avg. clustering coeff.	# Iterations	Execution time
Karate	34	78	.588	5	.63 s
Risk	42	165	.516	5	.47 s
Dolphin	62	159	.303	7	.94 s
Lesmis	77	254	.736	5	1.58 s
Polbooks	105	441	.488	6	1.93 s
Football	115	613	.403	6	1.14 s
Roget's Thesaurus	1022	5075	.160	10	6.83 s
Polblog	1490	19025	.361	7	4.97 min
Krogan's PPI	2708	7123	.277	11	8.64 min

The proposed algorithm identifies three communities on Polbooks network. The nodes were grouped into two main categories, liberals and conservatives. There were fewer misclassified nodes as compared to earlier methods [37, 38]. However, the books with no well-defined ideological proclivity (centrist) were roughly spread across three detected communities. The experiment on football network detected 10 communities with only 1 misclassification. However, 13 teams were not assigned to any community. These unassigned nodes consist of 7 teams of Sunbelt, 4 teams of Independents and 2 teams of Western Athletic. This is primarily because Sunbelt's and Independents' teams played as many games against the teams in other conferences as they did against the teams in their own conference. In similar cases, where the network structure does not truly correspond to the conference structure, our algorithm is unable to reveal the communities, which is consistent with our expectations. On the undirected version of Polblog network, algorithm generates two communities with 51 misclassifications.

The experiment on the Karate club network splits the network almost perfectly with no misclassification error. However, seven nodes have been classified in both the communities. This is in congruence with Zachary's observation that some of the individuals in the network were vacillated between the two sides, thus facilitating the information flow from one faction to another [33]. The experiment on the Dolphin network partitions the network into two communities with 21 and 41 nodes. Three sub-communities were detected in the larger community. The test on a network based on Victor Hugo's novel (Les Miserables), identified four overlapping communities. Fantine and Valjean were correctly detected as members of two different communities [39]. Other critical nodes were also correctly identified by the proposed algorithm as shown in Fig. 3.

Proposed Algorithm was also tested on a PPI network of the yeast Sacchromyces Cerevisiae. Since the ground truth community structure of this network is unknown, we compared results of the proposed algorithm with the results of a previous study on this network [36]. The protein complexes (communities) produced on PPI network were found to be very similar to that of Markov Clustering procedure. Further, we used an association network of English words and phrases from the 1879 edition of Roget's

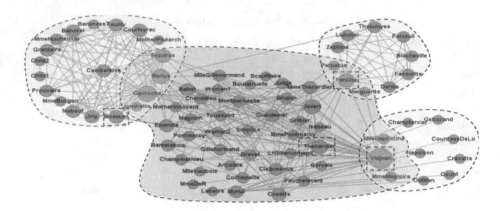

Fig. 3. Four overlapping communities detected in Les Miserables network

Thesaurus for experimental purpose. The detected community structure was found to be consistent with an algorithm based on bibliometric approach [40]. The visualization of results is shown in Appendix. In the interest of saving space, only a part of the community structure of PPI and Roget's Thesaurus network has been visualized.

5.1 Performance Comparison of the Proposed Algorithm

The performance of the proposed algorithm in detecting disjoint community structure has been compared with two algorithms which optimize conductance viz., PageRank based algorithm (PR) [4] and heat kernel based algorithm (HK) [5]. Since conductance is considered to be the most robust score when a network contains discrete communities, both the above algorithms achieve results closer to the ground truth. The average conductance of detected communities has been used as an evaluation metric.

Extant literature on overlapping community detection algorithms has commonly used partition density as one of the evaluation criteria [41]. Partition density emphasizes on the link density within the community rather than the conventional notion that intra-community density of links should be larger than inter-community density. For a network with M links, suppose $p = |C|$ communities are detected in a network, such that $C_1, C_2 ..... C_p \in G$. D refers to the partition density and it is defined as follows:

$$D = \frac{2}{M} \sum_i \frac{l_i \{ l_i - (n_i - 1) \}}{(n_i - 2)(n_i - 1)} \tag{5}$$

where l_i denotes the number of edges within C_i, n_i denotes number of nodes within C_i and i varies from 1 to p. We have compared the performance of the proposed algorithm with popular overlapping community detection algorithms: OSLOM [10], SBMF [12] and BNMF [13]. Partition density has been used as an evaluation metric.

Table 4. Comparison of average conductance on networks with disjoint ground truth

Networks	Ground truth	PR	HK	Proposed algorithm
Polbooks	0.322	0.571	0.127	**0.122**
Football	0.392	0.385	**0.244**	0.320
Polblogs	0.097	0.867	0.229	**0.113**

The best result for each dataset has been indicated in bold in Tables 4 and 5. The average conductance and partition density scores of the proposed algorithm are either best or second best amongst the other relevant algorithms used for performance comparison. The second best values in case of Football and Lesmis network are due to comparatively lesser number of communities detected in these networks.

Table 5. Comparison of partition density on networks with overlapping ground truth

Networks	OSLOM	SBMF	BNMF	Proposed algorithm
Karate	.002	.103	.131	**0.105**
Dolphin	.089	−.017	−.014	**0.113**
Lesmis	**.308**	.069	.049	0.266

6   Conclusion

In this study, we proposed a rough set based algorithm for community detection in complex networks. Our algorithm applies the concept of constrained connectedness and upper approximation to cluster the nodes of a network. Further, we introduced a relative connectedness metric which is used as the merging criteria during the iterative process. Experiments on various real world networks followed by comparative analysis show that the proposed approach effectively reveals the community structure in networks.

To this end, we have used unweighted networks to detect communities. One of the avenues of further research is to use the concept of rough membership function to detect communities in weighted networks. As a future work, a recommender system can be designed using the proposed approach to predict links between customers and products. In this case product reviews can be mined to compute the relative connectedness between customers and products.

Appendix

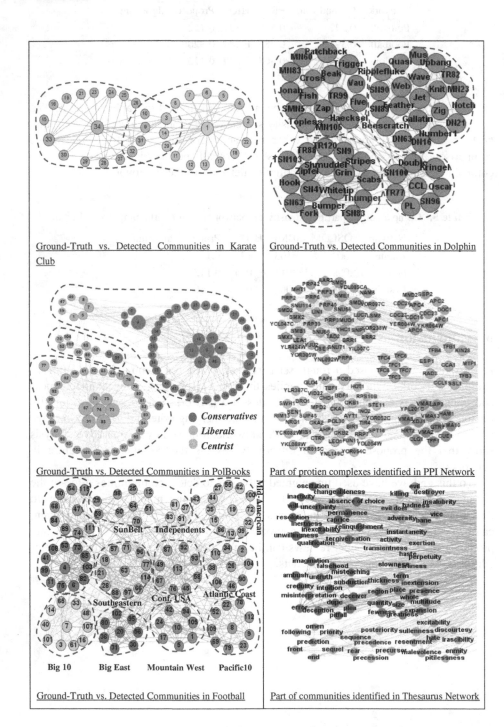

Ground-Truth vs. Detected Communities in Karate Club

Ground-Truth vs. Detected Communities in Dolphin

Ground-Truth vs. Detected Communities in PolBooks

- Conservatives
- Liberals
- Centrist

Part of protien complexes identified in PPI Network

Ground-Truth vs. Detected Communities in Football

Part of communities identified in Thesaurus Network

References

1. Steinhaeuser, K., Chawla, N.V.: Identifying and evaluating community structure in complex networks. Pattern Recogn. Lett. **31**, 413–421 (2010)
2. Fortunato, S.: Community detection in graphs. Phys. Rep. **486**, 75–174 (2010)
3. Newman, M.E.J., Girvan, M.: Finding and evaluating community structure in networks. Phys. Rev. E **69**, 26113 (2004)
4. Yang, J., Leskovec, J.: Defining and evaluating network communities based on ground-truth. Knowl. Inf. Syst. **42**, 181–213 (2015)
5. Kloster, K., Gleich, D.F.: Heat kernel based community detection. In: 20th International Conference on Knowledge Discovery and Data Mining, pp. 1386–1395. ACM (2014)
6. Palla, G., Derényi, I., Farkas, I., Vicsek, T.: Uncovering the overlapping community structure of complex networks in nature and society. Nature **435**, 814–818 (2005)
7. Gregory, S.: An algorithm to find overlapping community structure in networks. In: Kok, J. N., Koronacki, J., Lopez de Mantaras, R., Matwin, S., Mladenič, D., Skowron, A. (eds.) PKDD 2007. LNCS (LNAI), vol. 4702, pp. 91–102. Springer, Heidelberg (2007)
8. Shen, H., Cheng, X., Cai, K., Hu, M.-B.: Detect overlapping and hierarchical community structure in networks. Phys. Stat. Mech. Appl. **388**, 1706–1712 (2009)
9. Ball, B., Karrer, B., Newman, M.E.J.: Efficient and principled method for detecting communities in networks. Phys. Rev. E **84**, 36103 (2011)
10. Lancichinetti, A., Radicchi, F., Ramasco, J.J., Fortunato, S.: Finding statistically significant communities in networks. PLoS ONE **6**, e18961 (2011)
11. Lancichinetti, A., Fortunato, S., Kertész, J.: Detecting the overlapping and hierarchical community structure in complex networks. New J. Phys. **11**, 33015 (2009)
12. Zhang, Z.-Y., Wang, Y., Ahn, Y.-Y.: Overlapping community detection in complex networks using symmetric binary matrix factorization. Phys. Rev. E **87**, 62803 (2013)
13. Psorakis, I., Roberts, S., Ebden, M., Sheldon, B.: Overlapping community detection using bayesian non-negative matrix factorization. Phys. Rev. E **83**, 66114 (2011)
14. McDaid, A., Hurley, N.: Detecting highly overlapping communities with model-based overlapping seed expansion. In: 2010 International Conference on Advances in Social Networks Analysis and Mining (ASONAM), pp. 112–119. IEEE (2010)
15. Latouche, P., Birmelé, E., Ambroise, C.: Overlapping stochastic block models with application to the french political blogosphere. Ann. Appl. Stat. **5**(1), 309–336 (2011)
16. Gregory, S.: Finding overlapping communities in networks by label propagation. New J. Phys. **12**, 103018 (2010)
17. Xie, J., Szymanski, B.K., Liu, X.: SLPA: Uncovering overlapping communities in social networks via a speaker-listener interaction dynamic process. In: Data Mining Workshops (ICDMW), pp. 344–349. IEEE (2011)
18. Huang, J., Sun, H., Han, J., Feng, B.: Density-based shrinkage for revealing hierarchical and overlapping community structure in networks. Phys. Stat. Mech. Appl. **390**, 2160–2171 (2011)
19. Nepusz, T., Petróczi, A., Négyessy, L., Bazsó, F.: Fuzzy communities and the concept of bridgeness in complex networks. Phys. Rev. E **77**, 16107 (2008)
20. Zhang, S., Wang, R.-S., Zhang, X.-S.: Identification of overlapping community structure in complex networks using fuzzy c-means clustering. Phys. Stat. Mech. Appl. **374**, 483–490 (2007)
21. Xie, J., Kelley, S., Szymanski, B.K.: Overlapping community detection in networks: the state-of-the-art and comparative study. ACM Comput. Surv. **45**, 1–35 (2013)
22. Pawlak, Z.: Rough sets. Int. J. Comput. Inf. Sci. **11**, 341–356 (1982)

23. Kumar, P., Krishna, P.R., Bapi, R.S., De, S.K.: Rough clustering of sequential data. Data Knowl. Eng. **63**, 183–199 (2007)
24. Lingras, P., West, C.: Interval set clustering of web users with rough k-means. J. Intell. Inf. Syst. **23**, 5–16 (2004)
25. Kundu, S., Pal, S.K.: Fuzzy-rough community in social networks. Pattern Recogn. Lett. **67**, 145–152 (2015)
26. Zadeh, L.A.: Toward a theory of fuzzy information granulation and its centrality in human reasoning and fuzzy logic. Fuzzy Sets Syst. **90**, 111–127 (1997)
27. Gleich, D.F., Seshadhri, C.: Vertex neighborhoods, low conductance cuts, and good seeds for local community methods. In: 18th ACM SIGKDD International Conference on Knowledge Discovery and Data Mining, pp. 597–605. ACM (2012)
28. Grzymala-Busse, J.W.: LERS-a system for learning from examples based on rough sets. In: Słowiński, R. (ed.) Intelligent Decision Support. Theory and Decision Library, vol. 11, pp. 3–18. Springer, Netherlands (1992)
29. Dash, M., Liu, H., Scheuermann, P., Tan, K.L.: Fast hierarchical clustering and its validation. Data Knowl. Eng. **44**, 109–138 (2003)
30. Krebs, V.: Books about US Politics (2004). http://www.orgnet.com
31. Girvan, M., Newman, M.E.: Community structure in social and biological networks. Proc. Nat. Acad. Sci. **99**, 7821–7826 (2002)
32. Adamic, L.A., Glance, N.: The political blogosphere and the 2004 us election: divided they blog. In: 3rd International Workshop on Link Discovery, pp. 36–43. ACM (2005)
33. Zachary, W.W.: An information flow model for conflict and fission in small groups. J. Anthropol. Res. **33**(4), 452–473 (1977)
34. Lusseau, D., Schneider, K., Boisseau, O.J., Haase, P., Slooten, E., Dawson, S.M.: The bottlenose dolphin community of doubtful sound features a large proportion of long-lasting associations. Behav. Ecol. Sociobiol. **54**, 396–405 (2003)
35. Knuth, K.E.: The Stanford GraphBase: A Platform for Combinatorial Computing. Addison-Wesley, Reading (1993)
36. Krogan, N.J., Cagney, G., Yu, H., Zhong, G., Guo, X., Ignatchenko, A., Li, J., Pu, S., Datta, N., Tikuisis, A.P., et al.: Global landscape of protein complexes in the yeast Saccharomyces cerevisiae. Nature **440**, 637–643 (2006)
37. Newman, M.E.: Finding community structure in networks using the eigenvectors of matrices. Phys. Rev. E **74**, 36104 (2006)
38. Agarwal, G., Kempe, D.: Modularity-maximizing graph communities via mathematical programming. Eur. Phys. J. B **66**, 409–418 (2008)
39. He, D., Jin, D., Chen, Z., Zhang, W.: Identification of hybrid node and link communities in complex networks. Sci. Rep. **5**, 8638 (2015)
40. Balakrishnan, H., Deo, N.: Discovering communities in complex networks. In: 44th Annual Southeast Regional Conference, pp. 280–285. ACM (2006)
41. Ahn, Y.-Y., Bagrow, J.P., Lehmann, S.: Link communities reveal multiscale complexity in networks. Nature **466**, 761–764 (2010)

Applications of Big Data Mining I

Mining User Trajectories from Smartphone Data Considering Data Uncertainty

Yu Chi Chen[1]([✉]), En Tzu Wang[2], and Arbee L.P. Chen[3]

[1] Department of Computer Science, National Tsing Hua University,
Hsinchu, Taiwan
vickychen565@gmail.com
[2] Computational Intelligence Technology Center, Industrial Technology
Research Institute, Hsinchu, Taiwan
m9221009@em92.ndhu.edu.tw
[3] Department of Computer Science and Information Engineering,
Asia University, Taichung, Taiwan
arbee@asia.edu.tw

Abstract. Wi-Fi hot spots have quickly increased in recent years. Accordingly, discovering user positions by using Wi-Fi fingerprints has attracted much research attention. Wi-Fi fingerprints are the sets of Wi-Fi scanning results recorded in mobile devices. However, the issue of data uncertainty is not considered in the proposed Wi-Fi positioning systems. In this paper, we propose a framework to find user trajectories from the Wi-Fi fingerprints recorded in the smartphones. In this framework, we first discover meaningful places with the proposed Wi-Fi distance metric. Second, we propose two similarity functions to recognize the places and show the probabilities of the places where a user stayed in by the proposed uncertain data models. Finally, an algorithm on probabilistic sequential pattern mining is used for finding user trajectories. A series of experiments are performed to evaluate each step of the framework. The experiment results reveal that each step of our framework is with high accuracy.

1 Introduction

The number of mobile devices has quickly increased in recent years. Mobile devices are often equipped with position-aware modules such as Global Positioning System (GPS) and Wi-Fi system. The modules can help to collect mobile data of the corresponding devices. Several previous works use digital traces to understand the owners of the mobile devices. An algorithm is proposed in [2] to discovery places from the logs of Wi-Fi signals, collected by a mobile device. Fan et al. [3] propose a scheme for mapping the observed Wi-Fi signals to human-defined places. Li et al. [4] propose an algorithm to mine periodic behaviors for moving objects. Lee et al. [5] analyze customer behaviors in the shopping mall.

These position-aware modules have different characteristics. GPS provides a worldwide coverage excluding indoor and underground locations. The positioning system based on Wi-Fi provides rough location estimation when wireless access is available. Data uncertainty is inherent in the Wi-Fi positioning system. Many previous

© Springer International Publishing Switzerland 2016
S. Madria and T. Hara (Eds.): DaWaK 2016, LNCS 9829, pp. 51–67, 2016.
DOI: 10.1007/978-3-319-43946-4_4

works focus on discovering places by using the strength variation of Wi-Fi signals [2, 3], but the uncertainty of the places caused by the Wi-Fi positioning system is not considered. In addition, there are many different meanings of probabilistic sequential patterns [6, 7] because of different objectives. In Fan et al. [1], we proposed a data cleaning and information enrichment framework for enabling user preference understanding through collected Wi-Fi logs. Continuing this effort, in this paper, we build an uncertain data model from *Wi-Fi fingerprints*, which are the sets of Wi-Fi scanning results recorded in user mobile devices, to mine user trajectories. User trajectories, defined as the probabilistic sequential patterns in this paper, are the paths that the users frequently follow. From user trajectories, we understand user movement behaviors and are able to predict user movements. On the other hand, the trajectories of the same type of users are useful for recommendation. For example, when a traveler arrives in Hsinchu City for the first time, we can recommend her/him a traveler trajectory, mined from the travelers in Hsinchu City.

We propose an overall framework to mine user trajectories from the logs of Wi-Fi signals recorded in user smartphones. Four steps, including data collection, place detection, building uncertain models, and mining probabilistic sequential patterns are involved in this framework. First, we develop a monitoring app to collect the Wi-Fi scanning results, including the name (SSID), signal strength (RSSI), and the unique media access control address (BSSID) of the Wi-Fi access points and recruit participants to annotate places for building the Wi-Fi reference database. Next, we propose a novel distance metric based on the Wi-Fi fingerprints and then apply the K-Means clustering algorithm [8] to detect places. We then define two similarity functions to compute the similarity between a cluster and the place records in the Wi-Fi reference database and also design uncertain models using these similarities to compute the probabilities of the place where a user stayed in. Finally, we propose an algorithm based on PrefixSpan [9] to mine the probabilistic sequential patterns. To the best of our knowledge, it is the first work to build the uncertain models from real data rather than predefining them [6, 7], for mining user trajectories.

The remainder of this paper is organized as follows. The related works are reviewed in Sect. 2. The preliminaries including the Wi-Fi fingerprints, uncertain sequence model, probabilistic sequential patterns, and problem statement are described in Sect. 3. The proposed approach is presented in Sect. 4. Thereafter, the evaluation of each step of our framework and the probabilistic sequential patterns mined are reported in Sect. 5. Finally, Sect. 6 concludes this work.

2   Related Work

In recent years, lots of research [4, 5] use digital traces collected from position-aware sensors of mobile devices to understand user behaviors. To the best of our knowledge, we are the first approach focusing on building an uncertain model to mine user trajectories from the daily life data collected by mobile devices. There are many Wi-Fi access points in the cities. Due to different specifications of the Wi-Fi access points, we cannot get precise places. [2, 3] propose algorithms to detect the places by using the fingerprint-based algorithm, which is widely used for indoor localization. Kim et al. [2] present a place discovery algorithm, named PlaceSense, for the logs of Wi-Fi access points of a mobile device. PlaceSense discovers a place in two steps. The first step is to detect a stable Wi-Fi signal environment that indicates an arrival at a place. The second step is to detect the time of the Wi-Fi signal change that indicates a departure from a place. According to these two steps, PlaceSense needs to check the changes of Wi-Fi signal periodically. Therefore, PlaceSense collects GPS, Wi-Fi, and GSM traces every 10 s. Considering the consumption of smartphone battery, we collect Wi-Fi logs when the battery or the state of the screen changes rather than every 10 s. Therefore, the Wi-Fi logs are not recorded periodically and PlaceSense cannot be used to discover places from the Wi-Fi logs in this condition.

Fan et al. [3] propose a scheme by employing a crowdsourcing model to perform place name annotations by mobile participants to bridge the gap between Wi-Fi signals and human-defined places. This research uses Euclidean distance as the distance metric between Wi-Fi fingerprints and applies DBSCAN clustering algorithm to discover the places. However, they do not mention how to adjust two parameters of the DBSCAN clustering algorithm, i.e. ϵ (maximum radius of the neighborhood) and MinPts (minimum number of points in an ϵ-neighborhood). On the contrary, we apply the K-Means clustering algorithm [8] in our framework and propose a novel distance metric and two similarity functions for Wi-Fi fingerprints to discover the places. In the experiments, we use the purity, entropy and SSE (sum of squared error) to set the parameter K in K-Means clustering algorithm and perform a series of experiments to evaluate our distance metric and the proposed similarity functions. Considering the RFID location tracking system, [7] establishes a data model in which each location element in the traces is represented by a set of pairs (location l, probability p) where probability p is predefined. It focues on proposing an efficient and effective mining algorithm. Compared to [7], we propose an overall framework to mine user trajectories form smartphone data, including collecting data, building uncertain models and mining probabilistic sequential patterns.

3 Preliminaries

In this section, we describe the notations to be used and formally define the patterns we want to mine from the data. Basic Service Set IDentification (BSSID) is the MAC address of the Wi-Fi access point. Received Signal Strength Indicator (RSSI) is a measurement of the signal strength of a Wi-Fi access point. Accordingly, a Wi-Fi fingerprint in this paper is a set of tuples (Timestamp, BSSID, RSSI) recording all the Wi-Fi Access Points observed by a smartphone at the same timestamp.

Definition 1 (Wi-Fi Fingerprint Vector): Suppose that there are n different Wi-Fi access points. A smartphone will detect at most n access points at the same time. We represent a Wi-Fi fingerprint by a vector with a length of n, which lists the RSSI values of the n Wi-Fi access points. The Wi-Fi fingerprint vector is represented as $f = <f_1, f_2, f_3, ..., f_n>$ where f_i denotes the RSSI value of the i-th Wi-Fi access point. The range of f_i is within $(-100, 0)$. The greater value of f_i indicates a stronger signal and $f_i = -100$, if the i-th Wi-Fi access point cannot be detected. ∎

Table 1. A Wi-Fi fingerprint vector.

ID	BSSID
1	68:b6:fc:b9:59:a8
2	00:0b:86:3d:f6:50
3	28:10:7b:cf:79:4e
4	d8:fe:e3:d6:e0:45
5	00:0b:86:38:de:e0

Timestamp	BSSID	RSSI
2013/12/1 0:03:03	68:b6:fc:b9:59:a8	-72
2013/12/1 0:03:03	28:10:7b:cf:79:4e	-80
2013/12/1 0:03:03	d8:fe:e3:d6:e0:45	-74

(a) Wi-Fi access points   (b) Wi-Fi fingerprint f

Example 1: Assume there are 5 Wi-Fi access points. The Wi-Fi fingerprint f is shown in Table 1(b). Since the Wi-Fi access points 2 and 5 are not detected the Wi-Fi fingerprint is represented as $f = <-72, -100, -80, -74, -100>$. ∎

A user trajectory in this paper is represented by a sequence of meaningful places. Each place is determined by the Wi-Fi fingerprints of the smartphone held by a user. Due to the uncertainty of recognizing a place by the Wi-Fi fingerprints, each place is therefore associated with a probability.

Definition 2 (Uncertain Sequence Model): We build an uncertain data model to represent the probabilities of the places where a user stayed in. We transform the Wi-Fi data collected by user smartphones into an uncertain sequence model. An uncertain sequence is a sequence of place elements with probabilities. A probabilistic place element of a sequence is a set of pairs (related to a place element e and a probability p, where p denotes the probability of a user who stayed in e). The sum of the probabilities of the probabilistic place element must be equal to 1. ∎

Example 2: In Table 2, sequences s_1 and s_2 record the users partial trajectories of the first day and the second day, respectively. s_1 contains two uncertain location elements $s_1[1]$ and $s_1[2]$. $s_1[1]$ is with a probability of 70 % to be A and a probability of 30 % to be B. ∎

Table 2. Uncertain sequence model

SID	Probabilistic elements
S_1	$<s_1[1] = \{(A,0.7), (B,0.3)\}, s_1[2] = \{(B,0.8), (C,0.2)\}>$
S_2	$<s_2[1] = \{(A,1)\}, s_2[2] = \{(B,0.8), (C,0.2)\}, s_2[3] = \{(A,0.6), (C,0.4)\}>$

Definition 3 (Support of a Pattern): The support of a pattern α is the number of sequences in which α occurs, and denoted as $sup(\alpha)$. ∎

Example 3: As shown in Table 2, pattern AB appears in s_1 and s_2 and $sup(AB)$ is equal to 2. Pattern BA only appears in s_2 and $sup(BA)$ is equal to 1. ∎

Definition 4 (Probability of a Pattern): The probability of a pattern α in a sequence s_1 is the maximum probability of α occurring in s_1, and denoted as $Pr_{S1}(\alpha)$. ∎

Example 4: In Table 2, $Pr_{S1}(AB) = 0.7 \times 0.8 = 0.56$ and $Pr_{S2}(AB) = 0.8$. ∎

Problem Statement: Given a set of uncertain sequences, a support threshold τ_{sup}, and a probability threshold τ_{prob}, a probabilistic sequential pattern α has to satisfy both of the probability threshold and support threshold as follows:

$$\left|\{i|Pr_i(\alpha) \geq \tau_{prob}, \forall 1 \leq i \leq n\}\right| \geq \tau_{sup}$$

In this paper, we propose a novel framework to find the probabilistic sequential patterns from the Wi-Fi logs collected from user smartphones.   ∎

Example 5: As shown in Table 2, let the support threshold τ_{sup} be 2 and the probability threshold τ_{prob} be 0.5, pattern AB is a probabilistic sequential pattern.   ∎

4 The Proposed Approach

The flowchart of our approach is shown in Fig. 1. Four steps, including data collection, place detection, building uncertain models, and mining probabilistic sequential patterns are involved in this framework.

4.1 Data Collection

Wi-Fi Log. Raw data are collected from the participant Android smartphones installing a system monitoring app, which runs as a background service. This app records available Wi-Fi signals when the battery or the state of the screen changes and regularly uploads data to the server site. Raw data are a collection of Wi-Fi fingerprints composed of a set of tuples (Timestamp, BSSID, RSSI) and a participant ID. An example is shown in Table 3.

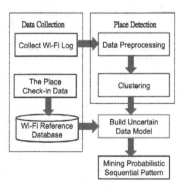

Table 3. A part of raw data

	Timestamp	Participant ID	BSSID	RSSI
1	2013/10/24 22:24:05	353567051353390	00:90:cc:f7:9d:68	−48
2	2013/10/24 22:24:05	353567051353390	20:c9:d0:1f:1e:e7	−92
3	2013/10/24 22:24:05	353567051353390	00:04:e2:dd:80:e8	−71
4	2013/10/24 22:24:05	353567051353390	00:a0:c5:80:06:b6	−85
5	2013/10/24 22:24:05	353567051353390	00:0b:86:3d:f6:51	−87
6	2013/10/24 22:24:14	353567051353101	00:90:cc:f7:9d:68	−40
7	2013/10/24 22:24:14	353567051353101	20:c9:d0:1f:1e:e7	−74

Fig. 1. Flowchart of our approach

The Place Checked-in Data. For recognizing the places where the participants stayed in, we build the Wi-Fi reference database to store the Wi-Fi fingerprints of the places visited by participants frequently. We recruit 10 participants in our campus to use the app for annotating and collecting the Wi-Fi fingerprints of 52 places, visited by participants frequently. The app records GPS coordinates and scans Wi-Fi signals every 15 s. The participants annotate where they stayed in by the pop-up window of the app, as shown in Fig. 2 and walk around in a place over 10 min. The participants annotate some Wi-Fi fingerprints as the same name, but the GPS coordinates of these Wi-Fi fingerprints do not point to the same location. As shown in Fig. 3, we mark the GPS coordinates of the Wi-Fi fingerprints annotated as "Make Lab" on the map. Since most places visited are indoor, the GPS positioning system only get rough coordinates, detecting places by the Wi-Fi fingerprints in this case is more precise.

4.2   Place Detection

Data Preprocessing. A Wi-Fi fingerprint is composed of a set of the Wi-Fi signals, observed at the same timestamp. In the procedure of data preprocessing, we transform the Wi-Fi logs to a set of Wi-Fi fingerprint vectors defined in Definition 1. For validating our approach, we collect one-month Wi-Fi logs of 10 participants to be the datasets in the experiments. There are a total of 25276 distinct Wi-Fi access points and 22610 Wi-Fi fingerprints in the Wi-Fi reference database. The numbers of Wi-Fi fingerprints are as shown in Table 4.

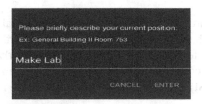

Fig. 2. The pop-up window of the app.

Fig. 3. The points annotated as MAKE Lab.

Table 4. The number of Wi-Fi fingerprints in the datasets

	Number of Wi-Fi fingerprints
Wi-Fi reference database	22610
Dataset 1	1802
Dataset 2	3344
Dataset 3	5890
Dataset 4	5444
Dataset 5	7402
Dataset 6	4857
Dataset 7	2154
Dataset 8	6186
Dataset 9	10529
Dataset 10	12653

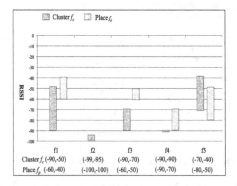

Fig. 4. The representatives of a cluster and a place

Clustering. Due to the properties of Wi-Fi, the Wi-Fi fingerprints of a place are similar rather than exactly the same. Accordingly, we apply clustering analysis to discover the places. Given a set of Wi-Fi fingerprints, the clustering analysis is modeled as follows. We first propose a novel Wi-Fi fingerprint distance. Given a Wi-Fi fingerprint $f_a = <f_{a1}, f_{a2}, f_{a3}, \ldots, f_{an}>$, S_{f_a} is defined to be the set $\{i | f_{ai} \neq -100, \forall : 1 \leq i \leq n\}$, denoting which Wi-Fi access points are detected in Wi-Fi fingerprint f_a. For two Wi-Fi fingerprints $f_a = <f_{a1}, f_{a2}, f_{a3}, \ldots, f_{an}>$ and $f_b = <f_{b1}, f_{b2}, f_{b3}, \ldots, f_{bn}>$, the distance $d(f_a, f_b)$ between f_a and f_b is computed by

$$d(f_a, f_b) = 1 - \frac{|S_{f_a} \cap S_{f_b}|}{|S_{f_a} \cup S_{f_b}|} \times \left(1 - \sqrt{\frac{\sum_{i \in S_{f_a} \cap S_{f_b}} (f_{ai} - f_{bi})^2}{|S_{f_a} \cap S_{f_b}| \times 100^2}} \right)$$

The first item $\frac{|S_{f_a} \cap S_{f_b}|}{|S_{f_a} \cup S_{f_b}|}$ denotes the size of the intersection of Wi-Fi access points which are detected both in f_a and f_b. The second item $1 - \sqrt{\frac{\sum_{i \in S_{f_a} \cap S_{f_b}} (f_{ai} - f_{bi})^2}{|S_{f_a} \cap S_{f_b}| \times 100^2}}$ denotes the depth of this intersection. If we want to use this Wi-Fi fingerprint distance, there are two properties must be satisfied: positivity and symmetry.

(1) **Positivity:** When $f_a = f_b$, $d(f_a, f_b) = 1 - 1 \times (1 - 0) = 0$. $\frac{|S_{f_a} \cap S_{f_b}|}{|S_{f_a} \cup S_{f_b}|}$ and $\left(1 - \sqrt{\frac{\sum_{i \in S_{f_a} \cap S_{f_b}} (f_{ai} - f_{bi})^2}{|S_{f_a} \cap S_{f_b}| \times 100^2}} \right)$ must be greater than or equal to 0 and less than or equal to 1. Above these, $d(f_a, f_b) \geq 0$ and $d(f_a, f_b) = 0$ only when $f_a = f_b$.

(2) **Symmetry:** $d(f_a, f_b) = 1 - \frac{|S_{f_a} \cap S_{f_b}|}{|S_{f_a} \cup S_{f_b}|} \times \left(1 - \sqrt{\frac{\sum_{BSSID_x \in f_a \cap f_b} (a_x - b_x)^2}{|S_{f_a} \cap S_{f_b}| \times 100^2}} \right) = 1 - \frac{|S_{f_b} \cap S_{f_a}|}{|S_{f_b} \cup S_{f_a}|} \times \left(1 - \sqrt{\frac{\sum_{BSSID_x \in f_b \cap f_a} (b_x - a_x)^2}{|S_{f_b} \cap S_{f_a}| \times 100^2}} \right) = d(f_b, f_a)$

After designing the distance metric, we are able to perform clustering analysis. Since the Wi-Fi signal range is a circle whose center is the location of the Wi-Fi access point, we apply a center-based clustering algorithm K-Means [8]. We adjust k by observing *entropy*, *purity*, and *SSE* (sum of squared error) of the clustering results. The clustering results are regarded as places frequently visited by participants.

4.3   Building Uncertain Models

In this step, we want to know the Wi-Fi fingerprint indicates which place in the Wi-Fi reference database and then denotes this answer as an uncertain model. We propose two methods, the centroid-based and ranged-based methods, to represent a cluster. We also propose two similarity functions for centroid-based and ranged-base method. We compute similarities between the representative of a cluster and the places in the Wi-Fi reference database. Similarity means how similar between a cluster and a place and we use similarity to build uncertain models.

The Centroid-Based Method. In this method, we use the centroid of the cluster that is the Wi-Fi fingerprint in to be the representative. Also, we use the average of the Wi-Fi fingerprint vectors to be the representative for each place in the Wi-Fi reference database. Given the centroid of cluster $f_a = <f_{a1}, f_{a2}, f_{a3}, \ldots, f_{an}>$ and the representative of *placeA* $f_A = <f_{A1}, f_{A2}, f_{A3}, \ldots, f_{An}>$, the similarity between f_a and f_A is computed by

$$sim(f_a, f_A) = \frac{|S_{f_a} \cap S_{f_A}|}{|S_{f_a} \cup S_{f_A}|} \times \left(1 - \sqrt{\frac{\sum_{i \in S_{f_a} \cap S_{f_A}} (f_{ai} - f_{Ai})^2}{|S_{f_a} \cap S_{f_A}| \times 100^2}} \right)$$

Example: Given the centroid of cluster $f_c = <-70, -97, -80, -90, -60>$ and the representative of *placeA* $f_A = <-50, -100, -55, -80, -70>$, the similarity between f_c and f_A is computed as follows.

$$sim(f_c, f_p) = \frac{4}{5 + 4 - 4} \times \left(1 - \sqrt{\frac{20^2 + 25^2 + 10^2 + 10^2}{4 \times 100^2}} \right) = 0.8 \times 0.825 = 0.66$$

∎

The Range-Based Method. In this method, we use the minimum and the maximum RSSI of each Wi-Fi access point in a set of Wi-Fi fingerprints to be a representative. According to above point, $f = <(f_{1\ min}, f_{1\ max}), (f_{2\ min}, f_{2\ max}), \ldots, (f_{n\ min}, f_{n\ max})>$ denotes the representative of a set of Wi-Fi fingerprints and we use this formula to represent a cluster and the places in the Wi-Fi reference database. $f_{i\ min}$ denotes the minimum RSSI of the i-th Wi-Fi access point in a set of Wi-Fi fingerprints. $f_{i\ max}$ denotes the maximum RSSI of the i-th Wi-Fi access point in a set of Wi-Fi fingerprints. Given a representative f_a of a set of Wi-Fi fingerprints, S_{f_a} is defined to be the set $\{i | f_{ai\ max} \neq -100, \forall : 1 \leq i \leq n\}$, denoting which Wi-Fi access points are detected in this set of Wi-Fi fingerprints. Given the representative f_c of a cluster that is the Wi-Fi fingerprint in and the representative f_p of a place which is in the Wi-Fi reference database, the similarity between f_c and f_p is computed by

$$sim(f_c, f_p) = \frac{|S_{f_c} \cap S_{f_p}|}{|S_{f_c} \cup S_{f_p}|} \times \frac{\sum_{i \in S_{f_a} \cap S_{f_A}} intersection(f_{ci}, f_{pi})}{\sum_{i \in S_{f_a} \cap S_{f_A}} \left(\max(f_{ci\ max}, f_{pi\ max}) - \min(f_{ci\ min}, f_{pi\ min}) + 1 \right)}$$

$$intersection(f_{ci}, f_{pi}) = \begin{cases} 0, & \min(f_{ci\ max}, f_{pi\ max}) < \max(f_{ci\ min}, f_{pi\ min}) \\ \min(f_{ci\ max}, f_{pi\ max}) - \max(f_{ci\ min}, f_{pi\ min}) + 1, & \min(f_{ci\ max}, f_{pi\ max}) \geq \max(f_{ci\ min}, f_{pi\ min}) \end{cases}$$

The first item $\frac{|S_{fc} \cap S_{fp}|}{|S_{fc} \cup S_{fp}|}$ of similarity function denotes the size of the intersection of Wi-Fi access points which are detected both in the cluster and the place. The second item denotes the overlap of two sets of Wi-Fi fingerprints.

Example: As shown in Fig. 4, there are a representative $f_c = <(-90, -50), (-99, -95), (-90, -70), (-90, -90), (-70, -40)>$ of the cluster and a representative $f_p = <(-60, -40), (-100, -100), (-60, -50), (-90, -70), (-80, -50)>$ of the place. The similarity between f_c and f_p is computed as follows.

$$sim(f_c, f_p) = \frac{4}{5 + 4 - 4} \times \frac{11 + 0 + 1 + 21}{51 + 41 + 21 + 41} = 0.8 \times 0.21 = 0.17$$

∎

Normalization. After computing the similarity between the clusters and the Wi-Fi reference database, we prune the places with related similarities lower than 0.05 and normalize the similarities to build an uncertain model of this Wi-Fi fingerprint. This uncertain model shows the probabilities of the places, which the Wi-Fi fingerprint indicates to. With the procedure of building the uncertain models, we can transform the Wi-Fi logs into the uncertain sequence models is defined in Definition 2.

4.4   Pattern Mining

After generating the uncertain sequence models, we want to mine the patterns defined in the problem statement. In this paper, we focus on an overall framework to mine user trajectories from smartphone data rather than the mining algorithm. We propose an approach based on PrefixSpan algorithm [9] to mine the patterns.

Table 5. Prune the elements with low probabilities

SID	Probabilistic elements
S_1	$<s_1[1] = \{(A,0.7)\}, s_1[2] = \{(B,0.8)\}>$
S_2	$<s_2[1] = \{(A,1)\}, s_2[2] = \{(B,0.8)\},$ $s_2[3] = \{(A,0.6)\}>$

Table 6. Find candidate patterns

Candidate pattern	SID
A	S_1, S_2
B	S_1, S_2
A -> B	S_1, S_2

Prune the Elements with Low Probabilities. If $Pr(\alpha) \geq \tau_{prob}$, all of the elements in α must have a probability greater than or equal to τ_{prob}. According to the above point, we prune the elements whose probabilities are smaller than τ_{prob}. This step can make sequences shorter and reduce amounts of computation in the following steps.

Example: Given an uncertain sequence model as shown in Table 2 and $\tau_{prob} = 0.5$, we prune the elements which probability is lower than 0.5. After this step, the uncertain model is shown as Table 5. ∎

Find Candidate Patterns. In this step, we apply the PrefixSpan algorithm [9] with a minimum support threshold τ_{sup} to find the candidate patterns and also record which sequences contain the candidate patterns. Candidate patterns are the patterns satisfying the support threshold, i.e. $sup(\alpha) \geq \tau_{sup}$.

Example: Suppose that τ_{sup}. is 1. Above previous example, we apply the PrefixSapn algorithm with $\tau_{sup} = 1$. The candidate patterns are as shown in Table 6. ∎

Check Probability Threshold. In this step, we consider the condition of the probability threshold based on candidate patterns. We compute the maximum probability of candidate patterns in sequences, which are recorded from the previous step. If the maximum probability is less than the probability threshold τ_{prob}, we remove this sequence out of support calculation. After this step, the candidate patterns that fix the condition of support threshold are the results of our problem.

Table 7. Check probability threshold

Candidate pattern	Probability
A	$Pr_{S1}(A) = 0.7$, $Pr_{S2}(A) = 1$
B	$Pr_{S1}(B) = 0.8$, $Pr_{S2}(B) = 0.8$
A -> B	$Pr_{S1}(AB) = 0.56$, $Pr_{S2}(AB) = 0.8$

Example: Above previous example, the procedure of this step is as shown in Table 7. The patterns, A, B, AB are the probabilistic sequential patterns with $\tau_{prob} = 0.5$ **and** $\tau_{sup} = 1$. ∎

The pseudo codes of the approach of pattern mining are shown as follows.

Algorithm of pattern mining

Input: an uncertain sequence model *uSeq*, a support threshold *τsup*, and
a probability threshold *τprob*
Output: Probabilistic Sequential Patterns
Prune the Elements with Low Probabilities (*uSeq, τprob*)
    For each sequence
        For each item
            If probability of item < *τprob*
                Drop this item
Find Candidate Patterns (*uSeq, τsup*)
    Apply the PrefixSpan algorithm with *τsup* to find candidate patterns in *Seq*.
    The support and the set of ID of the sequences related to a specific candidate pattern are kept.
    Return a set of candidate patterns *cp*
Check Probability Threshold (*cp, uSeq, τprob, τsup*)
    For each candidate pattern
        For each sequence contains this candidate pattern
            If max(probability of candidate pattern) < *τprob*
                Support of this candidate pattern minus one
    Return the candidate patterns whose support ≥ *τsup*

---

5 Experiments

A series of experiments are performed in this section to evaluate each step of our approach. We choose the place with the highest probability in the uncertain model to be a certain model. We focus on four parts, including clustering, certain models, uncertain models, and pattern mining.

5.1 Datasets and the Ground Truth

We choose one-month Wi-Fi logs of ten participants and ask them to label the places where they go on Wi-Fi fingerprints. Moreover, we ask them to write a diary for describing daily trajectories. The number of Wi-Fi fingerprints, number of distinct places, and percentage of unlabeled Wi-Fi fingerprints in each dataset are as shown in Table 8. We pick up 52 places into our Wi-Fi database in these experiments.

Table 8. 10 datasets

Dataset ID	# of Wi-Fi fingerprints	# of distinct place labels	Unlabeled Wi-Fi fingerprints
1	1802	11	10.38 %
2	3344	15	5.80 %
3	5890	10	3.84 %
4	5444	16	6.98 %
5	7402	17	6.76 %
6	4857	12	4.36 %
7	2154	12	5.90 %
8	6186	19	9.34 %
9	10529	17	10.64 %
10	12653	14	16.00 %

5.2 Evaluation on Clustering Results

In this section, we use entropy, purity, and SSE (sum of squared error) to evaluate the clusters and use them to adjust the parameter k in the K-Means clustering algorithm. Suppose that there are m Wi-Fi fingerprints and n distinct place labels in a dataset. m_i denotes the number of Wi-Fi fingerprints in the i-th cluster. m_{ij} denotes the number of Wi-Fi fingerprints labeled as $place_j$ in the i-th cluster. The entropy is defined as follows

$$entropy(i) = -\sum_{j=1}^{n} \frac{m_{ij}}{m_i} \log_n \frac{m_{ij}}{m_i} \quad entropy = \sum_{i=1}^{k} \frac{m_i}{m} \times entropy(i)$$

The purity is defined as follows.

$$purity(i) = \frac{\max(m_{ij})}{m_i} \quad purity = \sum_{i=1}^{k} \frac{m_i}{m} \times purity(i)$$

Table 9. k of the ten datasets

Dataset	K
Dataset 1	20
Dataset 2	15
Dataset 3	14
Dataset 4	15
Dataset 5	15
Dataset 6	15
Dataset 7	13
Dataset 8	17
Dataset 9	14
Dataset 10	15

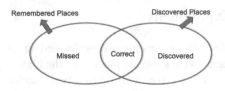

Fig. 5. *Remembered places* (recorded by participants) and *discovered places* (found by our framework)

C_i denotes the i-th cluster. M_i denotes the centroid of the cluster C_i. The SSE is defined as follows.

$$SSE = \sum_i^k \sum_{x \in C_i} d^2(M_i, x)$$

We run the k-Means clustering algorithm from $k = 1$ to $k = 15$, set entropy threshold to 0.2, and set purity threshold to 0.8. If there is no clustering result satisfying the entropy and purity threshold, we increase k to run the k-Means clustering algorithm until entropy and purity satisfy the conditions. The evaluation of the ten datasets clustering results is shown in Fig. 5. As shown in Fig. 5, the clustering results have good performance, with low entropy, low SSE, and high purity. k for each datasets is shown in Table 9 (Fig. 6).

5.3   The Certain Model Evaluation

In this section, we evaluate the certain models with two methods of representing clusters and compare with the certain models without clustering procedure. Suppose that a cluster is represented as a place, we choose the highest probability place in the uncertain model as a certain model. As shown in Fig. 5, places recorded in a participant's diary are called *remembered places* and places discovered by our framework are called *discovered places*. *Remembered places* that are not discovered are called *missed*. Places that are both *remembered places* and *discovered places* are called *correct places*. More *correct places* show better performance. Moreover, we define precision and recall as follows:

$$Precision = \frac{Correct}{Discovered}, Recall = \frac{Correct}{Remembered}$$

The evaluation of ten datasets certain models are shown in Table 7. As shown in Table 10 and Fig. 7, the certain model with clustering can discover more places, but the

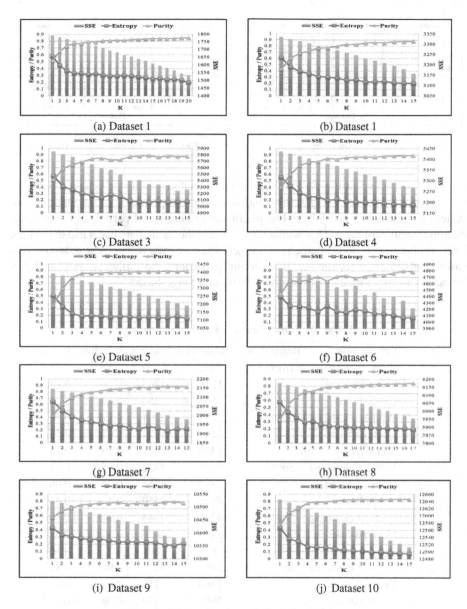

(a) Dataset 1

(b) Dataset 1

(c) Dataset 3

(d) Dataset 4

(e) Dataset 5

(f) Dataset 6

(g) Dataset 7

(h) Dataset 8

(i) Dataset 9

(j) Dataset 10

Fig. 6. The evaluation of clustering results

number of *correct* places does not increase, hence the precision of certain model without clustering is higher than certain model with centroid-based and range-based. The performance of certain models with range-based is worse than the other two because of the concept of range-based similarity function. We will demonstrate it in Sect. 5.4. Because the dataset 3, 6, 9 and 10 have good performances in clustering procedure, their recall of certain models with centroid-based is higher than the certain models without clustering

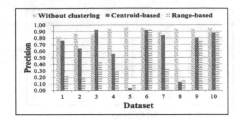

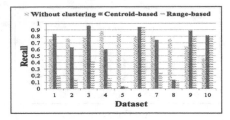

Fig. 7. The precision of bar chart   **Fig. 8.** The recall of bar chart

procedure, as shown in Fig. 8. This shows the procedure of clustering can effectively filter the noise of Wi-Fi fingerprints and discover the places. Certain models with clustering in dataset 5 and 8 have bad performance because of participants' behavior. The time duration of dataset 5 is winter vacation and the participant had a trip in the time duration of dataset 8, so the participant's trajectories are irregular in these datasets. This reason leads to bad performance in clustering procedure and further leads to bad performance in certain model with centroid-based and range-based.

Table 10. The evaluation of the certain model

Dataset	Certain model	Remembered	Discovered	Correct	Precision	Recall
1	Without clustering	1615	1492	1220	0.82	0.76
	Centroid-based		1776	1345	0.76	0.83
	Range-based		1602	356	0.22	0.22
2	Without clustering	3150	2775	2404	0.87	0.76
	Centroid-based		3101	1984	0.64	0.63
	Range-based		2456	471	0.19	0.15
3	Without clustering	5664	5201	4422	0.85	0.78
	Centroid-based		5890	5451	0.93	0.96
	Range-based		5519	2350	0.43	0.41
4	Without clustering	5064	4726	4444	0.94	0.88
	Centroid-based		5444	3025	0.56	0.6
	Range-based		2678	801	0.3	0.16
5	Without clustering	6899	5991	5736	0.96	0.83
	Centroid-based		7304	256	0.04	0.04
	Range-based		2999	240	0.08	0.03
6	Without clustering	4645	3863	3681	0.95	0.79
	Centroid-based		4762	4368	0.92	0.94
	Range-based		4695	4363	0.93	0.94
7	Without clustering	2027	1762	1605	0.91	0.79
	Centroid-based		1780	1501	0.84	0.74
	Range-based		1453	486	0.33	0.24

(Continued)

Table 10. (*Continued*)

Dataset	Certain model	Remembered	Discovered	Correct	Precision	Recall
8	Without clustering	5608	4515	4250	0.94	0.76
	Centroid-based		5976	734	0.12	0.13
	Range-based		4211	684	0.16	0.12
9	Without clustering	9409	6422	6019	0.94	0.64
	Centroid-based		10355	8314	0.8	0.88
	Range-based		10200	7628	0.75	0.81
10	Without clustering	10628	5135	4880	0.95	0.46
	Centroid-based		9811	8636	0.88	0.81
	Range-based		9440	8451	0.9	0.8

5.4 The Uncertain Model Evaluation

In this section, we evaluate the uncertain models with two methods of representing clusters and compare with the uncertain models without the clustering procedure. Given a Wi-Fi fingerprint f, we compute the cosine similarity between the uncertain model of Wi-Fi fingerprint f and the distribution of the place labels, which are labeled on the Wi-Fi fingerprints that are the top 3 % nearest to the Wi-Fi fingerprint f. The averages of the cosine similarity for ten datasets with two methods are shown in Fig. 9. Because dataset 3, 6, 9 and 10 have a good performance in clustering procedure, the uncertain models without clustering procedure perform better than the other two methods except dataset 3, 6, 9 and 10. The uncertain models with two methods of representing clusters have low value in dataset 5 and 8 because the trajectories in these two datasets are irregular and the performance of clustering results is not good enough. In some datasets, the uncertain models with range-based perform worse than centroid-based because of the defect of range-based similarity function. These datasets have the situation same as the example that is shown in Fig. 10. In Fig. 10, f_1 and f_3 are also detected in cluster f_c and Place f_p, but $intersection(f_{c1}, f_{p1})$ and $intersection(f_{c3}, f_{p3})$ are equal to 0. When we compute $sim(f_c, f_p)$, f_1 and f_3 make the second item of similarity function very small and then make the $sim(f_c, f_p)$ lower than a threshold (0.05). Therefore, there are many clusters considered as unknown place in these datasets. This is the reason for unstable performance of the uncertain models with range-based.

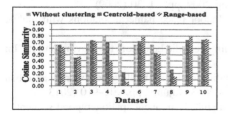

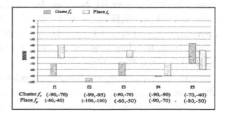

Fig. 9. The evaluation of uncertain model   **Fig. 10.** The concept of range-based method

When we implemented the step of uncertain model, we observe the first item $\frac{|S_{fc} \cap S_{fp}|}{|S_{fc} \cup S_{fp}|}$ of similarity function does not have performance that we expect. Because we use the average and the range of RSSI to represent a set of Wi-Fi fingerprints, if there have any Wi-Fi access point which RSSI is greater than -100 in a set and this Wi-Fi access point must be in the set S_f. Hence the definition of $S_f = \{i | f_i \neq -100, \forall : 1 \leq i \leq n\}$ is not good enough to describe Wi-Fi access points we detected in a representative of the set of Wi-Fi fingerprints. We can modify it as follows.

$$S_f = \{i | f_{ai} \geq min_{RSSI}, \forall : 1 \leq i \leq n\}$$

min_{RSSI} is a parameter in this definition. The greater value of min_{RSSI} makes S_f stricter.

6 Conclusion

In this paper, we propose an overall framework to mine user trajectories from smartphone data. We build an uncertain sequence model and use it to mine the probabilistic sequential patterns from real datasets. We propose the Wi-Fi fingerprint distance metric and use it to run the K-means clustering algorithm for discovering places. Two methods are proposed to represent a cluster and we define two similarity functions for these two methods. Then, we propose an algorithm based on PrefixSpan to mine the probabilistic sequential patterns. A series of experiments are performed to evaluate each step of our framework. In some datasets, based on the good performance of clustering results, the accuracy of the uncertain model with clustering is higher than the uncertain model without clustering. While considering efficiency, it need not do the procedure of clustering analysis. Because of the concept of the uncertain model with range-based, it has unstable performance. The experiment results show that our framework can deal with different kinds of datasets.

References

1. Fan, Y.C., Chen, Y.C., Tung, K.C., Wu, K.C., Chen, A.L.P.: A framework for enabling user preference profiling through Wi-Fi logs. IEEE Trans. Knowl. Data Eng. **28**(3), 592–603 (2016)
2. Kim, D.H., Hightower, J., Govindan, R., Estrin, D.: Discovering semantically meaningful places from pervasive RF-beacons. In: UbiComp 2009, pp. 21–30 (2009)
3. Fan, Y.C., Lee, W.H., Iam, C.T., Syu, G.H.: Indoor place name annotations with mobile crowd. In: ICPADS 2013, pp. 546–551 (2013)
4. Li, Z., Ding, B., Han, J., Kays, R., Nye, P.: Mining periodic behaviors for moving objects. In: KDD 2010, pp. 1099–1108 (2010)
5. Lee, S., Min, C., Yoo, C., Song, J.: Understanding customer malling behavior in an urban shopping mall using smartphones. In: UbiComp 2013 Adjunct, pp. 901–910 (2013)
6. Li, Y., Bailey, J., Kulik, L., Pei, J.: Mining probabilistic frequent spatio-temporal sequential patterns with gap constraints from uncertain databases. In: ICDM 2013, pp. 448–457 (2013)

7. Zhao, Z., Yan, D., Ng, W.: Mining probabilistically frequent sequential patterns in uncertain databases. In: EDBT 2012, pp. 74–85 (2012)
8. Hartigan, J.A., Wong, M.A.: A k-means clustering algorithm. Appl. Stat. **28**(1), 100–108 (1979)
9. Pei, J., Han, J., Mortazavi-asl, B., Pinto, H., Chen, Q., Dayal, U., Hsu, M.C.: PrefixSpan: mining sequential patterns by prefix-projected growth. ICDE **2001**, 215–224 (2001)

A Heterogeneous Clustering Approach for Human Activity Recognition

Sabin Kafle[✉] and Dejing Dou

Department of Computer and Information Science,
University of Oregon, Eugene, USA
{skafle,dou}@cs.uoregon.edu

Abstract. Human Activity Recognition (HAR) has a growing research interest due to the widespread presence of motion sensors on user's personal devices. The performance of HAR system deployed on large-scale is often significantly lower than reported due to the sensor-, device-, and person-specific heterogeneities. In this work, we develop a new approach for clustering such heterogeneous data, represented as a time series, which incorporates different level of heterogeneities in the data within the model. Our method is to represent the heterogeneities as a hierarchy where each level in the hierarchy overcomes a specific heterogeneity (e.g., a sensor-specific heterogeneity). Experimental evaluation on Electromyography (EMG) sensor dataset with heterogeneities shows that our method performs favourably compared to other time series clustering approaches.

Keywords: Time series · Heterogeneous clustering · Bayesian semiparametrics · Human Activity Recognition

1 Introduction

The widespread availability of sensors in everyday lives enables us to capture contextual information from underlying human behavior in real-time. This has led to the significant research focus on Human Activity Recognition (HAR) using sensor data [15]. Sensor data is used to determine the specific activity performed by the user at that instant, using either statistical or machine-learning approach. Despite a significant interest on HAR research, real-world performance variations across different sensors have been overlooked [15].

A significant research problem based on use of sensor networks is development of sensor-based automatic prosthetic limbs. The sensor network (usually an EMG sensor network) is used for detecting the intention of the user of the prosthetic limbs in order to provide a better control mechanism to the prosthetic limbs. The sensor network provides data related to the neural intent of the user, which is then interpreted by the prosthetic limb control mechanism to enable certain degree of freedom to the limb motion. For example, the control system is able to recognize whether the user is walking along a level ground or climbing up the

© Springer International Publishing Switzerland 2016
S. Madria and T. Hara (Eds.): DaWaK 2016, LNCS 9829, pp. 68–81, 2016.
DOI: 10.1007/978-3-319-43946-4_5

stairs based on the neural impulse of the user (inferred from sensor data using statistical and machine learning models), which then triggers an intent specific freedom on the prosthetic limbs; e.g., automated rising of the prosthetic limb when the user is climbing up the stairs. While significant progress has been made in the development of prosthetic limbs with such control mechanisms [5], most of the work focus on having a prosthetic limb trained to a specific user only. There is a distinct lack of research in unsupervised learning of user intent from such sensor data.

We focus on developing an unsupervised approach to recognize the user intent based on the sensor data. We treat the sensor data as a time series which is the most natural interpretation of such data. While time series clustering is a significant research area with many different approaches proposed, most of them are inapplicable to our current problem. Time series clustering usually cluster the data obtained from same or similar data source, which is not true for our case. Moreover, most of the approaches require the number of clusters (or activity) in the data to be predetermined which is not always feasible in sensor data. Another challenge lies in the interpretation of the sensor data itself. The sensor data comprises of additive noises and have been found to be inefficient in representing the user intent as raw data themselves [14]. Time and frequency domain features are extracted from sensor data which are then used in machine-learning models for intent interpretation.

In this work, we address the challenges of performing unsupervised learning approach on sensor datasets. We first introduce the heterogeneities in the dataset as a hierarchy with each level in the hierarchy representing a specific heterogeneity. Next, we perform clustering using Bayesian semiparametric approach to mitigate the problem of pre-specifying the number of clusters in the dataset. Our approach learns the number of clusters (or activities) present in the dataset as a parameter of the model, which is capped by some large number that is considered to be an upper limit on the possible number of clusters. Finally, we also develop a feature series clustering approach where we obtain features from the sensor dataset, which is then used to cluster the input data. For evaluation of our approach, we use an EMG sensor dataset collected while the subject performs a walking motion into different terrains. Our dataset consists of eight EMG sensors placed on different limb muscles of a single person during the data collection phase. We find that having a hierarchy to eliminate heterogeneity in the data helps in obtaining better clustering performance. Our method outperforms other approaches which treat the sensor data as a time series in unsupervised learning.

The paper is organized as follows. In Sect. 2, we review previous work on sensor data usage for activity recognition followed by brief description of time series clustering algorithms. We also introduce Hierarchical Normal Model, which is our approach for eliminating heterogeneities in sensor data. We describe our approach in Sect. 3 which includes description of our method and different parameters within the approach. We present experimental results in Sect. 4 and conclude in Sect. 5.

2   Background and Related Work

2.1   Activity Recognition Using EMG Sensors

EMG (Electromyographic) sensors measure electrical current generated in skeletal muscles during its contraction representing neuromuscular activity. The contraction of skeletal muscle is initiated by impulses in the neuron to the muscle and is usually under voluntary control, which are captured by surface EMG sensors. Such signals are significant for detection of gait events of individual.

A *gait* is defined as someone's manner of ambulation or locomotion, involving the total body [2]. The two main phases of gait cycle are the stance phase and the swing phase. A complete gait cycle comprises of - "Heel Strike" (HS), "Flat Foot" (FF), "Mid Stance" (MS), "Heel Off" (HO), "Toe Off" (TO) and "Mid Swing" (MS). Two phases of gait cycle have been found to be most effective in recognizing locomotion mode. One is Heel Strike (also called initial contact), a short period which begins the moment the foot touches the ground while the other is toe-off (also called pre-swing phase), a period when the toe begins to take stance. Activity recognition mechanism based on *gait cycles* involve extracting features from these two phases before classification. Each gait is classified as belonging to a particular activity; e.g., walking up the stairs or walking down the ramp.

Most of the work in EMG signal based terrain identification (also called locomotion mode identification or gait event detection) is based on using classification algorithms, which depend on having labelled training data. The earlier work for EMG signal analysis is based upon wavelet analysis [8] and auto-regressive models [1]. It was demonstrated by [13] that there is a difference in EMG signal envelope among level-ground walking and descending and ascending a ramp, with conclusion that EMG signals from hip-muscles could be used to classify the locomotion modes. The more recent approaches are based on using the features extracted from EMG signals for training a machine learning classification model.

The EMG signals by themselves are random signals with zero mean, but have significance during stages where the muscle contraction is maximum [4]. The features extracted from EMG signals are crucial for getting proper classification accuracy during prediction. The features extracted during the *150* ms phase before and after the "Heel Strike" and "Toe Off" is found to be most accurate for terrain identification [5]. The time domain features which are significant for gait event detection [3] are - *Mean, Variance, Mean Trend, Variance Trend, Windowed Mean Difference, Windowed Variance Difference and Auto-regressive coefficients.*

2.2   Time Series Clustering

Time series clustering is one of the most fundamental and complex task in data mining research. Time series clustering algorithms are usually applied by either converting the popular static clustering approaches to handle time series or by modifying time series to make static clustering methods applicable [9].

One of the most popular approach for static data clustering is k-means or k-mediods, which generate spherical-shaped cluster with a distance measure being considered for deciding cluster membership. Another popular approach for clustering is hierarchical clustering which generate clusters in agglomerative manner (assign each data as an individual cluster and proceed with merging to generate ideal cluster) or divisive manner (partition the data based on some metrics). This approach requires some cluster quality check metrics to determine the best cluster partition. Density-based clustering approach grows a cluster as long as the density of the "neighbourhood" exceeds some threshold value. Model-based clustering approach assumes a model for each cluster and attempts to best fit the data to the assumed model.

The most used approach for clustering time series data is based on computing the similarity measure between different time series and then using the similarity measure to obtain either a spherical cluster partition using *k-means* algorithm or a non-spherical cluster partition using *fuzzy k-means*. Another approach is to extract features from time series and then use those features to perform clustering, either by using a multinomial distribution (when the number of clusters is known *apriori*) or a Dirichlet Process (when the number of clusters is not known apriori). The detail survey and comparison of different time series approach can be found on [11].

2.3   Hierarchical Normal Model

Many different kinds of data, including observational data collected in human and biological sciences, have a hierarchical structure. Sensor signals such as EMG have a natural hierarchy where the measurement of each person is grouped under an individual person and each type of sensor is grouped under that particular sensor. This natural hierarchical tendency of data requires multi-level analysis, which can be incorporated using Hierarchical Normal Models (HNM). HNMs were first studied in the context of biological and human sciences where family, race, geographical location introduces a natural hierarchy in the data [10].

A hierarchy of normal distribution is considered in hierarchical normal model. The top-most level of hierarchy includes a prior for mean and variance of the model (joint prior or distinct prior). A mean value is sampled from the prior, which is then used to sample different means for Level 1 of hierarchy, with the variance obtained from variance prior. For each sub-hierarchy in Level 2, the mean is sampled from each parent sample separately. The variance at each level can be either estimated from the data of that group and kept fixed or obtained from Gibbs sampler step for variance.

An example representation of Hierarchical Normal Model (HNM) is given in Fig. 1. In the figure, the hierarchy moves from left to right. On each level of hierarchy, different components incorporate differences present in that level of hierarchy. We start with a base distribution and add heterogeneities as we move from left to right at each level. The existence of such hierarchies is the result of differentiation in all kind of activities (e.g., different gait events for different person and differing sensor metrics for different muscle activation).

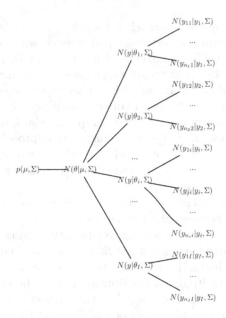

Fig. 1. A Hierarchical Normal Model (HNM).

3   Our Approach

In this section, we describe our model for clustering EMG sensor data. In our approach, we address three key challenges in clustering sensor data:

- Sensor data, especially EMG sensor data, by itself cannot be used for classification or clustering purposes, since it is a noisy time series with *zero mean*. We need to filter out noise from the sensor data before it can be used for activity recognition purposes.
- Heterogeneity is a key challenge and bottleneck for clustering EMG sensor data. Person and Sensor based heterogeneity is a significant impediment to clustering such sensor data.
- Usage of EMG sensor data for activity recognition is heavily dependent on features extraction. We need to incorporate the features from EMG signals for performing clustering, since they have been found to be more useful for activity recognition. Our experimental results also show that using features extracted from EMG sensor data gives much better clustering performance than using raw signal from EMG sensor as input.

We aim to address the above mentioned challenges with our approach. We explain our approach on three subsections each related to the above-mentioned aspects of sensor data analysis. We also borrow some ideas from statistical analysis of time series data and use them heavily in our approach.

3.1 Sensor Data Representation

We represent the sensor data as a time series of $T \times 1$ dimension, also called a vector y_i where i subscript is used to represent the ith time series in the sensor dataset of size N.

We break-down a single EMG sensor data into three distinct latent variables. We use the sampling model [12] to represent sensor data.

$$y_i = Z\alpha_i + X\beta_i + \theta_i + \epsilon_i, i = 1, 2, ..., n \qquad (1)$$

where, ϵ_i is $T \times 1$ dimensional random noise. The other three parameters given in the Eq. 1 represent three different latent variables:

- α_i is $p \times 1$ dimensional vector representing the non-clustering components of the sensor data. It is used to enhance the fit of the data to the cluster core to which that data belongs to. For example, mean of the sensor data cannot be significant aspect for clustering such data but can be represented by α_i to make the sensor data fit to its cluster better.
- β_i is the $d \times 1$ dimensional vector representing the clustering but non-autoregressive components of the sensor data. It is used to cluster the sensor data based on several components of time series including trends. d represents the number of components of time series that are considered for clustering. A polynomial trend would require the value of d to be 3.
- θ_i is the $T \times 1$ dimensional vector representing the Auto Regressive $AR(1)$ components of the sensor data assuming stationarity in the time series. We do not assume the sensor data to be stationary but only consider some components represented by θ_i as an auto-regressive component.

Our approach does not require explicit specification of different aspects of sensor data that are considered for clustering or not but they are learned during the training phase automatically. The matrices Z and X are design matrices of dimensions $T \times p$ and $T \times d$ respectively.

We provide Bayesian treatment to our approach. This is done by assuming that sensor data and latent variables are generated by multivariate Normal distributions. We, then represent a single sensor data as a function of multivariate Normal Distribution given below:

$$f(y_i) \propto N_T(Z\alpha_i + X\beta_i + \theta_i, \sigma^2_{\epsilon_i} I)$$
$$\alpha_i \sim N(0, \Sigma_\alpha)$$
$$\beta_i \sim N(\overline{\beta_{s,r,k}}, \Sigma_{\beta,s,r,k}) \qquad (2)$$
$$\theta_i \sim N(\overline{\theta_{s,r,k}}, \Sigma_{\theta,s,r,k})$$

where parameters with s, r, k as subscript represent the specific values obtained after incorporating the heterogeneity into the non-heterogeneous clustering parameters.

For simplicity, we assume covariance matrices to be diagonal matrices for Multivariate Normal Distribution and each diagonal elements obtained as a sample from Inverse-Gamma (IGa) prior. This completes the specification of our approach to separate noisy components from sensor data.

3.2 Incorporation of Heterogeneities in Sensor Data

We incorporate heterogeneities in EMG sensor data by using a Hierarchical Normal Model (HNM) where each level of hierarchy represents a specific heterogeneity in the data. We use two level of hierarchy to represent two different heterogeneities in EMG sensor datasets. The first level of hierarchy represents Sensor Level differences since the differences in calibration of different sensors and their positioning play a significant role in creating heterogeneity. The second level of hierarchy is used to represent person specific heterogeneities since for different person, a sensor with same calibrations and positioning will have different readings based on differences between the persons' biomechanics. We also incorporate heterogeneities for each activity (or clusters) separately. This is done to reduce the number of parameters since our approach enables us to consider heterogeneity only for clustering parameters and allow non-clustering parameters to fit independent of heterogeneity.

The top-most level of hierarchy for a specific cluster k is:

$$\gamma_{root} \sim N(0, \Sigma_\beta) \times N(0, \Sigma_\theta) \tag{3}$$

where γ_{root} is considered product of two clustering parameters i.e. $\beta_{root} \times \theta_{root}$. The prior for covariance of β parameter is diagonal matrix with IGa prior while for θ it is adapted from [12] to handle stationarity.

For the second level, which is the sensor specific heterogeneities incorporating level, we have R branches where R is the total number of sensor types present in the data. We sample a mean value from the parent which is $root$ and use the covariance matrix from that level attached to the specific component to sample the clustering components that incorporates sensor level of heterogeneities. Specifically, for a sensor r data belonging to cluster k, we obtain the clustering component as:

$$\gamma_{r,k} \sim N(\overline{\beta_{r,k}}, \Sigma_{\beta,r,k}) \times N(\overline{\theta_{r,k}}, \Sigma_{\theta,r,k})$$
$$\overline{\beta_{r,k}} \sim N(0, \Sigma_{\beta,k}) \tag{4}$$
$$\overline{\theta_{r,k}} \sim N(0, \Sigma_{\theta,k})$$

where $\Sigma_{\beta,r,k}$ and $\Sigma_{\theta,r,k}$ are covariances for β and θ parameters of cluster k, level 2 and branch r.

Similarly, for person specific heterogeneities incorporating level (which is level 3), we consider S persons for each sensor branch r in the previous level and obtain a person's heterogeneity incorporating clustering parameters as:

$$\gamma_{s,r,k} \sim N(\overline{\beta_{s,r,k}}, \Sigma_{\beta,s,r,k}) \times N(\overline{\theta_{s,r,k}}, \Sigma_{\theta,s,r,k})$$
$$\overline{\beta_{s,r,k}} \sim N(\overline{\beta_{r,k}}, \Sigma_{\beta,r,k}) \tag{5}$$
$$\overline{\theta_{s,r,k}} \sim N(\overline{\theta_{r,k}}, \Sigma_{\theta,r,k})$$

Here, k represents the kth cluster from K clusters, r represents the rth sensor from R sensors and s represents the sth person among S persons.

The block diagram of HNM for heterogeneities is given in Fig. 2 for more clarity.

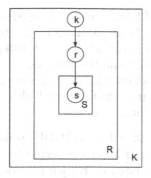

Fig. 2. Block HNMs for heterogencity. Each k cluster is represented by one such block which incorporates two level of heterogeneities. For Sensor data, two level of heterogeneities represent sensor (r) and person (s) based heterogeneities.

For clustering, we use the Generalised Dirichlet Process (GDD) based prior as explained in [6]. The selection of GDD based prior is for two main reasons - (i) It enables us to learn number of clusters from the data itself. (ii) The posterior of GDD is conjugate with multinomial sampling and thus we can easily run Gibbs Sampling for cluster inference. The number of clusters that can sufficiently represent the data is obtained by using Adequate Truncation Value method explained in [6] during the first few iterations of Gibbs Sampling.

The prior characterization of our approach is now complete. The posterior for every random variables introduced in our approach can be obtained analytically based upon the likelihood function given as:

$$f(y) = \prod_{i=1}^{N} N_T(Z\alpha_i + X\beta_i + \theta_i, \Sigma_y) \tag{6}$$

where, $\Sigma_y = \sigma_\epsilon^2 I$ and N is the total data count.

For the sake of clarity and brevity, we include posterior characterizations of our approach in Appendix A.

3.3   Inference and Feature Clustering

Gibbs Sampling algorithm is used for posterior inference, with Metropolis within Gibbs sampler being used for sampling ρ and σ_θ^2. The Gibbs Sampler algorithm used is same as given in [7] except for sampling from the hierarchical model and Metropolis steps for θ parameters. The hierarchical model's posterior sampling is done in bottom-up approach. The hierarchy is sampled beginning from the Sampling model until the top level is reached.

Features extraction from EMG sensor data is necessary to obtain better classification and clustering results. We extend our approach to multi-dimensional time series to handle EMG sensor data where we extract features for every 50 ms window. Each features extracted creates a series of their own (also called a vector).

We call such feature vector of EMG sensor data a feature series. We assume each feature series is independent of one another in order to reduce the complexity of the model. Then each feature is considered as an independent EMG sensor data and the above model is applied to all the features.

The most significant aspect during handling such multiple features is to consider how each feature series impacts the overall clustering aspect our approach. We propose two approaches for such cases:

- The Generalized Dirichlet Distribution (GDD) is used for combining different feature series. A single cluster label is selected for all the feature series. The parameters of GDD are updated with each feature series likelihood w.r.t. the data. Rest of the model is kept same as explained above.
- Different GDD is used for each feature series. The final cluster membership is based on majority voting of cluster assignments in individual feature series.

We perform experiments to evaluate each approach and find out that using a single GDD for combining different feature series works best for our dataset.

3.4   Cluster Selection

Each iteration of Gibbs Sampling produces a cluster assignment of data points, which is then filtered to select one cluster assignment as the best fit. One way of selecting a cluster membership used by [12] is Heterogeneity Measure (HM), which can be calculated as:

$$HM(G_1, .., G_m) = \sum_{k=1}^{m} \frac{2}{n_k - 1} \sum_{i<j \in G_k} \sum_{t=1}^{T} (y_{it} - y_{jt})^2 \qquad (7)$$

where, m is the number of clusters, G_k is k^{th} cluster, n_k is the number of data point belonging to cluster G_k, T is the length of time series representation of sensor data and y_{it} is the tth value in time series y_i.

The larger the value of HM, the more heterogeneous a clustering is. It is preferable to have a cluster with small HM and small m.

4   Experiments

Our dataset consists of 9 normal human subjects performing 5 different gait events which are measured by eight different sensors placed in their bodies. In this experiment, we attempt to cluster each sensor data into individual gait events (level ground walking, stair ascent, stair descent, ramp ascent and ramp descent). We consider a data point to be a sensor reading of a single gait cycle. We conduct several experiments with different configuration of hyper parameters in order to determine the best configuration for the dataset. The number of clusters is determined initially using the Adequate Truncation Value during the initial Gibbs Sampling phase but we found that using the same number of clusters as in original dataset gives the best result. We found out 15 is sufficient number of clusters for this data. In order to eliminate the disproportions of data in different classes, we use subsampling to get the equal number of input data for each class.

The dimension of design matrices $Z_{T \times p}$ and $X_{T \times d}$ is an important design decision. We use $p = 1$ since we found that the value of p doesn't play a significant role for our dataset. For X, we set $d = 7$ with first three columns representing the polynomial trend of degree 3, with remaining four columns being used as a latent trait indicator for four gait phases (Before Heel Strike, After Heel Strike, Before Toe Off, After Toe Off). We also set the inverse gamma prior fixed to $[2, 1]$ throughout the experiment. The results presented is based on the heterogeneity score of sampled cluster membership, with the lowest score being selected as best clustering assignment. The best cluster membership obtained from the sampler based on Heterogeneity Measure is then used to obtain a confusion matrix. The confusion matrix is used to compute the accuracy of the clustering approach with labelled examples. The Heterogeneity Measure for every result obtained is between 0.5 and 1.95, with random impact on the accuracy of the clustering. For all the experiments, we run Gibbs sampler upto 5000 iterations, with 3000 as burn-in phase and collect a sample every 200 iterations after the burn-in phase.

First we conduct experiments to determine the different aspects of the model. The results are presented in Fig. 3. The first result in Fig. 3 shows that EMG data by itself is not meaningful for classifying or clustering purposes.

We obtain best result with configurations given as second bar in Fig. 3 where we specify the number of clusters same as the number of labels. The comparison between third bar and fifth bar of Fig. 3 along with fourth and sixth bar of Fig. 3 illustrate that majority voting for feature series performs slightly worse than a single cluster membership for entire feature series approach. Also, the use of

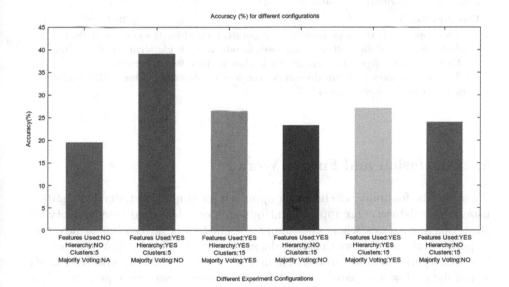

Fig. 3. Results obtained for different configuration of Model

hierarchy helps in obtaining the better performance compared to the non-usage of hierarchy as evident from third and fifth bar comparison.

Next, we compare the performance of our approach with other popular time series clustering algorithms. The result is presented in Table 1. It is evident that our approach outperforms other approaches for time series clustering in case of our dataset which consists of heterogeneity. For k-means based clustering experiment, we use TSclust package [11]. We refer the reader to [11] for more information about different distance metrices.

Table 1. Performance measure of different time series clustering approaches

Method	Accuracy (%)
Bayesian Nonparametrics Time Series Clustering (BNPTSclust)[1] [12]	26.0
Rest of the algorithms are k-means clustering algorithms implemented in [11]	
Autocorrelation based Dissimilarity (ACF)	26.0
Periodogram-based distances (PER)	25.3
Normalized Compression Distance (NCD)	23.3
Euclidean Distance (EUCL)	36.0
Compression-based dissimilarity measure (CDM)	24.7
Dynamic Time Warping (DTW) measure	31.3
Discrete Wavelet Transform (DWT)	30.7
Correlation Based Dissimilarity (COR)	29.3
Partial Autocorrelation based Dissimilarity (PACF)	28.0
Complexity Invariant Distance (CID)	30.7
Permutation Distribution Clustering (PDC)[2]	18.7
Our Approach[3]	**39.1**

[1] This approach is based on Bayesian non parametrics where the number of clusters is inferred from the data itself. This approach favours a single cluster most of the time.
[2] Used default configuration provided in TSclust package for clustering.
[3] The best accuracy is obtained when not considering Majority Voting, while specifying the number of clusters to be only 5.

5   Conclusion and Future Work

We study the feasibility of clustering approach for Human Activity Recognition using sensor dataset. Our approach introduces hierarchy-based heterogeneity for clustering time series where the number of clusters is not known in advance. Experimental result shows that introducing hierarchy helps in clustering sensor-based time series more accurately. Though the accuracy of our approach for EMG sensor data is low, comparison with other time series clustering approaches show that our method performs better than other approaches. The current method expresses the time series as a linear model only, future work will involve extension to non-linear models to handle more complex time series, along with using more datasets for further experiments.

Acknowledgements. This work is partially supported by the NIH grant R01GM1 03309. We acknowledge Deepak Joshi and Michel Kinsy for their inputs.

Appendix A Posterior Characterization

- α_i: The posterior for α_i is:

$$f(\alpha_i|rest) \propto N_P(\mu_a, \Sigma_a)$$
$$\Sigma_a = (\Sigma_\alpha^{-1} + Z^T \Sigma_y^{-1} Z)^{-1}$$
$$\mu_a = \Sigma_a Z^T \Sigma_y^{-1}(y_i - X\beta_i - \theta_i) \tag{8}$$

$$f(\sigma_{\alpha_j}^2|rest) = IGa(c_0^\alpha + \frac{n}{2}, c_1^\alpha + \frac{1}{2}\sum_{i=1}^{n}\alpha_{ij}^2), j = 1,..,p$$

- β_i: The posterior for β_i (or $\beta_{s,r,k}$) is:

$$f(\beta_i|rest) \propto N_D(\mu_b, \Sigma_b)$$
$$\Sigma_b = (\Sigma_{\beta,s,r,k}^{-1} + X^T \Sigma_y^{-1} X)^{-1}$$
$$\mu_b = \Sigma_b[X^T \Sigma_y^{-1}(y_i - Z\alpha_i - \theta_i) + \Sigma_{\beta,s,r,k}\overline{\beta}_{s,r,k}] \tag{9}$$

$$f(\sigma_{\beta,s,r,k,i}^2|rest) = IGa(c_0^{\beta,s,r,k,i} + \frac{m}{2}, c_1^{\beta,s,r,k,i} + \frac{1}{2}\sum_{j=1}^{m}\beta_{s,r,k,i}^2)$$
$$i = 1,..,p$$

where m is the number of data points belonging to that cluster.
- θ_i: The posterior for θ_i (or $\theta_{s,r,k}$) is:

$$f(\theta_i|rest) \propto N_T(\mu_c, \Sigma_c)$$
$$\Sigma_c = (\Sigma_{\theta,s,r,k}^{-1} + \Sigma_y^{-1})^{-1}$$
$$\mu_c = \Sigma_c[\Sigma_y^{-1}(y_i - Z\alpha_i - X\beta_i) + \Sigma_{\theta,s,r,k}\overline{\theta}_{s,r,k}] \tag{10}$$

$$f(\sigma_{\theta,s,r,k,i}^2|rest) = IGa(c_0^{\theta,s,r,k,i} + \frac{m}{2}, c_1^{\theta,s,r,k,i} + \frac{1}{2}\sum_{j=1}^{m}\theta_{s,r,k,i}^2)$$
$$i = 1,..,T$$

where m is the number of data points belonging to that cluster.
- $\sigma_{\epsilon_i}^2$: The posterior for $\sigma_{\epsilon_i}^2$ is:

$$f(\sigma_{\epsilon_i}^2|rest) \propto IGa(c_0^\epsilon + \frac{T}{2}, c_1^\epsilon + \frac{1}{2}M_i'M_i)$$
$$M_i = (y_i - Z\alpha_i - X\beta_i - \theta_i) \tag{11}$$

- *Level k posterior*: The posterior for any level of hierarchy except for top-most level consists of following updates:

$$f(\beta_k|rest) \propto N_D(\mu_g, \Sigma_g)$$
$$\Sigma_g = (\Sigma_{\beta,r,k}^{-1} + \Sigma_{\beta,k}^{-1})^{-1}$$
$$\mu_g = \Sigma_g(\Sigma_{\beta,k}\overline{\beta_k} + \Sigma_{\beta,r,k}^{-1}\beta_{r,k})$$
$$f(\sigma_{\beta_{k,i}}^2|rest) = IGa(c_0^{\beta_{k,i}}\frac{R}{2}, c_1^{\beta_{k,i}}\sum_{j=1}^{S}\beta_{k,i}^2)$$
$$f(\theta_k|rest) \propto N_D(\mu_h, \Sigma_h) \qquad (12)$$
$$\Sigma_h = (\Sigma_{\theta,r,k}^{-1} + \Sigma_{\theta,k}^{-1})^{-1}$$
$$\mu_h = \Sigma_h(\Sigma_{\theta,k}\overline{\theta_k} + \Sigma_{\theta,r,k}^{-1}\theta_{r,k})$$
$$f(\sigma_{\theta_{k,i}}^2|rest) = IGa(c_0^{\theta_{k,i}}\frac{R}{2}, c_1^{\theta_{k,i}}\sum_{j=1}^{R}\beta_{k,i}^2)$$

- *Top level posterior*: The posterior at top-most level is:

$$f(\beta|rest) \propto N_D(\mu_e, \Sigma_e)$$
$$\Sigma_e = (\Sigma_{\beta,k}^{-1} + \Sigma_{\beta}^{-1})^{-1}$$
$$\mu_e = \Sigma_e(\Sigma_{\beta,k}^{-1}\beta_k)$$
$$f(\sigma_{\beta_i}^2|rest) = IGa(c_0^{\beta_i}\frac{K}{2}, c_1^{\beta_i}\sum_{j=1}^{K}\beta_i^2)$$
$$f(\theta|rest) \propto N_D(\mu_f, \Sigma_f)$$
$$\Sigma_f = (\Sigma_{\theta,k}^{-1} + \Sigma_{\theta}^{-1})^{-1} \qquad (13)$$
$$\mu_f = \Sigma_f(\Sigma_{\theta,k}^{-1}\theta_k)$$
$$f(\sigma_{\theta}^2|rest) = IGa(\frac{KT}{2}, \frac{1}{2}\sum_{j=1}^{K}\theta_j'Q^{-1}\theta_j)$$
$$f(\rho|rest) \propto |Q|^{-K/2}\exp\frac{-1}{2\sigma_{\theta}^2}\sum_{j=1}^{K}\theta_j'Q^{-1}\theta_j\frac{\sqrt{1+\rho^2}}{1-\rho^2}$$

where $Q_{ij} = \rho^{|i-j|}$ for $i, j = 1, .., T$.
- *Posterior for GDD and p*: The posterior for GDD is conjugate with multinomial sampling. The probability p is updated based on the fit of the data with respect to the individual clusters lowest level mean using the likelihood function. The complete detail for GDD posterior characterization can be found in [6].

This completes the posterior characterization of our approach.

References

1. Graupe, D., Cline, W.K.: Functional separation of emg signals via arma identification methods for prosthesis control purposes. IEEE Trans. Syst. Man Cybern. **5**(2), 252–259 (1975)
2. Griffin, L.Y., Albohm, M.J., Arendt, E.A., Bahr, R., Beynnon, B.D., DeMaio, M., Dick, R.W., Engebretsen, L., Garrett, W.E., Hannafin, J.A., et al.: Understanding and preventing noncontact anterior cruciate ligament injuries: a review of the Hunt Valley II meeting, January 2005. Am. J. Sports Med. **34**(9), 1512–1532 (2006)
3. Gupta, P., Dallas, T.: Feature selection and activity recognition system using a single triaxial accelerometer. IEEE Trans. Biomed. Eng. **61**(6), 1780–1786 (2014)
4. Huang, H., Kuiken, T., Lipschutz, R.D., et al.: A strategy for identifying locomotion modes using surface electromyography. IEEE Trans. Biomed. Eng. **56**(1), 65–73 (2009)
5. Huang, H., Zhang, F., Hargrove, L.J., Dou, Z., Rogers, D.R., Englehart, K.B.: Continuous locomotion-mode identification for prosthetic legs based on neuromuscular-mechanical fusion. IEEE Trans. Biomed. Eng. **58**(10), 2867–2875 (2011)
6. Ishwaran, H., James, L.F.: Gibbs sampling methods for stick-breaking priors. J. Am. Stat. Assoc. **96**(453), 161–173 (2001)
7. Ishwaran, H., Zarepour, M.: Markov chain monte carlo in approximate dirichlet and beta two-parameter process hierarchical models. Biometrika **87**(2), 371–390 (2000)
8. Kumar, D.K., Pah, N.D., Bradley, A.: Wavelet analysis of surface electromyography. IEEE Trans. Neural Syst. Rehabil. Eng. **11**(4), 400–406 (2003)
9. Liao, T.W.: Clustering of time series data–a survey. Pattern Recogn. **38**(11), 1857–1874 (2005)
10. Lindley, D.V., Smith, A.F.: Bayes estimates for the linear model. J. Roy. Stat. Soc.: Ser. B (Methodol.) **34**, 1–41 (1972)
11. Montero, P., Vilar, J.A.: TSclust: An R package for time series clustering. J. Stat. Softw. **62**(1), 1–43 (2014). http://www.jstatsoft.org/v62/i01/
12. Nieto-Barajas, L.E., Contreras-Cristan, A.: A bayesian nonparametric approach for time series clustering. Bayesian Anal. **9**(1), 147–170 (2014)
13. Peeraer, L., Aeyels, B., Van der Perre, G.: Development of emg-based mode and intent recognition algorithms for a computer-controlled above-knee prosthesis. J. Biomed. Eng. **12**(3), 178–182 (1990)
14. Reaz, M., Hussain, M., Mohd-Yasin, F.: Techniques of EMG signal analysis: detection, processing, classification and applications. Biolog. Proc. Online **8**(1), 11–35 (2006)
15. Stisen, A., Blunck, H., Bhattacharya, S., Prentow, T.S., Kjærgaard, M.B., Dey, A., Sonne, T., Jensen, M.M.: Smart devices are different: assessing and mitigating mobile sensing heterogeneities for activity recognition. In: Proceedings of the 13th ACM Conference on Embedded Networked Sensor Systems, pp. 127–140 (2015)

SentiLDA — An Effective and Scalable Approach to Mine Opinions of Consumer Reviews by Utilizing Both Structured and Unstructured Data

Fan Liu$^{(\boxtimes)}$ and Ningning Wu

Information Science, University of Arkansas at Little Rock,
2801 S. University Ave., Little Rock, AR 72204, USA
{fxliu,nxwu}@ualr.edu

Abstract. With the help of Internet and Web technologies, more and more consumers tend to seek opinions online before making purchase decisions. However, with the ever-increasing volume of user generated reviews, people are overwhelmed with the amount of data they have. Thus there is a great need for a system that can summarize the reviews and produce a set of aspects being mentioned in the reviews together with the pros/cons being expressed to them. To address the need, this paper proposes a new probabilistic topic model, SentiLDA, for mining reviews (unstructured data) and their ratings (structured data) jointly to detect the product/service aspects and their corresponding positive and negative opinions simultaneously. A key feature of SentiLDA is that it is capable of mining positive and negative sub-topics under the same aspect without the need of sentiment seed words. Experiment results show that the performance of SentiLDA outperforms the other related state-of-the-art models in detecting product/service aspects and their corresponding sentiments in reviews.

Keywords: Opinion mining · Topic modeling · Big data

1 Introduction

The Internet has greatly changed our shopping experiences over the last decade. Coming to the big data era, user-generated content (UGC) becomes a rich information source on the Internet. Customer reviews can be found from all kinds of social media, from internet tycoons like amazon.com to personal blogs. Many online review platforms allow users to submit reviews in free text format to comment on pros and cons of a product or service, and to give numerical ratings on the overall satisfaction level. These reviews can help people seek information before making shopping decisions but they also bring problems to consumers. A survey shows 32 % of internet users have been confused by information they have found online during their shopping; 30 % have felt overwhelmed by the amount of information they found online [6]. It is impossible for a shopper to digest the huge volume of reviews available online without any post-processing of the data. Thus, some people just rely on the ratings, leave the

© Springer International Publishing Switzerland 2016
S. Madria and T. Hara (Eds.): DaWaK 2016, LNCS 9829, pp. 82–96, 2016.
DOI: 10.1007/978-3-319-43946-4_6

actual reviews aside. But ratings cannot tell it all. Besides, there is not a golden standard to guide how people give ratings. For example, someone may give a 5-star with minor defects, but others give a 5 only when they are 100 % satisfied. Thus, there are studies trying to provide an adjusted rating for the reviews [17]. However, most consumers need more than just a score. They actually need a system, which can digest big volume of reviews and produce a set of aspects being mentioned in the reviews together with the pros/cons being expressed to the aspects. Therefore, techniques for review summarization and integration are in great demand for the improvement of online shopping experience.

Many studies have been carried out to address the problem of review summarization. Some solely detect the aspects in reviews [7, 13], while others separate the opinions from the facts [20]. But they are still far from obtaining the opinion orientations associated with the aspects. Predicting the sentiments of the words requires extra knowledge about the words, and sometimes even the knowledge about the aspects being discussed. To conquer this difficulty, researchers developed different topic models incorporating prior information from a set of seed words with general deterministic sentiment orientations [8, 9, 18].

Only one study [2] tried to jointly model ratings and reviews of the movies. It applied an approach based on collaborative filtering and topic modeling, which had some limitations on the dataset due to the nature of collaborative filtering. To the best of our knowledge, no previous research has developed an approach to predict the sentiment orientations of the words in reviews by incorporating ratings information solely based on topic modeling.

The major contribution of this research is to propose a new probabilistic topic model, SentiLDA, for mining reviews (unstructured data) and their ratings (structured data) jointly to detect the product/service aspects and their corresponding positive and negative opinions simultaneously without using seed words. A key feature of SentiLDA is that it is capable of mining positive sub-topics and negative sub-topics under the same aspect topic without prior information of the sentiment seed words. Since it does not rely on domain knowledge, SentiLDA is general and can be applied to similar problems with unstructured text data and structured numerical data.

The second contribution is we implement SentiLDA in Spark [12] following the MapReduce paradigm by using variational inference. It takes the advantage of the parallel distributed in-memory computing environment to scale up and speed up the model inference. Experiment results show SentiLDA outperforms the other related state-of-the-art models in detecting product/service aspects and their corresponding sentiments in reviews.

In addition, the implementation avoids the scalability issue of the traditional Gibbs sampling technique, and thus makes it very suitable for big data analysis in distributed environment. SentiLDA could be used to support other research based on domain specific sentiment words, such as review rating prediction, opinions summarization and Integration.

The remaining of the paper is organized as follows. Section 2 introduces the related works. The proposed model is described in Sect. 3, followed by a brief explanation of the implementation in Sect. 4. Experiment setup, results and analysis are presented in Sect. 5. Finally, Sect. 6 concludes the paper and suggests the future work.

2 Related Work

Early studies in sentiment prediction mainly depended on using WordNet [3] or pointwise mutual information (PMI) [14] to determine the sentiment of a word. However, this approach has difficulties in predicting domain-specific sentiment words. Certain words may be positive in a domain, but negative in another domain, e.g. "big" is good for cell phone screen size, but bad for battery size of the cell phone.

Opinions are always expressed to objects. In order to perform review summarization and integration, it is desirable to know both the sentiment orientation of a word and what aspect it is talking about. Many researchers have been trying to solve this ultimate problem by using topic models. Tying-JST [9], TSM [11], ASUM [8], and JAS [18], are popular models proposed for this objective. Tying-JST modifies LDA by adding one variable to control the sentiment orientations of the words in the reviews. The sentiment variable is drawn from a document level sentiment distribution determined by a Dirichlet distribution. The approach of TSM is similar as Tying-JST. In additional, it introduces a background words variable in the model. ASUM adds some constraints on the basis of Tying-JST. It assumes the words from the same sentence are of the same topic and sentiment. JAS brings in more variables to control the subjectivities of the words and the sentiments of the subjective words. It also assumes each sentence in the review has two sentence-level sentiment distributions for opinion and fact respectively. The models mentioned above all have their drawbacks. Tying-JST and TSM extract topics-sentiments solely based on words co-occurrences, which loses the locality information of the words in the reviews. ASUM restricts the sentiments of the words in the same sentence to be the same, which is not held in many reviews. Rather than discovering T topics with positive and negative words separated, ASUM discovers T positive topics and T negative topics, which requires further post work to perform review summarization and integration. JAS introduces many latent variables, which increase the computational complexity of the model inference. Moreover, the generative process described by JAS is not intuitive as it assumes two sentiment distributions for each sentence in a review. Last but not least, in order to distinguish facts from opinions, and positive sentiment from negative sentiment, all these models heavily rely on a good set of sentiment seed words which is not always easy to obtain.

Review rating has been studied in some research recently. But most of them are trying to do rating prediction or justification based on review context [4, 16]. JMARS [2] is the closest one to our study. It also models aspects, ratings and sentiments jointly on movie reviews. The approach is based on collaborative filtering and topic modeling. In order to perform collaborative filtering, it requires each user writes more than one reviews on different movies. Then the model is able to construct user's expectations and movie's properties. For the reason of privacy protection and data accessibility, it is not easy to obtain multiple product reviews from the same user. Therefore, it prevents JMARS from being applied to the problem we are trying to solve in this research. Furthermore, product reviews are often shorter than movie reviews, the locality information of words are of great importance, but JAMRS fails to model it.

To the best of our knowledge, this is the first study to exploit review rating to predict the aspect-specific sentiment orientations of the words in reviews by solely using topic modeling and without using sentiment seed words.

3 Method

3.1 SentiLDA

In order to extract topics/aspects and their associated sentiment opinion bearing words in the reviews respectively, we proposed a probabilistic graphical model, which follows a hierarchical topic-vocabulary structure shown in Fig. 1. It is a two-level vocabulary hierarchy. The root node V stands for the whole vocabulary of the corpus. The vocabulary is virtually split into K child nodes in the first level, where each node contains all words for one topic/aspect of the product. In the second level each topic/aspect node is divided into three leaf nodes, which are the neutral, positive and negative words of the corresponding topic/aspect respectively. By definition, the neutral words mean purely descriptive ones that do not express any opinion, such as "hotel", "room", and "restaurant" etc. The positive and negative words stand for the ones that convey sentimental opinions, such as "excellent", "terrible", and so on. Negations will be detected and handled appropriately by using Stanford CoreNLP NLPT [10]. These words are not necessarily constrained to be adjectives and adverbs. Nouns and verbs can also bear sentiments, e.g., "noise" is negative and "recommend" express a positive opinion.

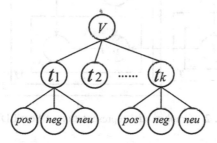

Fig. 1. Hierarchy of the topic-vocabulary structure

In LDA [1], a bag-of-words assumption is proposed, where the relative position of each individual word is neglected. Words located close to each other in the documents can be assigned to totally unrelated topics, which are inappropriate in many scenarios, especially when the model is applied to short documents like reviews. Thus, we made a stricter assumption that words co-occurring in the same sentence must be of the same topic. This assumption is similar to ASUM, but it has two main differences: (1) there is only one topic distribution for each review, compared to a positive and a negative distribution respectively in ASUM; (2) unlike ASUM, the sentiment orientations of the words in the same sentence are not constrained to be the same. Actually it is quite common that two sides of a coin are discussed in the same sentence of a review, e.g., "The restaurant in the hotel was great but fairly expensive." "Great" and "expensive" are two opposite sentiments of the topic restaurant in hotel reviews. The assumption in SentiLDA is more intuitive and logical, because people tend to discuss issues in a review topic by topic. For each topic there would be opinions of both positive and negative side, rather than setting a sentimental orientation first and then choose a topic to write. Moreover, this assumption also reduces the complexity of the model by introducing only one topic distribution instead of two. (3) We observe many narrative sentences in reviews, e.g. "I spend the Xmas with my family at hotel ABC this year", which express no sentiment, but just a fact. In ASUM, it has to be either positive or negative. But in SentiLDA, it could be neutral. Based on our observation and experiment results the proposed model outperforms ASUM. The graphical representation of SentiLDA is shown in Fig. 2. There are D reviews in the whole corpus, where each review consists of M_d sentences, and there are $N_{d,m}$ words in each sentence. The details of the model are described below.

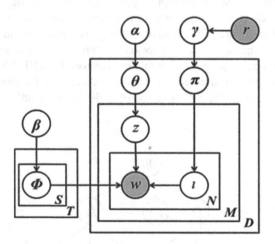

Fig. 2. Graph model representation of SentiLDA

Before actually writing any reviews, first draw three word distributions $\Phi_{t,s} \sim$ Dirichlet (β) for each topic t, in which s corresponds to neutral (facts-topic), positive, and negative sentiment topic respectively. When a reviewer writes a review d, the generative process for each word in a review is as following.

1. Draw a topic distribution $\theta_d \sim$ Dirichlet (α) for the review
2. Draw a sentiment distribution $\pi_d \sim$ Dirichlet (γ) for the review
3. For each sentence m in the review d,
   (a) Choose a topic $t \sim$ Multinomial (θ_d)
   (b) For each token n in the sentence m,
       (i) Choose a sentiment label $\iota_n \sim$ Multinomial (π_d)
       (ii) Choose a word $w \sim$ Multinomial (Φ_{t,ι_n}).

In the SentiLDA, π plays an important role in assigning a sentiment label to each word in the document, and it is generated from a Dirichlet distribution with a hyper parameter γ. For ease of use, it is suggested empirically to use a symmetric hyper parameter for a Dirichlet distribution; however, a symmetric γ means a random sentiment distribution in the proposed model [15]. Without a guidance of the overall sentiment distribution of the review, it is impossible to effectively separate the words into different sentimental orientations, because all the words are clustered solely based on co-occurrences. Fortunately, besides the unstructured text in the reviews, there is also a numerical overall rating being accompanied with the reviews in most of the online review platforms. It can be exploited to provide a clue of the sentiment distribution of the review. However, shown by previous study [17], the review ratings are inconsistent among different users, different review platforms. Therefore, generating a sentiment distribution π just based on the absolute value of the rating is not appropriate. But the ratings could be a very good clue for setting a prior γ for a Dirichlet distribution, which generates a sentiment distribution π. Then, the value of π could be further optimized in the parameters inference. Based on this assumption, SentiLDA is proposed with a variable r. Note that r is in a shadowed node, which means it is an observed value. In the case of modeling reviews, it is the review overall rating provided by the review writer. It determines the value of the prior γ, where $\gamma \in \Gamma$, and Γ is a set of possible priors corresponding to different ratings.

3.2   Model Inference

The key to solving the problem is to infer the latent variables in the proposed SentiLDA model. In practice, the latent variables are derived by maximizing the log-likelihood of the observed data. Given the hyper parameters α, γ, and word distributions over topics Φ, the joint distribution of the latent topic distribution θ, sentiment distribution π, topic assignments z, sentiment assignments ι, and observed words w is given by,

$$p(\theta, \pi, z, \iota, w | \alpha, \gamma, \Phi) = p(\theta|\alpha)p(\pi|\gamma) \prod_{m=1}^{M} \left\{ p(z_m|\theta) \prod_{n=1}^{N_m} [p(\iota_n|\pi_m)p(w_n|z_m, \iota_n, \Phi)] \right\} \tag{4.1}$$

If we integrate and sum over all the latent variables, then the marginal distribution of a review is obtained. After taking product of the marginal probability of every single review, we can obtain the likelihood of the whole set of reviews,

$$p(D|\alpha, \gamma, \Phi) = \prod_{d=1}^{D} \iint p(\theta_d|\alpha)p(\pi_d|\gamma) \Big(\prod_{m=1}^{M} \sum_{z_m}^{T} p(z_m|\theta_d) \\ \times \prod_{n=1}^{N_m} \sum_{\iota_n}^{S} p(\iota_n|\pi_d)p(w_n|z_m, \iota_n, \Phi) \Big) d\pi_d d\theta_d \tag{4.2}$$

There are two ways to find the values of the latent variables to maximize the probability of generating such a corpus, a collapsed Gibbs sampler based on Markov chain Monte Carlo (MCMC) and an inference technique based on variational methods. Due to its simplicity to be understood and implemented, the collapsed Gibbs sampler [5] dominates the research community in solving latent variables inference problem. However, it has several limitations that prevent it from being applied to big data scenario [19]. Therefore, we use variational method as an alternative technique to solve the variable inference problem in our proposed model. Compared to Gibbs sampling, variational inference has the following advantages: (1) there is clear convergence criterion for variational inference; (2) it does not require a shared state during each iteration; (3) it takes less number of iterations, usually 20 to 40, to converge, and thus reduces the communication overhead; and (4) it is able to optimize the hyper parameters due to its statistical nature.

Variational Inference. By introducing variational parameters δ, λ, ε, and η, shown in Fig. 3, the dependencies between θ and z, π and l are dropped. A family of distribution $q(\theta, \pi, z, \iota)$ on the latent variables is obtained. It is used to approximate the true posterior distribution of the latent variables in the proposed model. By minimizing the difference, Kullback-Leibler (KL) divergence, between these two distributions, the optimal values of variational parameters δ, λ, ε, and η can be derived.

$$q(\theta, \pi, z, \iota) = q(\theta|\delta)q(\pi|\lambda) \prod_{m=1}^{M} \left[q(z_m|\varepsilon_m) \prod_{n=1}^{N_m} q(l_n|\eta_n) \right] \qquad (4.3)$$

Minimizing the KL divergence between the variational distribution and true posterior distribution of the latent variables is equivalent to maximizing the evidence lower bound (ELBO) of the corpus.

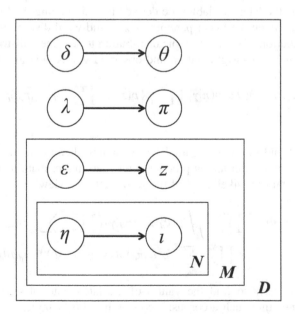

Fig. 3. Graphical Model Representation of the variational approximation of the posterior in SentiLDA

$$\mathcal{L} = E_q[\log p(\boldsymbol{\theta}, \boldsymbol{\pi}, z, \iota, w | \boldsymbol{\alpha}, \gamma, \boldsymbol{\Phi})] - E_q[q(\boldsymbol{\theta}, \boldsymbol{\pi}, z, \iota)] \tag{4.4}$$

Therefore, the updating equations for the variational parameters are obtained,

$$\delta_i = \alpha_i + \sum_{m=1}^{M} \varepsilon_{m,i} \tag{4.5}$$

$$\lambda_s = r_s + \sum_{m=1}^{M} \sum_{n=1}^{N_m} \eta_{m,n,s} \tag{4.6}$$

$$\varepsilon_{m,i} \propto \prod_{n=1}^{N_m} \prod_{s}^{S} \Phi_{i,s,w_n}^{\eta_{m,n,s}} \exp\left[\Psi(\delta_i) - \Psi\left(\sum_{i=1}^{T} \delta_i\right)\right] \tag{4.7}$$

$$\eta_{m,n,s} \propto \prod_{i=1}^{T} \Phi_{i,s,w_n}^{\varepsilon_{m,i}} \exp\left[\Psi(\lambda_s) - \Psi\left(\sum_{s}^{S} \lambda_s\right)\right] \tag{4.8}$$

The word distribution over topics is derived as follow,

$$\Phi_{i,s,j} \propto \sum_{d=1}^{D} \sum_{m=1}^{M} \sum_{n=1}^{N_m} \varepsilon_{d,m,i} * \eta_{d,m,n,s} * w_{d,m,n}^{j} \tag{4.9}$$

Incorporating POS Information. In the decoupled variational model η are the multinomial distributions for the sentiment labels of each word respectively. They will start from some random initial values, and to be updated iteratively during the inference process. However, a fully random initialization of these values may not be sufficient to separate the words of different sentiments effectively. Thus, we exploit the part of speech (POS) information from the reviews to initialize η, which has never been applied by previous researches. We use Stanford CoreNLP NLPT [10] to tag the reviews for the POS information. If the POS tag of a word is a noun or verb, η is initialized to a higher value to neutral, 0.6 in our experiment, and equal values to positive and negative, 0.2 in our experiment. Otherwise, it is initialized to a lower value to neutral (0.2), and 0.4 for both positive and negative.

4 Implementation

Spark is a popular big data processing engine that supports parallel distributed in-memory computing. Every document is independent to each other in the inference procedure, thus it adapts to the paradigm of MapReduce in Spark seamlessly. And the iterative model inference procedure requires the same set of data being processed many times. Spark's ability of doing in-memory computing could reduce the cost of I/O traffic of reading in the data significantly. Instead of reading in the data multiple times in Hadoop MapReduce, it only needs to read in the data once in Spark. Therefore, we implement the proposed model in Spark. The inference procedure can be implemented in two stages, map stage and reduce stage.

4.1   Map Stage: Document Level

There is a set of variational parameters δ, λ, ε, and η for each document. There is no dependency between sets of variational parameters of different documents. Thus, all the documents can be processed parallelly. Equations (4.5)–(4.8), and document level (4.9) are implemented in the map stage. The map stage emits document level word distribution for topics Φ' after processing each document.

4.2   Reduce Stage: Corpus Level Aggregation

Word distribution for topics Φ is a global variable. In order to update it, document level word distribution for topics Φ' has to be aggregated at the corpus level according to different key values, which consists of a topic index, a sentiment index and a word index. Corpus level Eq. (4.9) is implemented in the reduce stage.

5   Experiments

5.1   Data Set

We crawled a set of reviews covering 36 major hotels on the Strip in Las Vegas, USA from 4 websites: expedia.com, hotels.com, orbitz.com, and tripadvisor.com. It contains all the reviews in English and their numerical ratings from each source. We choose hotel reviews because it contains many different aspects, each has a lot domain specific sentiment words. SentiLDA is proposed to solve problems of such characteristics, but not limited to hotel reviews. All reviews were preprocessed through a pipeline consists of tokenization, sentences splitting, lemmatization, and POS tagging by using Stanford CoreNLP Toolkit [10]. Negations were also detected by Stanford NLP Toolkit. Words modified by negations were added with a prefix of "not_". Punctuations and stop-words were removed. Only nouns, verbs, adjectives and adverbs that carry actual meanings were kept. In order to reduce the sparsity of the vocabulary, we further removed words that appear less than 10 times in the corpus since they barely convey meaningful information. All the ratings are in the scale of 1 to 5, and are integers only. Table 1 shows the statistics of the resulting dataset in the experiment.

Table 1. Statistics of the corpus

Rating	# of reviews	# of sentences	# of words
1	25,490	210,719	1,429,640
2	40,000	311,828	2,093,742
3	80,213	575,936	3,814,216
4	157,389	1,026,331	6,679,301
5	185,268	1,097,737	6,959,797
Total	488,360	3,222,551	20,976,696

5.2 Experiment Setting

In this experiment, SentiLDA is compared with ASUM [8], Tying-JST [9] and JAS [18]. All of them are popular models in review aspect/sentiment discovery. However, in the original paper of the above models, they were all inferred by Gibbs sampling. In order make them run faster, we migrate them to Spark by using variational inference as well. We set the number of topics T for all the models to 35, since it could discover all the major features and has the least number of uninterpretable features. Hyper parameter α_i is set to 2 for each aspect in the Dirichlet distribution. $\Phi_{i,s,j}$ is randomly initialized and normalized for all words in an aspect-sentiment distribution. Since the proposed model exploits the review ratings to indicate the prior of sentiment distribution in a review, there is a set of γ corresponding to different ratings. Table 2 shows the different configurations of γ. All the other models do not exploit review rating information and use the default symmetric setting of γ instead. The sum of the elements in all configurations of γ is kept to be 1.

Table 2. Different γ settings by different ratings

Rating	Neutral	Positive	Negative
1	0.75	0.02	0.23
2	0.75	0.07	0.18
3	0.75	0.12	0.13
4	0.75	0.18	0.07
5	0.75	0.23	0.02
Symmetric	0.33	0.33	0.33

5.3 Qualitative Analysis

Tables 3, 4 and 5 show the top 20 words of each sentiment for the customer service aspect obtained from the proposed SentiLDA model. For comparison purpose, the top words of the sentiments for the same aspects derived from JAS and ASUM are also shown in the tables. We have also compared with Tying-JST. Since its result is similar to JAS, we do not include it here due the length limit of this paper. ASUM doesn't extract neutral sentiment words directly, but we can extract them by looking for the common words in positive and negative sentiment.

Table 3. Top 20 words of customer service obtained by SentiLDA for each sentiment

neutral	staff, friendly, helpful, hotel, service, room, great, desk, nice, clean, front, check, good, always, courteous, excellent, pleasant, housekeeping, stay, extremely
positive	concierge, professional, attentive, greet, welcome, make, name, spa, smile, warm, special, reception, level, feel, doorman, gracious, hotel_3, outstanding, impeccable, efficient
negative	rude, customer, unfriendly, unhelpful, attitude, not_helpful, poor, manager, horrible, terrible, bad, management, lack, not_friendly, less, not_care, unprofessional, dirty, worst, employee

Table 4. Top 20 words of customer service obtained by JAS for each sentiment

neutral	make, hotel, stay, help, feel, go, staff, check, need, time, ask, more, get, room, guest, take, way, question, treat, say
positive	staff, service, friendly, helpful, great, hotel, room, nice, excellent, good, customer, concierge, housekeeping, always, professional, pleasant, courteous, polite, best, wonderful
negative	desk, staff, front, rude, customer, people, hotel, employee, manager, work, check, attitude, management, person, kind, guest, poor, speak, extremely, member

Table 5. Top 20 words of customer service obtained by ASUM for each sentiment

positive	staff, friendly, helpful, hotel, clean, nice, room, great, courteous, service, pleasant, extremely, polite, professional, always, casino, check, stay, accommodate, well
negative	Service, hotel, staff, room, customer, poor, top, rude, notch, bad, experience, food, horrible, restaurant, terrible, lack, overall, housekeeping, cleanliness, casino

The comparison shows SentiLDA model captures most of the neutral words that are discovered by JAS and ASUM, such as "staff", "front", "desk", "service", "room", "check", "housekeeping", "stay" in customer service aspect. SentiLDA discovers more aspect-specific sentiment words than both JAS and ASUM, such as "attentive", "greet", "welcome", "smile", "warm", "gracious", "outstanding", "impeccable", and "efficient" in the positive side, "unfriendly", "unhelpful", "not_friendly", "not_care", "not_helpful", and "unprofessional" in the negative side.

Most of the sentiment words discovered by JAS and ASUM are from sentiment seed words set, or closely related to them. Because sentiment seed words are also the words used frequently by people, such as "great", "excellent", "rude", "horrible", they tend to dominate the high possibility words in an aspect sentiment. However, instead of relying on sentiment seed words, SentiLDA exploits review ratings as sentiment distribution prior. It evens out the frequent words to all sentiments according to the sentiment prior. Thus, aspect-specific sentiment words have higher probabilities in the correct sentiments.

Furthermore, SentiLDA is capable of detecting non-adjective sentiment bearing words. Table 6 shows the top 20 words of each sentiment for the aspect of bathroom. Words like "slipper", "robe", "toiletries" are all neutral if mentioned not in the bathroom. But when people talk about amenity in hotel bathrooms, they are definitely good to have. Thus they convey positive sentiment is this situation. On the contrast, the

Table 6. Top 20 words of bathroom aspect obtained by SentiLDA for each sentiment

neutral	room, shower, bathroom, water, bed, floor, towel, clean, day, get, tub, dirty, toilet, hair, carpet, sink, sheet, leave, stain, take
positive	slipper, robe, toiletries, chocolate, lotion, kit, amenity, gel, provide, product, bath, body, shave, cotton, toothbrush, razor, spa, cream, polish, steam
negative	stain, dirty, filthy, carpet, blood, sheet, bug, mold, black, cover, look, disgusting, dirt, wall, gross, break, notice, nasty, foot, find

existence of "stain", "mold", "bug", and even "blood" in the bathroom is obviously a negative sign. Models solely rely on sentiment seed words, such as ASUM, JAS and JST are not able to discover this kind of sentiment aspects.

5.4   Quantitative Analysis

Convergence Test. We study the convergence speed of different models on the training data set. The iterative updates process stops when the improvement of log likelihood is less than 0.01 %. We test on the proposed SentiLDA, ASUM, Tying-JST, and JAS. Figure 4 shows the result. From the plot, we can observe that SentiLDA achieve slightly better log likelihood than ASUM. All of them are much better than the other two. Regarding to the iterations take to convergence, SentiLDA and ASUM are quite similar with 33, 37 respectively. JAS only takes 18 iterations to converge, but with a much worse log likelihood. It may be caused by being trapped in a local optimal. Tying-JST takes the most iteration (78) to converge to the worst result of them. One of the possible reasons would due to the number of variables of Tying-JST. It has a topic variable and sentiment variable for each word in the review. Thus, it needs more iterations to update them all to a stationary state. Another possible reason for the low log likelihood may due to the lack of constraints of Tying-JST. The words in a review can be of any topics and sentiments, the combinations of the values of the variables are much larger than the other models, and then results in low log likelihood.

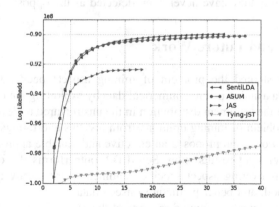

Fig. 4. Comparison of Log Likelihood convergence

Perplexity Test. The ability to predict unseen data is another important metric to evaluate the fitness of a topic model. We divided the data set in this experiment into two parts. We use two thirds of the data as a training set to train a model, and then use remaining one thirds of data as a held-out test set to evaluate the training models. The perplexity of the held-out test set is computed for comparison. The perplexity is calculated by take the inverse of the geometric mean per-word likelihood. A lower perplexity indicates better generalization ability of a model. Table 7 shows the results of the perplexity comparison. SentiLDA obtained the best predictive perplexity on held-out data set. It indicates that the proposed SentiLDA is not only good in extracting aspects-specific sentiment topics from seen data, but also performs well on unseen data. More complex models, such as JAS, and Tying-JST, tend to suffer from over-fitting problem.

Table 7. Perplexities of the held-out test set by applying different models

Model	SentiLDA	ASUM	JAS	JST
Perplexity	646.99	664.40	760.94	991.83

5.5   Result Analysis

From the qualitative analysis and quantitative analysis, it shows the proposed SentiLDA outperforms the other state-of-art opinion mining topic models. It takes less iterations to converge to a higher log likelihood on the training data, and performs better generalization ability in unseen testing data. SentiLDA is capable of discovering aspect specific sentiment words without using sentiment seed words. However, some general sentiment words, such as "nice", "excellent", "great" appear in neutral side after being modeled by SentiLDA. The possible reason might be without the hard constraints of sentiment seed words, some common sentiment strong words that frequently used by people will be detected as common fact(neutral) words by SentiLDA. But the bottom line is they have never been detected as the opposite sentiment.

6   Conclusion and Future Work

In this paper, we studied the problem of mining aspect specific sentiments from unstructured text data and structured numerical data by exploiting the numerical review rating as a prior for the sentiment distribution in the unstructured review. In specific, we defined a novel problem of mining opinions from reviews and ratings without relying on sentiment seed words and proposed an effective and scalable approach to solve this problem. The experiment results show SentiLDA outperforms the other state-of-art topic models in discovering aspect specific sentiment words, converging faster to a higher log likelihood, and better predicting unseen data.

There are some interesting future directions of this study. First, we have not tried to optimize the hyper parameters according to different ratings. It would be interesting to study how the performance of the model would improve if the hyper parameters are optimized. Second, sentiment seed words are not incorporated in this model, in the future we would like to study how we can incorporate the seed words in the model to improve the performance.

References

1. Blei, D.M., Ng, A.Y., Jordan, M.I.: Latent dirichlet allocation. J. Mach. Learn. Res. **3**, 993–1022 (2003)
2. Diao, Q., Qiu, M., Wu, C.Y., Smola, A.J., Jiang, J., Wang, C.: Jointly modeling aspects, ratings and sentiments for movie recommendation (jmars). In: Proceedings of the 20th ACM SIGKDD International Conference on Knowledge Discovery and Data Mining, pp. 193–202. ACM, August 2014
3. Fellbaum, C.: WordNet. Blackwell Publishing Ltd. (1998)
4. Ganu, G., Elhadad, N., Marian, A.: Beyond the stars: improving rating predictions using review text content. In: WebDB, vol. 9, pp. 1–6, June 2009
5. Griffiths, T.L., Steyvers, M.: Finding scientific topics. Proc. Natl. Acad. Sci. **101**(suppl. 1), 5228–5235 (2004)
6. Horrigan, J.A.: Online shopping. In: Pew Internet & American Life Project Report (2008)
7. Hu, M., Liu, B.: Mining opinion features in customer reviews. In: AAAI, vol. 4, No. 4, pp. 755–760, July 2004
8. Jo, Y., Oh, A.H.: Aspect and sentiment unification model for online review analysis. In: Proceedings of the Fourth ACM International Conference on Web Search and Data Mining, pp. 815–824. ACM, February 2011
9. Lin, C., He, Y.: Joint sentiment/topic model for sentiment analysis. In: Proceedings of the 18th ACM Conference on Information and Knowledge Management, pp. 375–384. ACM, November 2009
10. Manning, C.D., Surdeanu, M., Bauer, J., Finkel, J., Bethard, S.J., McClosky, D.: The Stanford CoreNLP natural language processing toolkit. In: Proceedings of 52nd Annual Meeting of the Association for Computational Linguistics: System Demonstrations, pp. 55–60, June 2014
11. Mei, Q., Ling, X., Wondra, M., Su, H., Zhai, C.: Topic sentiment mixture: modeling facets and opinions in weblogs. In: Proceedings of the 16th International Conference on World Wide Web, pp. 171–180. ACM, May 2007
12. http://spark.apache.org/
13. Titov, I., McDonald, R.: Modeling online reviews with multi-grain topic models. In: Proceedings of the 17th International Conference on World Wide Web, pp. 111–120. ACM, April 2008
14. Turney, P.D.: Thumbs up or thumbs down?: semantic orientation applied to unsupervised classification of reviews. In: Proceedings of the 40th Annual Meeting on Association for Computational Linguistics, pp. 417–424. Association for Computational Linguistics, July 2002
15. Wallach, H.M., Mimno, D.M., McCallum, A.: Rethinking LDA: why priors matter. In: Advances in Neural Information Processing Systems, pp. 1973–1981 (2009)
16. Wang, H., Lu, Y., Zhai, C.: Latent aspect rating analysis on review text data: a rating regression approach. In: Proceedings of the 16th ACM SIGKDD International Conference on Knowledge Discovery and Data Mining, pp. 783–792. ACM, July 2010
17. Wu, N., Liu, F., Zhang, J.: A study on consistency of cross-site online reviews. In: The 10th IEEE International Conference on Pervasive, Intelligence and Computing, December 2013
18. Xu, X., Tan, S., Liu, Y., Cheng, X., Lin, Z.: Towards jointly extracting aspects and aspect-specific sentiment knowledge. In: Proceedings of the 21st ACM International Conference on Information and Knowledge Management, pp. 1895–1899. ACM, October 2012

19. Zhai, K., Boyd-Graber, J., Asadi, N., Alkhouja, M.L.: Mr. LDA: a flexible large scale topic modeling package using variational inference in mapreduce. In: Proceedings of the 21st International Conference on World Wide Web, pp. 879–888. ACM, April 2012
20. Zhai, Z., Liu, B., Xu, H., Jia, P.: Clustering product features for opinion mining. In: Proceedings of the Fourth ACM International Conference on Web Search and Data Mining, pp. 347–354. ACM, February 2011

Mining Big Data II

Mining Data Streams with Dynamic Confidence Intervals

Daniel Trabold[1]([✉]) and Tamás Horváth[1,2]

[1] Fraunhofer IAIS, Schloss Birlinghoven, Sankt Augustin, Germany
{daniel.trabold,tamas.horvath}@iais.fraunhofer.de
[2] Department of Computer Science, University of Bonn, Bonn, Germany

Abstract. We consider data streams of transactions that are generated independently with some non-stationary distribution and regard an itemset to be interesting if its average success probability in the data stream reaches a user specified threshold. We propose an algorithm approximating the family of all interesting itemsets in a data stream. Using Chernoff bounds, our algorithm dynamically adjusts the confidence intervals of the candidate itemsets' probabilities. Though the method proposed assumes the itemsets to be independent Poisson trials, our extensive empirical evaluations on synthetic and real-world benchmark datasets clearly demonstrate that it can be applied also to frequent itemset mining from data streams. In addition, the transactions are not necessarily independent. In fact, the experimental results show the superiority of our algorithm over state-of-the-art frequent itemset mining algorithms in data streams if high F-measure and short processing time per transaction are crucial requirements at the same time.

1 Introduction

We consider the following problem of mining data streams: Suppose we receive a sequence $T_1, T_2, \ldots$ of transactions one by one, where each transaction T_i is a subset of some finite ground set I of items and has been generated *independently* with some probability distribution D_i over 2^I. The distributions D_i are *unknown* for the mining algorithm and non-stationary, i.e., they may change over time. For any itemset $X \subseteq I$, this gives rise to the success probability p_i of the event $X \subseteq T_i$ $(i = 1, 2, \ldots)$. Given a sequence $\langle T_1, \ldots, T_t \rangle$ of transactions for some $t > 0$, our goal is to generate *all* itemsets X with average probability at least θ for some user specified threshold $0 < \theta \leq 1$. This problem is of high relevance to a wide range of practical applications including e.g. automatic detection of concept drifts in market basket analysis, payment monitoring and fraud detection. As we will discuss in Sect. 3, it strongly relates also to the problem of mining *frequent* itemsets in data streams.

In many real-world problem settings, including the ones mentioned above, the transactions arrive continuously, typically at a high rate. To process them in a feasible way, streaming algorithms have to address the following challenges: (1) As data streams are typically regarded to be of unbounded length and the

© Springer International Publishing Switzerland 2016
S. Madria and T. Hara (Eds.): DaWaK 2016, LNCS 9829, pp. 99–113, 2016.
DOI: 10.1007/978-3-319-43946-4_7

memory available is limited, some *compact* synopsis is needed to capture the essential characteristics. (2) Another consequence of limited space is that we cannot store *all* the transactions received, implying that the data stream must be processed in a *single pass*. (3) Finally, the data structure devised must allow for a *quick* answering of user queries at any time. To fulfil these requirements, the outputs produced are typically approximate only. Solutions of different algorithms differ from each other in their approximation quality, synopsis size, time per update, and stability.

We present a probabilistic algorithm, called Dynamic Confidence Interval Miner (DCIM), that only *approximates* the exact solution of the pattern mining problem outlined above. Our algorithm relies on probabilistic reasoning, which, in turn, is based on *dynamic* confidence intervals. The confidence intervals are calculated by *Chernoff bounds* and become tighter as more evidence is collected from the stream. They are used to associate the itemsets with one of the following three states: *interesting*, *uninteresting*, and *undecided*. The algorithm stores only all singleton and interesting itemsets, as well as a typically small subset of the family of undecided itemsets. Thus, the space complexity of our algorithm is linear in the size of the family of interesting itemsets.

We first empirically evaluated the performance of our algorithm in terms of F-measure, memory consumption, and running time on artificial datasets. In particular, we generated data streams on simulated Poisson trials with various parameters. The results of these experiments clearly demonstrate the excellent performance of our algorithm: For each dataset, the output was produced in less than 5 ms average transaction processing time and its F-measure was close to 1.

Motivating by Poisson's theorem, we have then studied the suitability of our algorithm also for mining *frequent* itemsets in data streams. The transactions in these experiments were not necessarily independent. Using artificial and real-world benchmark datasets, we have compared the performance of our algorithm, again in terms of F-measure, memory consumption, and running time, to state-of-the-art algorithms mining frequent itemsets from data streams. In particular, we compared DCIM to Lossy Counting [1], FDPM [2], Stream Mining [3], S_Apriori [4], and EStream [5]. While there was no algorithm consistently outperforming all other algorithms in *all* of the above three aspects on *all* datasets, our algorithm performed very well in all settings. In fact, the results of the experiments show that the only algorithm that is in competition with DCIM is the FDPM algorithm [2] which uses also Chernoff bounds. However, FDPM was for some datasets several orders of magnitude slower. The differences to FDPM are discussed in the next section. In summary, DCIM is the algorithm of choice, if F-score and processing time are both crucial properties.

The rest of the paper is organized as follows. We present related work in Sect. 2, formalize the problem setting in Sect. 3, and describe our algorithm in Sect. 4. We report and discuss our experimental results in Sect. 5 and conclude in Sect. 6.

2   Related Work

A large variety of algorithms has been proposed for mining frequent itemsets from data streams. Some of them work under the landmark (e.g. [1–7]) and some under the sliding window model (e.g. [8,9]). For space limitations, we only discuss some selected algorithms working in the landmark model. Besides the frequency threshold parameter θ, some algorithms take also additional parameters. If so, we mention them and describe their purpose.

Stream Mining [3] processes 1- and 2-itemsets immediately and buffers transactions to process larger itemsets later. Processing of larger itemsets is delayed until a certain condition holds; when the process starts, it counts all itemsets level-wise from the buffer from length 3 onwards. After a level has been processed all counters of this level are reduced by 1 and all itemsets with counter 0 are removed. At the end, all itemsets with counter at least $\theta t - r$ will be output, where t is the stream length and r the number of reduction steps until t. This counting and pruning strategy, as we will see in Sect. 5, is however infeasible for most real-world datasets.

S_Apriori [6] takes an error and confidence parameter ϵ and δ, respectively. It divides the stream into blocks of size $2\theta \log(2/\delta)/\epsilon^2$ and uses each block for mining frequent itemsets of some particular length. While, on the one hand, the algorithm is extremely fast, on the other hand it requires the largest transaction buffer size and must process several buffers in order to produce reasonable F-scores. DCIM can work with much smaller buffers and still return very good results from the beginning, in contrast to S_Apriori.

EStream [5] is a complete, but not sound mining algorithm, that has an error and a maximum pattern length parameter $0 < \epsilon < \theta/2$ and L, respectively. It does not use any buffer. For each itemset length l, it specifies a minimum frequency threshold as a function of l and L, and stores all l-itemsets X as candidates if all $(l-1)$-subsets of X reach the minimum threshold calculated for $l-1$. Whenever $\lceil 2^{L-2}\epsilon^{-1} \rceil$ transactions have been processed the algorithm prunes all but the 1-itemsets. Since it uses no buffer, EStream stores many infrequent itemsets. The number of counted itemsets of our algorithm is up to two orders of magnitude smaller than that of EStream. Furthermore, the output depends on the specification of L. Unlike EStream we do not require such a user specified parameter.

Lossy Counting [1] is perhaps the most recognized algorithm. Using an additional parameter $0 < \epsilon < \theta$, typically with $\epsilon \ll \theta$, it provides the following guarantees: (i) The algorithm is complete, i.e., it returns all frequent itemsets, (ii) no itemset with frequency below $(\theta - \epsilon)t$ is generated, where t is the length of the data stream, and (iii) all estimated frequencies are less than the true frequencies by at most ϵt. Lossy Counting first buffers the transactions in as many blocks as memory available, each of size $\lceil \frac{1}{\epsilon} \rceil$ and processes then all blocks together. The algorithm keeps all itemsets with frequency at least ϵ in a data structure and updates them regularly. Since the gap between ϵ and θ is relatively large, a huge amount of unnecessary information is stored by the algorithm. In contrast to the *static* error bound ϵ, we use *dynamic* error bounds derived from Chernoff bounds that become tighter with increasing stream length.

FDPM [2] takes two user defined parameters to control reliability (δ) and the number of blocks in memory (k). The algorithm processes k blocks, each of size $(2 + 2\ln(2/\delta))/\theta$. A larger k reduces the runtime, at the expense of space. For a data stream $\langle T_1, \ldots, T_t \rangle$, all itemsets with frequency at least $\theta - \epsilon_t$ with $\epsilon_t = \sqrt{(2\theta \ln(2/\delta))/t}$ are kept in memory, where ϵ_t is derived from Chernoff bounds. FDPM is sound, but incomplete. Though it is based on the same idea of using dynamic confidence intervals calculated by Chernoff bounds, DCIM has three main features distinguishing it from FDPM:

(1) DCIM has three itemset states instead of two,
(2) it applies a different probabilistic reasoning, and
(3) its output is not necessarily sound.

As we will empirically show in Sect. 5, these differences result in an algorithm that is significantly more stable than FDPM, without any loss in F-measure.

3   The Problem Setting

In this section we formally define the mining problems considered in this work. A data stream of transactions over some finite set I of items (in what follows, simply a data stream) is a sequence $\langle T_1, T_2, \ldots, T_m \rangle$ with $\emptyset \neq T_i \subseteq I$ for all $i \in [m]$, where $[m] = \{1, \ldots, m\}$. For an integer $t \in [m]$ and data stream $\langle T_1, T_2, \ldots, T_m \rangle$, $\mathcal{S}_t$ denotes the sequence $\langle T_1, \ldots, T_t \rangle$. In the problem setting considered in this paper we assume that each T_i is generated *independently* by some *unknown* probability distribution $\mathcal{D}_i$ over the space $2^I \setminus \{\emptyset\}$ of all transactions. For any non-empty itemset $X \subseteq I$ and $i \in [m]$, this random generation of the transactions gives rise to the probability $p_i(X)$ that X is a subset of T_i, i.e., to

$$p_i(X) = \mathbf{Pr}_{T \in \mathcal{D}_i}[X \subseteq T]$$

where the subscript on the right-hand side indicates that the probability is taken w.r.t. the random generation of T according to $\mathcal{D}_i$. For any $X \subseteq I$ and $i \in [m]$, let X_i be the binary random variable indicating whether or not $X \subseteq T_i$. It follows from the definitions above that the X_is are independent Poisson trials with success probability $\Pr(X_i = 1) = p_i(X)$ that is *unknown* for the mining algorithm.

Given a data stream $\langle T_1, T_2, \ldots, T_m \rangle$, a threshold $\theta \in [0, 1]$ specified by the user, and some $t \in [m]$, an itemset $X \subseteq I$ is called *interesting* w.r.t. $\mathcal{S}_t$ if $p^{(t)}(X) \geq \theta$, where $p^{(t)}(X)$ is the average success probability of X in $\mathcal{S}_t$, i.e.,

$$p^{(t)}(X) = \frac{p_1(X) + \ldots + p_t(X)}{t};$$

X is called *frequent* w.r.t. $\mathcal{S}_t$ if $f^{(t)}(X) \geq \theta$, where $f^{(t)}(X)$ is the relative frequency of X in $\mathcal{S}_t$, i.e.,

$$f^{(t)}(X) = \frac{X_1 + \ldots + X_t}{t}.$$

We note that the above notion of *interestingness* strongly relates to that of *frequency*. Indeed, according to Poisson's theorem, for any $\epsilon > 0$,

$$\lim_{t \to \infty} \Pr\left(\left|f^{(t)}(X) - p^{(t)}(X)\right| < \epsilon\right) = 1,$$

i.e., the probability that the (relative) frequency $f^{(t)}(X)$ of X in $\mathcal{S}_t$ arbitrarily approximates to the average success probability $p^{(t)}(X)$ tends to 1 when $t \to \infty$.

Our goal in this paper is to answer interestingness and frequency queries for any $X \subseteq I$, at any time point $t > 0$. More precisely, we consider the following two problems in this paper:

Problem 1: *Given* a single pass over a data stream $\langle T_1, \ldots, T_m \rangle$ with transactions generated independently at random according to some unknown distributions $\mathcal{D}_1, \ldots \mathcal{D}_m$, an integer $t \in [m]$, a threshold $\theta \in [0,1]$, and a query of the form *"Is X interesting w.r.t. $\mathcal{S}_t$?"* for some $X \subseteq I$, *return* TRUE along with $p^{(t)}(X)$ if $p^{(t)}(X)$ is interesting w.r.t. $\mathcal{S}_t$; FALSE otherwise.

Problem 2: *Given* a single pass over a data stream $\langle T_1, \ldots, T_m \rangle$ of transactions, an integer $t \in [m]$, a threshold $\theta \in [0,1]$, and a query of the form *"Is X interesting w.r.t. $\mathcal{S}_t$?"* for some $X \subseteq I$, *return* TRUE along with $f^{(t)}(X)$ if $f^{(t)}(X)$ is interesting w.r.t. $\mathcal{S}_t$; FALSE otherwise.

We assume that such queries are received by the algorithm at time t, the data stream is traversed in one pass, and that the full set of transactions we have already seen cannot be stored by the algorithm.

In the next section we give a probabilistic algorithm approximating the solution of Problem 1. In Sect. 5 we then report extensive empirical results on artificial and real-world datasets which clearly demonstrate that our algorithm is of good approximation quality not only for Problem 1, but also for Problem 2.

4   The Mining Algorithm

In this section we present our algorithm for approximating the solution of Problem 1 defined above in Sect. 3. At any time of the algorithm, each itemset is associated with one of the following three states: *interesting*, *uninteresting*, and *undecided*. The state of an itemset may change over time. The transitions among the states are defined by dynamic confidence intervals calculated by Chernoff bounds. The algorithm stores all interesting and singleton itemsets (independently of their current state), as well as a typically small subset of the family of undecided itemsets. Similarly to other related stream mining algorithms, we assume that the algorithm is provided a buffer that can be used to store B transactions for some $B > 0$ integer.

The main steps of the algorithm are given in Algorithm 1. It has four parameters: A threshold $\theta \in [0,1]$ as defined in Problem 1, two confidence parameters $\delta_1, \delta_2 \in (0,1)$ for defining the dynamic confidence intervals, and the buffer size $B \in \mathbb{N}$. For each itemset $X \subseteq I$ stored by the algorithm, we record the index τ

Algorithm 1. MAIN

Parameters: threshold $\theta \in [0,1]$, confidences $\delta_1, \delta_2 \in (0,1)$, and buffer size $B \in \mathbb{N}$

Initialization:
   for all $i \in I$ **do** $\{i\}.\tau = 1$, $\{i\}.count = 0$, $\{i\}.state = $ INFREQ

Processing transaction T_t at time t
   set Buffer$[t \bmod B] = T_t$
   if $t \bmod B = B - 1$ **then** process and empty the buffer   // see Alg. 2

Processing query X at time t
   if $(X.state = $ FREQ$) \vee (X.state = $ UNDECIDED $\wedge X.count/(t - X.\tau + 1) \geq \theta)$ **then**
       return "interesting" with $X.\hat{p} = X.count/(t - X.\tau + 1)$
   else
       return "uninteresting"

---

of the transaction since X has (again) been stored by the algorithm $(X.\tau)$, the number of occurrences of X in the transactions from transaction T_τ $(X.count)$, the current state of X $(X.state)$, and an upper bound on the number of occurrences of X $(X.oe)$. When a new transaction arrives, we add it to the buffer. If the buffer becomes saturated we first process (Algorithm 2) and then empty it. Finally, when the algorithm receives an interestingness query for an itemset X, it returns "uninteresting", if X is not among the itemsets stored. Otherwise, it checks its current state and returns "interesting" if its state is interesting or its state is undecided and its relative frequency since $X.\tau$ is at least θ.

We now turn to the description of processing the buffer (see Algorithm 2). All itemsets are of state uninteresting initially. We do not store uninteresting itemsets, except for singletons. The state of an itemset X w.r.t. a data stream $\mathcal{S}_t = \langle T_1, \ldots, T_t \rangle$ is based on the following probabilistic reasoning: Let $p^{(t)}(X) = \mu > 0$. Since

$$p^{(t)}(X) = \frac{\mathbb{E}[X_1 + \ldots + X_t]}{t},$$

where $\mathbb{E}$ denotes the expected value, applying Chernoff bounds to Poisson trials we have

$$\mathbf{Pr}\left(\frac{X_1 + \ldots + X_t}{t} \leq (1 - \epsilon)\mu\right) \leq e^{-t\mu\epsilon^2/2}$$

for any $0 < \epsilon < 1$ (see, e.g., Chap. 4 of [10]). Bounding the right-hand side by δ_1, we get

$$\epsilon \geq \sqrt{\frac{-2 \log \delta_1}{t\mu}}.$$

Thus, for any $t > \frac{-2 \log \delta_1}{\mu}$, we have

$$\mathbf{Pr}\left(\frac{X_1 + \ldots + X_t}{t} \leq \left(1 - \sqrt{\frac{-2 \log \delta_1}{t\mu}}\right)\mu\right) \leq \delta_1.$$

Algorithm 2. Dynamic Confidence Interval Miner (DCIM)

Input: frequency threshold $\theta \in [0,1]$, confidences $\delta_1, \delta_2 \in (0,1)$, buffer size $B \in \mathbb{N}$

1: **function** OPTIMISTICESTIMATE(non-empty itemset X)
2:     **return** $\min\limits_{Y \prec X} Y.oe$

Process the buffer at time t:
3: LHE $= \max\left(0, \left(1 - \sqrt{\frac{-2\log\delta_1}{t\theta}}\right)\theta\right)$, RHE $= \min\left(1, \left(1 + \sqrt{\frac{-3\log\delta_2}{t\theta}}\right)\theta\right)$
4: **for all** $\emptyset \neq X \subseteq I$ in increasing cardinality s.t. $\nexists Y \subset X$ with $Y.state = $ UNINT **do**
5:     c = support of X in the buffer
6:     **if** $|X| = 1$ **then**
7:         $X.count = X.count + c$
8:         **if** $X.count/t \geq \theta$ **then** $X.state = $ INT, $X.oe = X.count/t$
9:         **else** $X.state = $ UNINT
10:    **else**   // $|X| > 1$
11:        **if** $X.state = $ UNINT **then**   // i.e., X is not in the data structure
12:            $M = (OE(X) \cdot (t - B) + c)/t$
13:            **if** $c/t \geq$ RHE **then**
14:                $X.state = $ INT, $X.\tau = t - B + 1$, $X.count = c$, $X.oe = M$
15:            **else if** $(c = \min\limits_{Y \prec X} Y.count \vee \forall Y \prec X : Y.state = $ INT$) \wedge M \geq$ LHE **then**
16:                $X.state = $ UNDECIDED, $X.\tau = t - B + 1$, $X.count = c$, $X.oe = M$
17:        **else**   // $X.state \in \{$INT, UNDECIDED$\}$
18:            $X.count = X.count + c$
19:            $X.oe = (OE(X) \cdot (X.\tau - 1) + X.count)/t$
20:            **if** $X.count/t \geq$ RHE **then** $X.state = $ INT
21:            **else if** $X.oe \geq$ LHE **then** $X.state = $ UNDECIDED
22:            **else** $X.state = $ UNINT
    Prune
23: **for all** X with $|X| > 0$ in increasing cardinality **do**
24:    **if** $X.state == $ UNDECIDED $\wedge X.count/(t - X.\tau + 1) < \theta$ **then**
25:        **for all** $Y \supseteq X$ with $Y.state \neq $ UNINT **do**
26:            $Y.state = $ UNINT

Now consider the case that

$$\frac{X_1 + \ldots + X_t}{t} \leq \left(1 - \sqrt{\frac{-2\log\delta_1}{t\theta}}\right)\theta. \tag{1}$$

One can easily check that if X was interesting, i.e., $\mu \geq \theta$, then we had

$$\left(1 - \sqrt{\frac{-2\log\delta_1}{t\mu}}\right)\mu \geq \left(1 - \sqrt{\frac{-2\log\delta_1}{t\theta}}\right)\theta$$

whenever

$$t = O\left(\frac{-\log\delta_1}{\theta}\right).$$

Thus, the probability that (1) occurs and X is *interesting* is bounded by δ_1, implying that X is *not* interesting with probability at least $1 - \delta_1$.

In a similar way, one can show that if $t = O\left(\frac{-\log \delta_2}{\theta}\right)$ and

$$\frac{X_1 + \ldots + X_t}{t} \geq \left(1 + \sqrt{\frac{-3 \log \delta_2}{t\theta}}\right)\theta \tag{2}$$

then X is *interesting* with probability at least $1 - \delta_2$.

Using the probabilistic reasoning sketched above, an itemset X is regarded by the algorithm as uninteresting (UNINT) for a data stream S_t if (1) holds, interesting (INT) if (2) is satisfied; undecided o/w. We process the itemsets in increasing cardinality for two reasons (Line 4 of Algorithm 2): First, we thus only need to enumerate itemsets which satisfy the Apriori property and, in addition, have all subsets in our data structure. Second, we can only compute an upper bound on the relative frequency of an itemset given its subsets, if the subsets have already been processed. All singleton itemsets are processed and stored by the algorithm (see Lines 6–9). For an itemset X of state uninteresting we first calculate an optimistic estimate of its relative frequency (Lines 12 and 1–2). If its relative frequency in the current buffer satisfies condition (2), then we change its state to interesting and store it (Lines 13–14). Otherwise, we change its state to undecided and store it if it does not satisfy (1) and one of the following two conditions holds (Lines 15–16):

(i) The absolute frequency of X in the buffer is identical with the minimum of that of Y over all $Y \prec X$, where $Y \prec X$ holds iff Y is a subset of X obtained by removing one item.
(ii) All proper subsets of X are interesting.

Note that if (i) holds then there is no $Y \prec X$ with state uninteresting; this is guaranteed by the Apriori condition in Line 4. Furthermore, X must be added in this case to the family of undecided itemsets, as it always occurs with its direct descendant Y that has the smallest $Y.count$. Finally, in Lines (18–22) we update the state of interesting and undecided itemsets according to the probabilistic reasoning described above.

Once all itemsets have been processed, for all stored undecided itemsets X with a relative frequency less than θ we delete all undecided proper supersets of X from the data structure (Lines 23–26).

5   Evaluation

In this section we empirically evaluate our algorithm on artificial and real-world datasets. We first experimentally investigate the performance of the DCIM algorithm in terms of F-measure, memory, and runtime for Problem 1 on artificial data streams, where the transactions are generated independently and with different probability distributions. In the second part of this section we then show

Table 1. Algorithms for frequent itemsets mining from data streams using the landmark model.

Algorithm	Error type	Chernoff based	Buffer	Threshold
Lossy Counting [1]	complete, but not sound	no	yes	static
FDPM [2]	sound, but incomplete	yes	yes	dynamic
Stream Mining [3]	sound, but incomplete	no	yes	static
S_Apriori [6]	not sound, incomplete	yes	yes	static
EStream [5]	complete, but not sound	no	no	static
DCIM	not sound, incomplete	yes	yes	dynamic

on artificial and real-world benchmark data streams that our algorithm is suitable not only for Problem 1 but also for Problem 2. In particular, we empirically demonstrate that it outperforms many state-of-the-art algorithms for mining frequent itemsets in data streams both in F-measure and runtime.

The list of the algorithms we have used for the comparison is presented in Table 1 (cf. Sect. 2). For completeness, in the last row of the table we show our algorithm DCIM as well. Column 2 of the table gives the error type of the algorithm and columns 3 and 4 specify whether the algorithm is based on Chernoff bounds and whether it uses a buffer. Finally, Column 5 indicates whether the decision function predicting interestingness is static or may change during the mining process. We implemented our and the other algorithms in Java. Whenever possible we used similar data structures to avoid that one implementation is favoured over another. For all algorithms which use a buffer we process the buffer consistently if it is not full at query time.

To evaluate the performance of our algorithm on Problem 1, we have generated random data streams of length $10\,M$. The transactions in the data streams have been generated independently as follows: Let the parameters n, p_{min}, and p_{max} denote the cardinality of the set I of possible items, the minimum, and the maximum success probabilities of the items, respectively. For each transaction T_t, $(1 \leq t \leq 10\,M)$, we first generated a probability $p_{t,i} \in (p_{min}, p_{max})$ of item i uniformly at random, for all $i \in I$. We then tossed a biased coin with probability $p_{t,i}$ of head, and added i to T_t if and only if the outcome was head. In this way, the transactions were generated independently with different probability distributions. We have used various values for n, p_{min}, and p_{max} resulting in the following four datasets, each of length $10\,M$:

Setting	n	p_{min}	p_{max}
a	50	0.4	0.6
b	100	0	0.4
c	100	0	0.1
d	1000	0	0.05

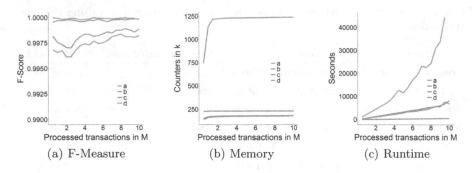

(a) F-Measure   (b) Memory   (c) Runtime

Fig. 1. The influence of the stream length on (a) F-Measure, (b) Memory, and (c) Runtime on the Poisson trial datasets a-d with θ corresponding to 20 k frequent itemsets, $\delta = 0.001$ and buffer size $= 100\,k$.

In Fig. 1 we first present our results for the data streams obtained. The thresholds in these experiments have been selected in a way that they resulted in roughly 20 k frequent itemsets. In particular, in Fig. 1(a) we plot the F-score, in Fig. 1(b) the memory consumption, and in Fig. 1(c) the runtime in seconds, all for different number of processed transactions (axis x). The results clearly demonstrate that for Problem 1, the solutions of our algorithm are of high F-measure, the memory consumption stabilizes after a certain stream length, and the runtime linearly scales with the stream length. One can observe that the F-score increases as the stream gets longer (see Fig. 1(a)). This is conform with Poisson's theorem (cf. Sect. 3). Figure 1(b) indicates that up to a certain number of processed transactions, the space required by our algorithm increases with the length. From a certain stream length, however, the memory required will stabilize. This property can also be explained by Poisson's theorem. Finally, the runtime scales linearly with the stream length in most of these experiments (see Fig. 1(c)), implying that the time needed to process a transaction is nearly constant; an exception can be observed for data stream a beyond 8 M transactions. As this stream needs more memory than the other settings, we attribute the increase in runtime to Java garbage collection.

In the experiments reported in Fig. 1, δ_1 was set to 0.001 and δ_2 to $1 - \delta_1$. We have empirically investigated other values of δ_1 and δ_2 as well and studied their influence on the F-measure, memory, and runtime of our algorithm. The results for various values of δ_1 are presented in Fig. 2, for data stream lengths 1 M, 5 M, and 10 M. One can see that if δ_1 is close to zero then, as expected, the F-measure and the memory increase. However, the change is not significant, indicating that our algorithm is stable with regard to δ_1. Somewhat surprisingly, the runtime seems less effected by the particular choice of δ_1. We have observed a similar tendency for δ_2 on the F-measure and runtime. However, in contrast to δ_1, the memory increases with δ_2. For space limitations, we omit the plots for δ_2.

In the second part of our experiments we have investigated the potential of our algorithm on Problem 2, i.e., on mining frequent itemsets in data streams.

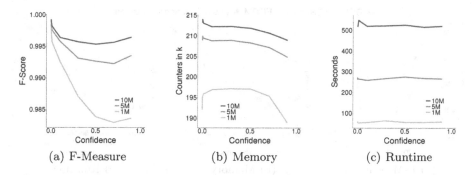

Fig. 2. The influence of δ_1 on (a) F-Measure, (b) Memory, and (c) Runtime on the Poisson dataset c with θ corresponding to 20 k frequent itemsets.

To this extent we have compared the DCIM algorithm to the algorithms listed in Table 1 in terms of F-measure, memory and runtime. We use F-measure for performance evaluation, as our algorithm has two-sided error. The memory is measured by the number of itemsets stored by the algorithm. As this number may increase and decrease, we report the maximum number of counters during the mining process. We have run all experiments on the same hardware and software platform and measured the runtime in seconds.

For the comparison we have used artificial [11] and real-world benchmark datasets. The real-world datasets were taken from the UCI machine learning repository [12]. As the algorithms work with different parameters, we have used the following parameter settings for the algorithms in our experiments, regardless of the dataset at hand, unless otherwise specified:

Algorithm	Parameter(s)
Lossy Counting	$\epsilon = 0.1\theta$
FDPM	$k = 5; \delta = 0.001$
Stream Mining	$\epsilon = 0.75$
S_Apriori	$\epsilon = 0.1\theta; \delta = 0.001$
EStream	$\epsilon = 0.1\theta; k =$ as from ground truth

Most of these parameters were used/recommended in the original papers by the authors, except for δ and k for FDPM. For δ we have used the same value as for δ_1 in our algorithm. In case of FDPM, the value $k = 5$ seemed a natural choice.[1] For each dataset, we used three data dependent frequency thresholds, referred to as *high*, *mid*, and *low*, which correspond to approximately 2 k, 20 k and 200 k frequent itemsets, respectively.

---

[1] In contrast to [2], we wanted to avoid the use of some large k which allows to buffer half the stream.

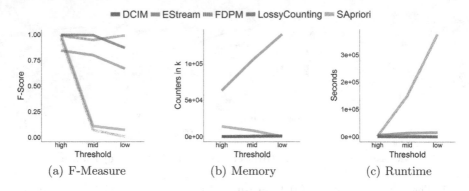

(a) F-Measure   (b) Memory   (c) Runtime

Fig. 3. The impact of θ on F-Measure, Memory and Runtime on *T40I10D100k*. Stream Mining is missing, as it was unable to produce any result in reasonable time. DCIM, FDPM, and S_Apriori required nearly the same space. DCIM, EStream, and S_Apriori had very similar runtimes.

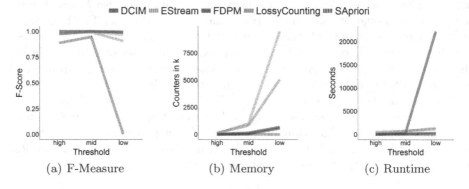

(a) F-Measure   (b) Memory   (c) Runtime

Fig. 4. The impact of θ on F-Measure, Memory and Runtime on the *poker-hand* dataset. Stream Mining is missing, as it was unable to produce any result in reasonable time. DCIM and FDPM required nearly the same space. DCIM and S_Apriori had very similar runtimes.

We first discuss the results obtained for the benchmark datasets T40I10D100k [11] (see Fig. 3) and poker-hand (see Fig. 4). The runtime results of these experiments clearly demonstrate one of the strengths of our algorithm over FDPM. In particular, while our DCIM algorithm and FDPM have roughly the same performance in F-measure and memory (cf. Figs. 3(a), 4(a) and 3(b), 4(b)), our algorithm is significantly faster than FDPM on both datasets (cf. Figs. 3(c) and 4(c)). For T40I10D100k one can observe that both EStream and S_Apriori reach very poor F-scores for the thresholds "mid" and "low" (Fig. 3(a)). Note also that Lossy Counting requires several orders more memory than any other algorithm for this dataset (Fig. 3(b)). For the dataset poker-hand one can observe that all algorithms but S_Apriori reach very high F-scores (Fig. 4(a)). Regarding memory, Lossy Counting and EStream store the most itemsets (Fig. 4(b)); our algorithm records far less itemsets.

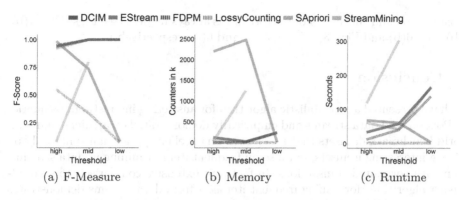

(a) F-Measure   (b) Memory   (c) Runtime

Fig. 5. The impact of θ on F-Measure, Memory and Runtime on the dataset *retail*. The result for threshold "low" could not be calculated for Stream Mining in reasonable time. DCIM, FDPM, and Lossy Counting achieved very similar F-scores and required nearly the same space for thresholds "mid" and "low".

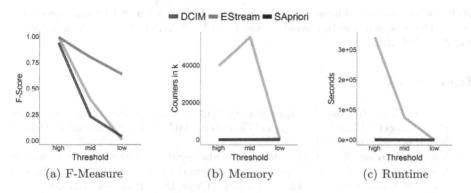

(a) F-Measure   (b) Memory   (c) Runtime

Fig. 6. The impact of θ on F-Measure, Memory and Runtime on the dataset *kosarak*. FDPM, Lossy Counting, and Stream Mining were unable to produce any result in reasonable time. DCIM and S_Apriori used nearly the same space and runtimes.

On the retail dataset (see Fig. 5), DCIM, FDPM, and Lossy Counting reached high F-scores (Fig. 5(a)). Few itemsets were counted by DCIM, FDPM, Lossy Counting and S_Apriori (Fig. 5(b)) and all algorithms were fast except for Stream Mining (cf. Fig. 5(c)).

Finally, for the dataset kosarak (see Fig. 6) one can see that only three algorithms terminated in acceptable time. Our algorithm has by far the best F-measure (see Fig. 6(a)). Together with S_Apriori it also requires significantly less memory and time than EStream (see Fig. 6(b) and 6(c)).

In summary, our algorithm works well on different datasets under varying frequency thresholds and parameters. Its F-score is amongst the best, ranking first or second depending on the dataset and threshold (see Figs. 3(a), 4(a), 5(a), and 6(a)). This is especially remarkable because some algorithms require way more memory and/or time for some of the settings than our algorithm. The memory consumption and runtime of our algorithm is always modest (see Figs. 3(b), 4(b), 5(b), and 6(b) and Figs. 3(c), 4(c), 5(c), and 6(c), respectively).

6   Conclusion

We have presented a probabilistic algorithm for answering interestingness queries on Poisson trial data streams and empirically demonstrated on artificial and real-world benchmark datasets that it achieves an excellent performance not only on mining interesting itemsets, but also on frequent itemset mining in data streams, even with correlated transactions. In fact, our extensive comparison to state-of-the-art algorithms for mining frequent itemsets from data streams demonstrates its superiority if high F-measure and short runtime are both crucial requirements.

We plan to generalize the algorithm to *distributed* data streams using limited communication amongst the sites, without any loss in the approximation quality.

Acknowledgments. The authors thank Michael Mock for useful discussions on the topic. This research was supported by the EU FP7-ICT-2013-11 project under grant 619491 (FERARI).

References

1. Manku, G.S., Motwani, R.: Approximate frequency counts over data streams. In: Proceedings of the 28th International Conference on Very Large Data Bases (VLDB 2002), pp. 346–357 (2002)
2. Yu, J.X., Chong, Z., Lu, H., Zhou, A.: False positive or false negative: mining frequent itemsets from high speed transactional data streams. In: Proceedings of the 30th International Conference on Very Large Data Bases (VLDB 2004), pp. 204–215 (2004)
3. Jin, R., Agrawal, G.: An algorithm for in-core frequent itemset mining on streaming data. In: Fifth IEEE International Conference on Data Mining, pp. 210–217 (2005)
4. Wang, E., Chen, A.: A novel hash-based approach for mining frequent itemsets over data streams requiring less memory space. Data Min. Knowl. Disc. **19**(1), 132–172 (2009)
5. Dang, X.H., Ng, W.K., Ong, K.L.: Online mining of frequent sets in data streams with error guarantee. Knowl. Inf. Syst. **16**(2), 245–258 (2007)
6. Sun, X., Orlowska, M.E., Li, X.: Finding frequent itemsets in high-speed data streams. In: SIAM Conference on Data Mining Workshop and Tutorial Proceedings (2006)
7. Li, C.W., Jea, K.F.: An approach of support approximation to discover frequent patterns from concept-drifting data streams based on concept learning. Knowl. Inf. Syst. **40**(3), 639–671 (2014)
8. Jiang, N., Gruenwald, L.: Cfi-stream: mining closed frequent itemsets in data streams. In: Proceedings of the 12th ACM SIGKDD International Conference on Knowledge Discovery and Data Mining, pp. 592–597. ACM (2006)
9. Cheng, J., Ke, Y., Ng, W.: Maintaining frequent itemsets over high-speed data streams. In: Ng, W.-K., Kitsuregawa, M., Li, J., Chang, K. (eds.) PAKDD 2006. LNCS (LNAI), vol. 3918, pp. 462–467. Springer, Heidelberg (2006)

10. Mitzenmacher, M., Upfal, E.: Probability and Computing. Randomized Algorithms and Probabilistic Analysis. Cambridge University Press, New York (2005)
11. Agrawal, R., Srikant, R.: Fast algorithms for mining association rules. In: Proceedings of the 20th VLDB Conference, pp. 487–499 (1994)
12. Lichman, M.: UCI machine learning repository (2013)

Evaluating Top-K Approximate Patterns
via Text Clustering

Claudio Lucchese[1]([⊠]), Salvatore Orlando[1,2], and Raffaele Perego[1]

[1] ISTI-CNR, Pisa, Italy
claudio.lucchese@isti.cnr.it
[2] DAIS - Università Ca' Foscari Venezia, Venice, Italy

Abstract. This work investigates how approximate binary patterns can be objectively evaluated by using as a proxy measure the quality achieved by a text clustering algorithm, where the document features are derived from such patterns. Specifically, we exploit approximate patterns within the well-known FIHC (Frequent Itemset-based Hierarchical Clustering) algorithm, which was originally designed to employ exact frequent itemsets to achieve a concise and informative representation of text data. We analyze different state-of-the-art algorithms for approximate pattern mining, in particular we measure their ability in extracting patterns that well characterize the document topics in terms of the quality of clustering obtained by FIHC. Extensive and reproducible experiments, conducted on publicly available text corpora, show that approximate itemsets provide a better representation than exact ones.

1 Introduction

Clustering is one of the most studied fields in text mining. Text data are characterized by high dimensionality, and this greatly influences the scalability and the effectiveness of clustering algorithms. One of the most common approaches to reduce dimensionality is to exploit stemming and removal of stop words, and, finally, use a vector representation of documents, where the presence of each term in a vector is weighted by using, for example, TF-IDF. However, stemming and removal of stop words only alleviate the curse of dimensionality, due to the large vocabulary of terms mentioned in a typical document corpus. A technique pursued by seminal papers (see [1,12]) is to exploit an algorithm for mining Frequent Itemset Mining (FIM) to extract a reduced set of more meaningful features, with the aim of shrinking the vectorial representation of documents. In particular, in this paper we focus on the FIHC algorithm (Frequent Itemset-based Hierarchical Clustering), proposed by Fung et al. in their influential paper [3].

The idea of FIHC is to first mine the frequent itemsets, namely *word-sets*, that co-occur in the corpus with a frequency not less than a pre-determined *minimum support* threshold, thus leading to a set of top-k most frequent word-sets. FIHC exploits the extracted frequent word-set to determine an initial clustering of documents, based on the documents that contain or overlap those patterns. The collection of mined word-sets naturally identifies a vocabulary of frequent

© Springer International Publishing Switzerland 2016
S. Madria and T. Hara (Eds.): DaWaK 2016, LNCS 9829, pp. 114–127, 2016.
DOI: 10.1007/978-3-319-43946-4_8

terms, which determines the dimensionality of the vectors used to represent each document. This representation is then used to recursively merge the initial clusters until the desired number of clusters is obtained.

The quality of the top-k word-sets exploited by FIHC impacts on the quality of the resulting clustering. For instance, changing the minimum support threshold has the effect of changing the set of top-k most frequent word-sets, and thus impacts on the feature space into which documents are mapped, as well as on the seed clusters initially identified by each word-sets. Specifically, the smaller the minimum support, the larger the set of the word-sets extracted. In fact, a large support threshold may result in a reduced set of word-sets that occur in many documents and thus having a loose discriminative power. On the other hand, a small threshold may result in too many patterns and, consequently, a large vocabulary of words used in the vectorial representation of documents, again suffering from the curse of dimensionality during clustering.

We propose to use the clustering quality as a *proxy measure* to quantitatively evaluate the quality of different kinds of patterns that are used to feed the FIHC algorithm. In particular, we study *approximate patterns* that identify itemsets that are approximately included in the corresponding sets of transactions [7,9,13]. This means that, given an approximate pattern, some *false positives* are allowed, i.e., some of the items included in the patterns may not occur in a few transactions supporting the pattern. While the *exact* patterns are commonly ranked according to the popular concept of frequency in the collection, the alternative *approximate* patterns we study in this paper are ranked according to different definitions of importance.

To limit the number of patterns used to model documents and identity the initial clusters, for both the exact and approximate cases we select the top-k one. Whereas for frequent ones these top-k patterns are simply the most frequent ones, an approximate pattern algorithm aims at discovering the set of k patterns that best *describes/models*, the input dataset. State-of-the-art algorithms differ in the formalization of the above concept of *dataset description*. For instance, in [9] the goodness of the description is given by the number of occurrences in the dataset incorrectly modeled by the extracted patterns, while shorter and concise patterns are promoted in [7,13]. The goodness of a description is measured with some cost function, and the top-k mining task is casted into an optimization of such cost. In most of such formulations, the problem is proved to be NP-hard, and greedy strategies are therefore adopted. At each iteration, the pattern that best optimizes the given *cost function* is added to the solution. This is repeated until k patterns have been found or until it is not possible to improve the cost function.

In this paper we study the quality of document clustering achieved by exploiting the approximate top-k patterns extracted by three state-of-the-art algorithms: ASSO [9], HYPER+ [13] and PANDA$^+$ [8], where the cost functions adopted by ASSO [9] and HYPER+ [13] share important aspects that can be generalized into a unique formulation. The PANDA$^+$ framework can be plugged with such generalized formulation, which makes it possible to greedily mine approximate patterns according to several cost functions, including the ones proposed in [7,10]. PANDA$^+$ also allows to include maximum noise constraints [2].

Concerning the evaluation methodology of these patterns, we adopt the quality of the clustering obtained by FIHC as a proxy of the quality of the patterns extracted by the various algorithms. Specifically, we use the aforementioned mining algorithms to extract top-k approximate patterns sets, which are then used to fed FIHC in order to cluster the input documents. Several commonly used "external" measures are used to evaluate the goodness of the pattern-based clusters with respect to the true classes of the documents. Moreover, we also compared such methods with a couple of baselines, i.e., K-MEANS and a version of FIHC exploiting classical frequent word-sets (exact patterns).

The main contribution of this paper is an extensive evaluation of approximate patterns. Our investigation shows that approximate patterns provide a better representation of the given dataset than exact patterns, and that PANDA$^+$ generates patterns of better quality than other state-of-the-art algorithms. In other words, our experiments shows that PANDA$^+$ seems to be able to better capture the patterns/features characterizing the most salient topics being discussed in the given corpus of documents.

The rest of the paper is organized as follows. Section 2 discusses exact and approximate pattern mining, and briefly introduces some algorithms for top-k approximate pattern miming. Section 3 discusses the clustering algorithm FIHC, and the possible exploitation within FIHC of either frequent or approximate patterns. In Sect. 4 we describe the experimental setting and the quality of document clustering identified by the various versions of FIHC and the baselines. Finally, Sect. 5 draws some concluding remarks.

2   Approximate and Exact Patterns

The *binary representation* of a transactional dataset, indeed a multi-set of item-set where each itemset is a subset of a given collection of items $\mathcal{I}$, is convenient to introduce pattern mining extracted from textual datasets. A *transactional dataset* of N transactions and M items – which is analogous to representing a corpus of N documents with a vocabulary of M terms as a collections of "sets of words", thus ignoring the positions and the number of occurrences of each term in a document – can be represented by a *binary* matrix $\mathcal{D} \in \{0,1\}^{N \times M}$, where $\mathcal{D}(i,j) = 1$ if the jth item occurs in the ith transaction, and $\mathcal{D}(i,j) = 0$ otherwise.

An *pattern* P is thus identified by a set of items, along with the set of transactions where the items occur. In terms of text documents, P is a word-set occurring in a given set of documents. We represent these two sets as binary vectors $P = \langle P_I, P_T \rangle$, where $P_I \in \{0,1\}^M$ and $P_T \in \{0,1\}^N$ are the *indicator vectors* of two subsets of items and transactions, respectively. The outer product $P_T \cdot P_I^\mathsf{T} \in \{0,1\}^{N \times M}$ identifies a sub-matrix of $\mathcal{D}$. These patterns are also called *hyper-rectangles* [13]: each pattern can be visualized as a rectangle if we properly reorder rows (transactions) and columns (items) to make them contiguous.

If a pattern is *exact*, the sub-matrix only covers 1-bits in $\mathcal{D}$, where $\|P_I\|$ is the *length* of the pattern and $\|P_T\|$ is its *support*, with $\|\cdot\|$ being the L^1-norm (or

Hamming norm) that simply counts the number of 1 bits in each binary vector. Conversely, in case a pattern is approximate, it only approximately covers 1-bits in $\mathcal{D}$ (*true positives*), but it may also cover a few 0-bits too (*false positives*). Still we have that $\|P_I\|$ is the *length* of the patten, and $\|P_T\|$ is its *approximate support*.

2.1   Exact Closed Patterns

Let $\Pi^\sigma = \{P_1, \ldots, P_{|\Pi^\sigma|}\}$ be a set of exact frequent patterns, where σ is the *minimum support ratio*. These patterns may *overlap*, since they may share items or transactions. Therefore, $\forall P \in \Pi^\sigma$, $P = \langle P_I, P_T \rangle$, we have that $\frac{\|P_T\|}{N} \geq \sigma$, where N is the number of documents in the corpus, represented as $\mathcal{D}$.

In this paper, we exploit the popular concept of *closed* frequent patterns, by removing from Π^σ some redundant patterns, since this also prevents the creation of redundant initial seed clusters used by FIHC. Specifically, a pattern $P_i \in \Pi^\sigma$, $P_i = \langle P_I^i, P_T^i \rangle$, is said closed *iff* $\nexists P_j \in \Pi^\sigma$, $P_j = \langle P_I^j, P_T^j \rangle$, such that $set(P_I^j) \subset set(P_I^i)^1$ and $P_T^i = P_T^j$. In other words, we maintain in Π^σ only the frequent itemsets such that there is no other super-itemset occurring in exactly the same set of transactions.

The number of frequent closed item sets may be orders of magnitudes smaller than all the frequent ones, still providing the same information: frequent itemsets can be in fact derived from closed ones. We denote by $\widehat{\Pi}^\sigma$, where $\widehat{\Pi}^\sigma \subseteq \Pi^\sigma$, the set of closed patterns given a minimum support σ. Several frequent closed itemsets mining algorithm [5,14] can be used to mine $\mathcal{D}$.

Since we need to limit the set of (closed) patterns to the top-k most frequent ones, we first select the largest σ_k such that:

$$\sigma_k = \operatorname*{argmax}_\sigma |\widehat{\Pi}^\sigma| \geq k \tag{1}$$

and then select the top-k in $\widehat{\Pi}^{\sigma_k}$, denoted by $\widehat{\Pi}_k^{\sigma_k}$, where the patterns in $\widehat{\Pi}_{\sigma_k}$ are first sorted in decreasing order of support (and then of pattern length). This minimum support σ_k used to identify $\widehat{\Pi}_k^{\sigma_k}$, is then employed by FIHC within specific similarity measures.

In this work, we thus exploit such top-k closed frequent itemsets as a baseline of the possible pattern-based features used to model the text documents that feed the FIHC algorithm.

2.2   Approximate Patterns

Let $\Pi = \{P_1, \ldots, P_{|\Pi|}\}$ be a set of approximate overlapping patterns that aim at best describing/modelling the input dataset $\mathcal{D}$. This means that Π *approximately cover* the 1's in dataset $\mathcal{D}$, except for some noisy item occurrences, identified by matrix $\mathcal{N} \in \{0, 1\}^{N \times M}$:

---

[1] $set(\cdot)$ takes an indicator vector and returns the corresponding subset.

$$\mathcal{N} = \bigvee_{P \in \Pi} (P_T \cdot P_I^\mathsf{T}) \;\veebar\; \mathcal{D}. \tag{2}$$

where $\vee$ and $\veebar$ are respectively the element-wise *logical or* and *xor* operators. Note that some 1-bits in $\mathcal{D}$ may not be covered by any pattern in Π (*false negatives*).

Indeed, our formulation of noise (matrix $\mathcal{N}$) models both *false positives* and *false negatives*. If an occurrence $\mathcal{D}(i,j)$ corresponds to either a false positive or a false negative, we have that $\mathcal{N}(i,j) = 1$.

We define the top-k approximate pattern discovery problem as an optimization one, where the goal is to minimize a given cost function $J(\Pi_k, \mathcal{D})$:

$$\overline{\Pi}_k = \operatorname*{argmin}_{\Pi_k} J(\Pi_k, \mathcal{D}) \tag{3}$$

A general formulation of the cost function J is the following:

$$J(\Pi_k, \mathcal{D}) = \gamma_{\mathcal{N}}(\mathcal{N}) + \rho \cdot \sum_{P \in \Pi_k} \gamma_P(P) \tag{4}$$

where $\mathcal{N}$ is the noise matrix defined by Eq. 2, $\gamma_{\mathcal{N}}$ and γ_P are user defined functions measuring the cost of encoding noise and pattern descriptions, respectively. Constant $\rho \geq 0$ works as a regularization factor weighting the relative importance of the patterns cost. It is worth noting that such cost J is directly proportional to the complexity of the pattern set and the amount of noise, respectively.

The various algorithms for top-k patterns greedily optimize a specialization of the function of Eq. 4. In addition, they exploit some specific *parameters*, whose purpose is to make the pattern set Π_k subject to particular *constraints*, with the aim of (1) reducing the algorithm search space, or (2) possibly avoiding that the greedy generation of patterns brings to local minima. As an example of the former type of parameters, we mention the frequency of the pattern. Whereas, for the latter type of parameters, an example is the amount of false positives we can tolerate in each pattern. Table 1 summarizes the specialization of the generalized cost function.

In the following, we briefly discuss some state-of-the-art algorithms for top-k approximate pattern mining, in turn used to select a significant set of features modelling documents to be clustered by FIHC.

Asso [9] is a greedy algorithm that minimizes function J_A in Table 1, which only measures the amount of noise in describing the input data matrix $\mathcal{D}$. Note that this noise, namely $\gamma_{\mathcal{N}}(\mathcal{N}) = \|\mathcal{N}\|$, is measured as the L^1-norm $\|\mathcal{N}\|$ (or Hamming norm), which simply counts the number of 1 bits in matrix $\mathcal{N}$. Indeed, Asso aims at finding a solution for the *Boolean matrix decomposition problem*, thus identifying two low-dimensional factor binary matrices of rank k, such that their *Boolean product* approximates $\mathcal{D}$. The authors of Asso called this matrix decomposition problem the Discrete Basis Problem (DBP). It can be shown that the DBP problem is equivalent to the approximate top-k pattern mining problem when optimizing J_A. Asso works as follows. First, it creates a set of candidate item sets by extending each item with every other item having correlation grater

Table 1. Objective functions for Top-k Pattern Discovery Problem.

Cost function	Specialization	Description
$J_A(\Pi_k, \mathcal{D})$	$\gamma_\mathcal{N}(\mathcal{N}) = \|\mathcal{N}\|$	Minimize noise [9]
	$\gamma_P(P) = 0$	
	$\rho = 0$	
$J_H(\Pi_k, \mathcal{D})$	$\gamma_\mathcal{N}(\mathcal{N}) = 0$	Minimize pattern set complexity [13]
	$\gamma_P(P) = \|P_T\| + \|P_I\|$	
	$\rho = 1$	
$J_P(\Pi_k, \mathcal{D})$	$\gamma_\mathcal{N}(\mathcal{N}) = \|\mathcal{N}\|$	Minimize noise and pattern set complexity [6,7]
	$\gamma_P = \|P_T\| + \|P_I\|$	
	$\rho = 1$	
$J_P^{\bar{\rho}}(\Pi_k, \mathcal{D})$	$\gamma_\mathcal{N}(\mathcal{N}) = \|\mathcal{N}\|$	Extend J_P to leverage the trade-off between noise and pattern set complexity
	$\gamma_P(P) = \|P_T\| + \|P_I\|$	
	$\rho = \bar{\rho}$	
$J_E(\Pi, \mathcal{D})$	$\gamma_\mathcal{N}(\mathcal{N}) = \mathsf{enc}(\mathcal{N})$	Minimize the encoding length [11] of the pattern model according to [10]
	$\gamma_P(P) = \mathsf{enc}(P)$	
	$\rho = 1$	

than a given parameter τ. Then ASSO iteratively selects a pattern from the candidate set by greedily minimizing the J_A.

HYPER+ [13] is a two-phase algorithm aiming at minimizing function J_H in Table 1, which only considers the pattern set complexity. Specifically, the complexity of each pattern in $P \in \Pi_k$ is measured by $\gamma_P(P) = \|P_T\| + \|P_I\|$. In the first phase, the algorithm aims to cover in the best way all the items occurring in $\mathcal{D}$, with neither false negatives nor positives, and thus without any noise. The rationale is to promote the simplest description of the whole input data $\mathcal{D}$, without any constraint on the amount k of patterns. For this first phase HYPER+ uses a collection of frequent item sets, for a given minimum support parameter σ. In the second phase, pairs of patterns previously extracted are recursively merged as long as a new collection of approximate patters can be obtained without generating an amount of *false positive* occurrences larger than a given budget β. Finally, since the pattern set produced by HYPER+ is ordered (from most to least important), we can simply select Π_k as the top-listed k patterns, as done by the algorithm authors in Sect. 7.4 of [13]. Note that this also introduces false negatives, corresponding to all the occurrences $\mathcal{D}(i, j) = 1$ in the dataset that remain uncovered after selecting only the top-k patterns.

Finally, we considered PANDA$^+$, a pattern mining framework [8] that can be plugged in with all the cost functions in Table 1, including the last three functions J_P, $J_P^{\bar{\rho}}$, and J_E, which can fully leverage the trade-off between patterns description cost and noise cost. In particular, J_E, originally proposed in [10], realizes the MDL principle [11]. The regularities in $\mathcal{D}$, corresponding to the discovered approximate patterns Π_k, are used to *lossless compress* the whole $\mathcal{D}$, expressed as pattern model and noise, as in Eq. 2. Hence, the best pattern set Π_k is the one that induces the smallest encoding of $\mathcal{D}$, namely J_E. PANDA$^+$ adopts a greedy strategy by exploiting a two-stage heuristics to iteratively select each pattern: (a) discover a noise-less pattern that covers the yet uncovered 1-bits of $\mathcal{D}$, and (b) extend it to form a good approximate pattern, thus allowing some false positives to occur within the pattern. Finally, in order to avoid the greedy search strategy accepting too noisy patterns, PANDA$^+$ supports two maximum noise thresholds $\epsilon_r, \epsilon_c \in [0, 1]$, inspired by [2], aimed at bounding the maximum amount of noise along the *rows* and *columns* of each pattern.

3   Frequent Itemset-Based Hierarchical Clustering

In this section we discuss the FIHC framework [3] for document clustering. FIHC implements 3 steps:

1. frequent item sets are mined and transformed in a set of initial grouping of transactions;
2. these groups are refined to produce a partitional clustering;
3. these clusters are recursively merged until the desired number of clusters is obtained.

In the first step, FIHC transforms the input corpus of documents in a binary representation suitable for frequent pattern mining algorithms, where the vector dimensions state the presence/absence of a vocabulary term in a document. According to the framework of Sect. 2, each mined frequent (closed) pattern $P \in \widehat{\Pi}_k^{\sigma_k}$, $P = \langle P_I, P_T \rangle$, trivially identifies a *group* of documents – i.e., the dataset documents corresponding to P_T that support the word-set identified by P_I. Since a transaction may support several frequent itemsets, by construction these document groups may overlap. We focus on closed frequent itemsets [5, 14] as they are a succinct representation of *all* the frequent itemsets, avoid redundancies by definition: this is because they are maximal with respect to the set of supporting transactions, and therefore there are no two closed itemsets supported by the same set of transactions.

We call *candidate* seed clusters the resulting groups of transactions supporting the the various frequent (closed) itemsets extracted, used in the subsequent step of FIHC.

In the second step, FIHC enforces a *partitional* clustering, where each transaction is assigned to only one of the *candidate* clusters. To this end, FIHC uses a function Score($\cdot$), which measures how well a given transaction fits within a candidate cluster. Each transaction is then assigned to the best fitting cluster. The Score($\cdot$) function is defined in terms of the items' *global frequency* and *local frequency*.

Given an item $x \in \mathcal{I}$, the global frequent $\phi_{\mathcal{D}}(x)$ is the ratio $supp_{\mathcal{D}}(x)/|\mathcal{D}|$ that considers all the transactions in the input dataset $\mathcal{D}$ supporting item x, while the local frequency $\phi_{\mathcal{C}}(x)$ is the ratio $supp_{\mathcal{C}}(x)/|\mathcal{C}|$ limited to transactions associated with a given cluster $\mathcal{C}$ under consideration.

On the basis of the extracted pattern set $\widehat{\Pi}_k^{\sigma_k}$, we first prune from $\mathcal{I}$ the infrequent items, thus obtaining $\mathcal{I}' = \{x \in \mathcal{I} \mid \phi_{\mathcal{D}}(x) \geq \sigma_k\}$.

Besides the minimum *global* frequency threshold σ_k, the same used to extract the top-k frequent (closed) patterns, FIHC also defines a minimum *local* frequency threshold σ_{loc}, used to identify the *set of locally frequency items* of cluster $\mathcal{C}$, defined by $LF_{\mathcal{C}} = \{x \in \mathcal{I}' \mid \phi_{\mathcal{C}}(x) \geq \sigma_{loc}\}$.

It is worth recalling that in order to compute $\phi_{\mathcal{C}}(x)$ and $\phi_{\mathcal{D}}(x)$, in this phase we ignore possible multiple occurrences of term x in each transaction/document. Indeed, FIHC combines this concept of global/local frequency with a typical word *weighting* scheme, where the term frequency is instead taken into account. Given a transaction t, which simply represents the presence/absence of the various words in a document, the algorithms builds an associated vector $\overrightarrow{w}_t$, where $\omega_t(x)$ weights the *importance* of term x in the original document, measured by the usual $TF \cdot IDF$ statistics. The matching of a transaction t to a cluster $\mathcal{C}$ is thus defined as a function of such weight vector, that only consider the items in the pruned set $\mathcal{I}'$:

$$\text{Score}(\mathcal{C} \leftarrow \overrightarrow{w}_t) = \sum_{x \in \mathcal{I}', x \in t, x \in LF_{\mathcal{C}}} \omega_t(x) \cdot \phi_{\mathcal{C}}(x) - \sum_{x \in \mathcal{I}', x \in t, x \notin LF_{\mathcal{C}}} \omega_t(x) \cdot \phi_{\mathcal{D}}(x) \quad (5)$$

where the first term of the function *rewards* cluster $\mathcal{C}$ if word x is locally frequent in $\mathcal{C}$, whereas the second term penalizes the same cluster for all the items of t that not locally frequent. The last term encapsulates the concept of dissimilarity into the score.

Intuitively, a cluster $\mathcal{C}$ is *good* for t if there are relatively many items in t that appear in many other transactions assigned to $\mathcal{C}$, and this happens when t is similar to these transactions because they share many common frequent items. Finally, terms with larger $TF \cdot IDF$ values have a larger impact.

According to such scoring function, each transaction in the dataset is associated with one and only one of the *candidate* clusters identified in the previous step. At the end of this second stage, a *partitional clustering* is thus attained, where only a few of the original candidate clusters survived by attracting other transactions.

During the third and last phase, FIHC merges similar pairs of clusters, using an ad-hoc similarity measure. Merging is performed recursively until the desired number of final clusters is reached. Indeed, the *inter-cluster similarity* is defined on top of the above Score($\cdot$) function. Given two clusters $\mathcal{C}_i$ and $\mathcal{C}_j$, all the transactions in the latter are combined and matched to the former. Let $\omega_{\mathcal{C}_j}$ be this combined weight vector, obtained by summing up the weight vectors of all the transactions in $\mathcal{C}_j$, i.e. $\overrightarrow{w}_{\mathcal{C}_j} = \sum_{t \in \mathcal{C}_j} \overrightarrow{w}_t$. Thus, the following cluster similarity is defined:

$$\mathsf{Sim}(\mathcal{C}_i \leftarrow \mathcal{C}_j) = \frac{\mathsf{Score}\left(\mathcal{C}_i \leftarrow \vec{\omega}_{\mathcal{C}_j}\right)}{\Omega} + 1 \tag{6}$$

where Ω in a normalization factor. The whole similarity computed in Eq. 6 is asymmetric and normalized between 0 and 2. It is finally made symmetric by taking the geometric mean of $\mathsf{Sim}(\mathcal{C}_i \leftarrow \mathcal{C}_j)$ and $\mathsf{Sim}(\mathcal{C}_j \leftarrow \mathcal{C}_i)$:

$$\mathsf{Inter\_Sim}(\mathcal{C}_i \leftrightarrow \mathcal{C}_j) = \sqrt{\mathsf{Sim}(\mathcal{C}_i \leftarrow \mathcal{C}_j) \cdot \mathsf{Sim}(\mathcal{C}_j \leftarrow \mathcal{C}_i)} \tag{7}$$

At each step of the recursive merging, the two most similar clusters $\mathcal{C}_i$ and $\mathcal{C}_j$ are replaced by a new cluster $\mathcal{C}_{ij} = \mathcal{C}_i \cup \mathcal{C}_j$.

3.1   Exploiting Approximate Patterns in FIHC

The FIHC framework can be easily adapted to produce a clustering starting from the approximate patterns extracted by algorithms such as ASSO, HYPER+, and PANDA$^+$. Indeed, they return patterns of the form $P = \langle P_I, P_T \rangle$, where each pattern identifies not only a set of items (namely vector P_I), but also a set of related transactions (namely vector P_T). Therefore, these patterns are analogous to those returned by a frequent (or frequent closed) itemset mining algorithm. The only difference is that, due to noise, the item set identified by P_I may be only *approximatively* supported by the set of transactions corresponding to P_T.

In order to apply the score function in Eq. 5, we need to redefine the concept of globally frequent item, since both PANDA$^+$ and ASSO do not use any frequency threshold to extract the patterns.

The original FIHC uses the minimum support threshold σ_k, which works as a sort of *a priori* filter, by limiting the selected features to only the frequent items and disregarding the infrequent ones. Conversely, the patterns returned by PANDA$^+$ or ASSO are not derived on the basis of any frequency threshold, even if both the algorithms extract significant patterns, even if the single items occurring in each approximate pattern are likely to be well supported in the dataset.

In analogy with frequent itemsets, where the single items that occur in any patterns are globally frequent by definition, we consider all the items occurring in the various approximate patterns $P \in \overline{\Pi}_k$ as the ones to be included in the pruned set of items $\mathcal{I}'$, $\mathcal{I}' \subseteq \mathcal{I}$. Thus, in order to permit FIHC to exploit an approximate pattern set, we need to replace the concept of global frequency of an item by the concept of occurrence of the same items in the pattern set.

We argue that a high quality pattern set should boost the quality of the generated clustering by FIHC. This is confirmed by our experimental results, where the clustering quality obtained by using top-k approximate patterns are better than using exact frequent (closed) ones.

4   Experimental Evaluation of Approximate Patterns

We compared the quality of the approximate patterns extracted by PANDA$^+$, ASSO, and HYPER+ by using as a proxy the quality of the clustering obtained by FIHC, which in turn uses such top-k patterns as described in Sect. 3. We run our experiments on four categorized text collections (R52 and R8 of Reuters 21578, WebKB [2], and Classic-4 [3]). The main characteristics of the datasets used are reported in Table 2. As expected, these datasets have a very large vocabulary with up to 19,241 distinct terms/items. The binary representation of those datasets, after class labels removal, was used to extract patterns. The number L of the class labels varies from 4 to 52.

Table 2. Datasets.

Dataset	L	M	N	avg. doc. len
Classic-4	4	5896	7094	34.8
R8	8	17387	7674	40.1
R52	52	19241	9100	42.1
WebKB	4	7770	4199	77.2

During the cluster generation step, the usual $TF \cdot IDF$ scoring was adopted to instantiate $\vec{w}_t$, and $\sigma_{loc} = 0.25$ was used. We forced FIHC to produce a number of clusters equal to L. Even if the goal of this work is to evaluate different solutions for pattern-based clustering, we also reported as a reference the results obtained with the K-Means clustering algorithm, by still setting parameter K of K-Means equal to the number L of classes in the datasets. Finally, cosine similarity was used to compare documents. This baseline is used only to make sure that the generated clustering is of good quality. All the pattern-based algorithms evaluated perform better than K-Means.

The quality of the clusters generated by each algorithm was evaluated with 5 different measures: Jaccard index, Rand index, Fowlkes and Mallows index (F-M), Conditional Entropy (the conditional entropy H_K of the class variable given the cluster variable), and average F-measure (denoted F_1) [4]. For each measure the higher the better, but for the conditional entropy H_K where the opposite holds. The quality measures reflect the matching of the generated clusters with respect to the true documents' classification.

Tables 3,4 report the results of the experiments conducted on the four text categorization collections.

In order to evaluate the benefit of approximate patterns over exact frequent patterns, we also investigated the clustering quality obtained by FIHC with the 50 and 100 most frequent closed item sets. As shown in Table 3, closed patterns provide a good improvement over K-Means. The best F_1 is achieved when 100 patterns are extracted, with an improvement of 13% over K-Means, and similarly for all other measures. This validates the hypothesis of pattern-based text clustering algorithms, according to which frequent patterns provide a better feature space than raw terms.

---

[2] http://www.cs.umb.edu/~smimarog/textmining/datasets/index.html.
[3] http://www.dataminingresearch.com/index.php/2010/09/classic3-classic4-datasets/.

Table 3. Pattern-based clustering evaluation. Best results are highlighted in boldface.

Algorithm	# Patt.	Dataset	F_1 ↑	Rand↑	Jaccard↑	F-M↑	H_K ↓	Avg.Len.	Avg.Supp.
K-Means	L	Classic-4	0.397	0.525	0.214	0.358	1.668	–	–
		R52	0.394	0.687	0.271	0.428	2.630	–	–
		R8	0.523	0.464	0.377	0.596	1.843	–	–
		WebKB	0.508	0.617	0.287	0.453	1.364	–	–
		avg.	*0.455*	*0.573*	*0.288*	*0.459*	*1.876*	–	–
Closed	50	Classic-4	0.470	0.633	0.250	0.400	1.461	1.000	852.220
		R52	0.407	0.769	0.177	0.360	1.894	1.860	2977.580
		R8	0.624	0.671	0.384	0.555	1.404	1.940	2675.980
		WebKB	0.432	0.331	0.277	0.506	1.880	2.020	1828.500
		avg.	*0.483*	*0.601*	*0.272*	*0.455*	*1.660*	*1.705*	*2083.570*
Closed	100	Classic-4	0.472	0.585	0.249	0.401	1.539	1.100	660.890
		R52	0.495	0.819	0.355	0.557	1.888	2.240	2442.170
		R8	0.648	0.692	0.423	0.596	1.364	2.360	2215.390
		WebKB	0.435	0.318	0.281	0.516	1.879	2.400	1534.810
		avg.	*0.512*	*0.603*	***0.327***	***0.517***	*1.668*	*2.025*	*1713.315*
Asso	L	Classic-4	0.452	0.628	0.217	0.357	1.537	1.000	1456.250
		R52	0.300	0.761	0.098	0.272	1.574	4.692	976.385
		R8	0.446	0.680	0.222	0.401	1.258	6.500	1656.875
		WebKB	0.436	0.627	0.200	0.333	1.700	9.000	1142.750
		avg.	*0.409*	***0.674***	*0.184*	*0.341*	***1.517***	*5.298*	*1308.065*
Asso	50	Classic-4	0.519	0.633	0.256	0.407	1.406	1.040	844.300
		R52	0.287	0.762	0.106	0.283	1.669	4.640	995.920
		R8	0.693	0.762	0.454	0.630	1.116	5.040	800.980
		WebKB	–	–	–	–	–	–	–
		avg.	–	–	–	–	–	–	–
Hyper+	L	Classic-4	0.452	0.628	0.217	0.357	1.537	1.000	1456.250
		R52	0.352	0.749	0.117	0.264	1.953	4.558	132.404
		R8	0.368	0.599	0.156	0.283	1.667	6.375	236.000
		WebKB	0.410	0.422	0.248	0.433	1.831	7.500	185.500
		avg.	*0.396*	*0.599*	*0.185*	*0.335*	*1.747*	*4.858*	*502.538*
Hyper+	50	Classic-4	0.480	0.596	0.255	0.409	1.509	1.040	854.700
		R52	0.357	0.749	0.118	0.265	1.962	4.580	136.660
		R8	0.668	0.733	0.404	0.581	1.191	4.840	116.940
		WebKB	0.436	0.313	0.283	0.520	1.883	5.940	70.100
		avg.	*0.485*	*0.598*	*0.265*	*0.444*	*1.636*	*4.100*	*294.600*
Hyper+	100	Classic-4	0.511	0.675	0.271	0.427	1.345	1.010	656.930
		R52	0.480	0.803	0.313	0.511	1.955	3.930	86.190
		R8	0.639	0.665	0.376	0.547	1.315	4.180	75.160
		WebKB	0.437	0.305	0.284	0.525	1.884	5.460	48.960
		avg.	***0.517***	*0.612*	*0.311*	*0.503*	*1.625*	*3.645*	*216.810*

For all the *approximate* pattern mining algorithms, we evaluated the clusters generated by feeding FIHC with L, 50, or 100 patterns.

The Asso algorithm has a minimum correlation parameter τ which determines the initial patterns candidate set. We reported results of $\tau = 0.6$, for which

Table 4. Pattern-based clustering evaluation. Best results are highlighted in boldface.

Algorithm	# Patt.	Dataset	F_1 ↑	Rand↑	Jaccard↑	F-M↑	H_K ↓	Avg.Len.	Avg.Supp.
PANDA$^+$ ($\epsilon = 0.75$)	L	Classic-4	0.439	0.621	0.215	0.354	1.637	3.250	401.000
		R52	0.347	0.771	0.133	0.334	1.653	6.712	578.692
		R8	0.479	0.658	0.214	0.377	1.354	6.250	1468.250
		WebKB	0.361	0.557	0.192	0.324	1.886	14.500	1261.500
		avg.	*0.406*	*0.652*	*0.188*	*0.347*	*1.632*	*7.678*	*927.361*
PANDA$^+$ ($\epsilon = 1.00$)	L	Classic-4	0.471	0.639	0.228	0.373	1.528	3.750	356.500
		R52	0.314	0.765	0.115	0.299	1.730	5.962	558.942
		R8	0.529	0.697	0.266	0.452	1.297	5.250	1676.000
		WebKB	0.351	0.576	0.179	0.305	1.885	22.750	1111.000
		avg.	*0.416*	***0.669***	*0.197*	*0.357*	*1.610*	*9.428*	*925.611*
PANDA$^+$ ($\epsilon = 0.75$)	50	Classic-4	0.498	0.633	0.238	0.384	1.436	2.560	193.120
		R52	0.352	0.769	0.126	0.320	1.624	7.380	566.920
		R8	0.672	0.756	0.435	0.614	1.172	8.220	457.320
		WebKB	0.433	0.331	0.279	0.509	1.886	33.100	325.940
		avg.	*0.489*	*0.622*	*0.269*	*0.457*	*1.530*	*12.815*	*385.825*
PANDA$^+$ ($\epsilon = 1.00$)	50	Classic-4	0.468	0.573	0.242	0.393	1.525	2.320	209.020
		R52	0.320	0.768	0.125	0.316	1.746	12.120	561.400
		R8	0.643	0.698	0.421	0.593	1.277	9.700	486.200
		WebKB	0.426	0.372	0.268	0.479	1.885	11.120	362.580
		avg.	*0.464*	*0.603*	*0.264*	*0.445*	*1.608*	*8.815*	*404.800*
PANDA$^+$ ($\epsilon = 0.75$)	100	Classic-4	0.510	0.647	0.251	0.402	1.444	2.710	158.645
		R52	0.554	0.827	0.376	0.581	1.642	5.700	372.340
		R8	0.704	0.769	0.467	0.642	1.055	6.140	326.360
		WebKB	0.435	0.320	0.282	0.517	1.886	14.190	252.520
		avg.	***0.551***	*0.641*	***0.344***	***0.535***	***1.507***	*7.185*	*277.466*
PANDA$^+$ ($\epsilon = 1.00$)	100	Classic-4	0.490	0.644	0.239	0.387	1.420	3.910	151.110
		R52	0.564	0.826	0.378	0.580	1.649	5.000	374.710
		R8	0.645	0.702	0.420	0.592	1.280	6.790	320.990
		WebKB	0.432	0.337	0.277	0.504	1.885	21.180	229.990
		avg.	*0.533*	*0.627*	*0.329*	*0.516*	*1.558*	*9.220*	*269.200*

we observed the best average results after fine-tuning in the range $[0.5, 1.0]$. We always tested the best performing variant of the algorithm which is named ASSO + *iter* in the original paper. Unfortunately, we were not able to include all ASSO results, since this algorithm was not able to process the four datasets (we stopped the execution after 15 h). We highlight that ASSO is however able to provide good performance on the datasets with a limited number of classes. The results on the other datasets are not as high quality as those obtained by PANDA$^+$.

To get the best performance of HYPER+, we used a minimum support threshold of $\sigma = 10\%$ and we fine-tuned its β parameter on every single dataset by choosing the best β in the set $\{1\%, 10\%\}$. The results obtained with only L patterns are poorer than the K-MEANS baseline, and 50 HYPER+ patterns do not improve over the most frequent 50 closed item sets. However, some improvement is visible with 100 HYPER+ patterns. Both F_1 and Rand index exhibit some improvement over closed item sets, and an improvement over K-MEANS of 14% and 26% respectively.

Finally, we report quality of PANDA$^+$ patternsin Table 4. We tested several settings for PANDA$^+$, and we achieved the best results with the J_P cost function and varying the noise tolerance, namely $\epsilon = \epsilon_r = \epsilon_c$. For the sake of space, we report only results for $\epsilon \in \{0.75, 1.0\}$. Even in this case, L patterns are insufficient to achieve results at least as good as K-MEANS, and 50 patterns provide similar results as the other algorithms tested. The best results are observed with the top-100 patterns extracted. In this case, PANDA$^+$ patterns are significantly better, achieving an improvement over the K-MEANS baseline in terms of F_1 and Rand index of 21% and 41% respectively. In fact, PANDA$^+$ patterns with $\epsilon = 0.75$ provide a better clustering with all of the measures adopted. We thus highlight, that imposing noise constraints $\epsilon < 1$ generally provides better patterns.

Tables 3,4 also report the average length and support of the patterns extracted by the various algorithms (see the last two columns). As expected, the most frequent closed itemsets are also very short, with at most 2.4 items on average. HYPER+ is better able to group together related items, mining slightly longer patterns up to an average length of 5.4 for the WebKB dataset. Unlike all other algorithms, PANDA$^+$ provides much larger patterns, e.g., of length 14.19 for WebKB in the best setting. We conclude that PANDA$^+$ is more effective in detecting items correlations, even in presence of noise, thus providing longer and more relevant patterns which are successfully exploited in the clustering step.

5   Conclusion

This paper analyzes the performance of approximate binary patterns for supporting the clustering of high-dimensionality text data within the FIHC framework. The result of reproducible experiments conducted on publicly available datasets, show that the FIHC algorithm fed with approximate patterns outperforms the same algorithm using *exact* closed frequent patterns. Moreover, we show that the approximate patterns extracted by PANDA$^+$ performs better than other state-of-the-art algorithms in detecting, even in presence of noise, correlations among items/words, thus providing more relevant knowledge to exploit in the subsequent FIHC clustering phase. From our tests, one of the motivation is the higher quality of the patterns extracted by PANDA$^+$, which are longer than the ones mined by the other methods. These patterns are fundamental for FIHC, which exploits them for the initial document clustering which is then refined in the following steps of the algorithm.

Acknowledgments. This work was partially supported by the EC H2020 Program INFRAIA-1-2014-2015 *SoBigData: Social Mining & Big Data Ecosystem* (654024).

References

1. Beil, F., Ester, M., Xiaowei, X.: Frequent term-based text clustering. In: Proceedings of the Eighth ACM SIGKDD International Conference on Knowledge Discovery and Data Mining, pp. 436–442. ACM (2002)

2. Cheng, H., Yu, P.S., Han, J.: Ac-close: Efficiently mining approximate closed itemsets by core pattern recovery. In: Sixth International Conference on Data Mining, 2006, ICDM 2006, pp. 839–844. IEEE (2006)
3. Fung, Benjamin C. M Wang, K., Ester, M.: Hierarchical document clustering using frequent itemsets. In: Proceedings of SIAM International Conference on Data Mining (SDM), pp. 59–70 (2003)
4. Jain, A.K., Dubes, R.C.: Algorithms for Clustering Data. Prentice-Hall, Dubes (1988)
5. Lucchese, C., Orlando, S., Perego, R.: Fast and memory efficient mining of frequent closed itemsets. IEEE Trans. Knowl. Data Eng. **18**, 21–36 (2006)
6. Lucchese, C., Orlando, S., Perego, R.: A generative pattern model for mining binary datasets. In: Proceedings of the 2010 ACM Symposium on Applied Computing, pp. 1109–1110. ACM (2010)
7. Lucchese, C., Orlando, S., Perego, R.: Mining top-k patterns from binary datasets in presence of noise. In: Proceedings of SIAM International Conference on Data Mining (SDM), pp. 165–176. SIAM (2010)
8. Lucchese, C., Orlando, S., Perego, R.: A unifying framework for mining approximate top-k binary patterns. IEEE Trans. Knowl. Data Eng. **26**, 2900–2913 (2014)
9. Miettinen, P., Mielikainen, T., Gionis, A., Das, G., Mannila, H.: The discrete basis problem. IEEE Trans. Knowl. Data Eng. **20**(10), 1348–1362 (2008)
10. Miettinen, P., Vreeken, J.: Model order selection for boolean matrix factorization. In: Proceedings of the 17th ACM SIGKDD International Conference on Knowledge Discovery and Data Mining, pp. 51–59 (2011)
11. Rissanen, J.: Modeling by shortest data description. Automatica **14**(5), 465–471 (1978)
12. Wang, K., Chu, X., Liu, B.: Clustering transactions using large items. In: International Conference on Information and Knowledge Management, CIKM-99, pp. 483–490 (1999)
13. Xiang, Y., Jin, R., Fuhry, D., Dragan, F.F.: Summarizing transactional databases with overlapped hyperrectangles. Data Min. Knowl. Discov. **23**(2), 215–251 (2011)
14. Zaki, M.J., Hsiao, C.J.: Efficient algorithms for mining closed itemsets and their lattice structure. IEEE Trans. Knowl. Data Eng. **17**(4), 462–478 (2005)

A Heuristic Approach for On-line Discovery of Unidentified Spatial Clusters from Grid-Based Streaming Algorithms

Marcos Roriz Junior[1(✉)], Markus Endler[1], Marco A. Casanova[1], Hélio Lopes[1], and Francisco Silva e Silva[2]

[1] Department of Informatics, Pontifical Catholic University of Rio de Janeiro, Rio de Janeiro, Brazil
{mroriz,endler,casanova,lopes}@inf.puc-rio.br
[2] Department of Informatics, Federal University of Maranhão, São Luís, Brazil
fssilva@deinf.ufma.br

Abstract. On-line spatial clustering of large position streams are useful for several applications, such as monitoring urban traffic and massive events. To rapidly and timely detect in real-time these spatial clusters, algorithms explored grid-based approaches, which segments the spatial domain into discrete cells. The primary benefit of this approach is that it switches the costly distance comparison of density-based algorithms to counting the number of moving objects mapped to each cell. However, during this process, the algorithm may fail to identify clusters of spatially and temporally close moving objects that get mapped to adjacent cells. To overcome this answer loss problem, we propose a density heuristic that is sensible to moving objects in adjacent cells. The heuristic further subdivides each cell into inner slots. Then, we calculate the density of a cell by adding the object count of the cell itself with the object count of the inner slots of its adjacent cells, using a weight function. To avoid collateral effects and detecting incorrect clusters, we apply the heuristic only to transient cells, that is, cells whose density are less than, but close to the density threshold value. We evaluate our approach using real-world datasets and explore how different transient thresholds and the number of inner slots influence the similarity and the number of detected, correct and incorrect, and undetected clusters when compared to the baseline result.

Keywords: Answer loss on-line spatial clustering · On-line spatial clustering · On-line grid-based spatial clustering · On-line clustering

1 Introduction

Advances in mobile computing enabled the popularization of portable devices with Internet connectivity and location sensing [1]. Collectively, these devices can generate large position data streams, which can be explored by applications to extract patterns. A mobility pattern that is particularly relevant to many applications is clustering [2],

© Springer International Publishing Switzerland 2016
S. Madria and T. Hara (Eds.): DaWaK 2016, LNCS 9829, pp. 128–142, 2016.
DOI: 10.1007/978-3-319-43946-4_9

a concentration of mobile devices (*moving objects*) in some area, *e.g.*, a mass street protest, a flash mob, a sports or music event, a traffic jam, etc.

A fast and on-line (continuous) detection of spatial clusters from position data streams is desirable in numerous applications [1, 3]. For example, for optimizing traffic flows (*e.g.*, rapidly detecting traffic jams), or ensuring safety (*e.g.* detection of suspicious or anomalous movements of rioters).

Traditionally, such clusters were detected using off-line techniques [3], where data are processed in batch (*e.g.*, from a dataset or a database). Most of these techniques are based on the classic DBSCAN algorithm [4, 5], which provides a density-based definition for clusters. DBSCAN uses two thresholds: an ε distance and the minimum density *minPts* of moving objects to form a cluster. A moving object p that has more than *minPts* moving objects in its ε-*Neighborhood* is a *core moving object*, where the ε-*Neighborhood* of p is the set $N_\varepsilon(p) = \{q \in D \,|\, distance(p, q) \leq \varepsilon\}$ and D is the set of all moving objects. The main idea of DBSCAN is to recursively visit each moving object in the neighborhood of a dense moving object to discover other dense objects. By such, a cluster is (recursively) expanded until no further objects are added to it.

However, to timely detect such clusters, the assumptions of traditional clustering algorithm can become troublesome. For example, to obtain the neighborhood set $N_\varepsilon(p)$ of a moving object p [6], DBSCAN compares the distance between p and the remainder moving object in order to select those that are within ε distance. Since these techniques were designed for off-line scenarios, they can employ spatial data structures, such as R-Trees and Quad-Trees, which provide efficient indexes for spatial data. However, for continuous-mode cluster detection in data streams, this optimization becomes troublesome, due to the high cost of continuously maintaining the spatial tree balanced. To overcome this issue, some approaches employ [5] incremental algorithms to reduce the number of distance comparisons. However, even these approaches become problematic in data stream scenarios due to the difficulty in accessing and modifying the spatial tree in parallel [7, 8], *e.g.*, to handle several data items at once. For example, to avoid inconsistencies in the tree index, these approaches sequentially execute the algorithm for each data item, which does not scale to large data streams (*e.g.*, with thousands of items per seconds).

Algorithms [8–11] that combine grid and density-based clustering techniques have been proposed as means to address such challenges. Such approaches handles this issue by dividing the monitored spatial domain into a grid G of $\frac{\varepsilon}{\sqrt{2}} \times \frac{\varepsilon}{\sqrt{2}}$ cell segments such that the maximum distance between any two moving objects inside a cell is ε. Then, rather than measuring the distance between each pair of moving objects, it counts the number of moving objects mapped to each cell. Cells that contain more than a given threshold (*minPts*) of moving objects are further clustered. This process triggers an expansion step that recursively merges a dense cell with its adjacent neighbor cells. Since cells are aligned in a grid, the recursive step is straightforward. With this approach, the main performance bottleneck is not anymore the distance comparison, but the number of grid cells. A primary benefit of grid-based algorithms is that they transform the problem semantics from distance comparison to counting.

Although the counting semantics enables grid-based approaches to scale and provides faster results over other approaches, they may fail to identify some clusters, a problem known as *answer loss* (or *blind spot*) [12–14]. Answer loss is a problem that happens in any grid-based approach, such as [8–11], due to the discrete subdivision of the space domain into grid cells, which can lead to spatial and temporally close moving objects being mapped to different cells and, thus, not contributing for a cell density that exceeds the *minPts* even though the objects are closer than ε to each other. For example, consider the clustering scenario shown in Fig. 1. Suppose that the threshold for a dense cell is *minPts* = 4. Then, although the moving objects are close to each other *w.r.t.* ε, the grid-based clustering process will not detect such cluster since none of the cells is considered dense.

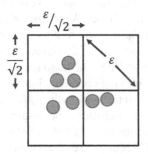

Fig. 1. Answer loss (blind spot) issue in grid-based clustering approaches (*minPts* = 4).

To address the *answer loss* problem, in this paper we propose a counting density heuristic that is sensitive to the number of moving objects in adjacent cells. The idea is to further subdivide cells into "logical" slots and to consider the distribution of moving objects inside these slots on adjacent cells when calculating the cell density. We consider two discrete functions (linear and exponential) to weight the adjacent cells inner slot distributions and combine the result with the cell's own density. However, since the heuristic considers moving objects in adjacent cells when calculating the cell density, it can wrongly detect a cluster that does not exists as a collateral effect. Thus, we propose that the heuristic should be used only when the evaluated cell has a *transient density*, *i.e.*, the number of its objects is less than the required *minPts*, but larger than a lower threshold. We evaluate the tradeoff between the transient cell threshold with the similarity and the number of detected and undetected clusters of a grid-based cluster result using the heuristic, when compared to the baseline (DBSCAN) off-line clustering algorithm. We also evaluate if and how the number of slots impacts the clustering result. Hence, the main contributions of this paper are:

- A counting heuristic, with linear and exponential weights, for mitigating the answer loss problem of on-line grid-based clustering approaches in spatial data streams;
- An extensive evaluation of the proposed heuristic using a real-word data stream.

The remainder of the paper is structured as follows. Section 2 overviews the concepts used throughout the paper. Section 3 presents our approach, the density heuristic that considers the moving objects in adjacent cells. Section 4 extensively evaluates and discusses the proposed technique using a real world dataset. Section 5 summarizes related work, while Sect. 6 states the concluding remarks and our plans for future work.

2 Basic Concepts

As mentioned before, on-line grid-based clustering algorithms [8–11] have been proposed as a means of scaling the clustering process [3]. The overall idea of these approaches is to cluster the grid cells, rather than moving objects, by counting the moving objects in each cell. Since cells are aligned in a grid, the expansion step is straightforward. To do this process, a spatial domain $S = \left(\left[lat_{min}, lat_{max} \right], \left[lng_{min}, lng_{max} \right] \right)$, defined by a pair of latitude and longitude intervals, is divided into a grid G of $\frac{\varepsilon}{\sqrt{2}} \times \frac{\varepsilon}{\sqrt{2}}$ cells. As said, the choice for an $\frac{\varepsilon}{\sqrt{2}} \times \frac{\varepsilon}{\sqrt{2}}$ partition is to guarantee that the maximum distance between any moving object inside a cell is ε. Each cell G_{ij} contains objects in the latitude and longitude intervals respectively defined as:

$$\left(lat_{min} + j \times \frac{\varepsilon}{\sqrt{2}}, \; lat_{min} + (j+1) \times \frac{\varepsilon}{\sqrt{2}} \right] \text{ and } \left(lng_{min} + i \times \frac{\varepsilon}{\sqrt{2}}, \; lng_{min} + (i+1) \times \frac{\varepsilon}{\sqrt{2}} \right]$$

Then, the moving objects in the data stream are mapped to these cells. The number of moving objects mapped to a cell within a given period is called the *cell density*. Cells that have a density greater than or equal to the *minPts* threshold are further clustered (expanded). Due to the grid alignment, this expansion is simpler than the pairwise distance comparison in DBSCAN. The condition for a grid cell to be part of a cluster is to have a density higher than the *minPts* threshold or else to be a border cell of some dense cell, analogously to the core and border objects in DBSCAN. A cluster output is thus the set of cells, core and border, reached from the triggered dense cell.

However, the discrete space subdivision into grid cells can lead to answer loss (or blind spot) problems [12–15]. Due to this division, spatially and temporally close moving objects may be mapped to adjacent cells. Although the moving objects are close *w.r.t. ε*, such cluster will not be detected since no cell is dense *w.r.t. minPts*.

To address this issue, we propose a density heuristic function that considers moving object in adjacent cells when evaluating the *cell density*. The main goal of the heuristics is to discover and output clusters that would not be discovered using only the cell density. We aim at providing a clustering result as close as possible to that of DBSCAN, which we take as ground truth. We will show that our heuristics is capable of on-line detecting a higher number of clusters to the off-line DBSCAN ground-truth result.

3 Proposed Approach

To address the *answer loss* problem while retaining the clustering counting semantic, the proposed heuristic logically divides each adjacent cell into s inner slots, in both dimensions (horizontal and vertical). Then, the density function counts the number of mapped objects in those slots in a way that slots closer to G_{ij} have higher weight than those that are more distant. The exact process is detailed below.

First, each cell is logically divided into vertical and horizontal inner slots (strips) of equal width and height. Each object mapped to a cell G_{ij} is further associated with one of these slots. This operation can be done in constant time during the cell mapping step of a grid algorithm by comparing the location update position with the size of each slot.

The next step is to update the clustering density function by considering the inner density of the neighboring cells, as illustrated by Fig. 2. Not only the number of mapped objects in a cell G_{ij} is considered, but also the distribution of moving objects in the inner slots of each neighboring cell in $N_\varepsilon(G_{ij})$. To do this, we propose a discrete decay weight function that counts the number of moving objects inside each of the inner slots of each neighbor cell in such way that slots closer to G_{ij} receive a higher weight. The closer slots indexes vary according to the position of the neighboring cell, as shown by the darker shades of gray in Fig. 2. Thus, to avoid such specific cases when computing the density function, we "normalize" the neighboring cells' inner slots distribution. To do so, we reorder the neighbor cells $\rho \in N_\varepsilon(G_{ij})$ inner slots $n_{\rho,i}$ in such a way that the first slot, $i = 0$, is closer to the evaluated cell, while the last slot, $i = s - 1$, is the farthest, to enable them to be handled as if they were aligned in the same position.

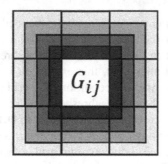

Fig. 2. Density neighborhood of a given cell. Note that the neighbor's closer inner slots vary according to the position of the neighbor.

After normalization, the density function can be described as:

$$density\left(G_{ij}\right) = \left|G_{ij}\right| + \sum\nolimits_{\rho \in N_\varepsilon(G_{ij})}\left(\sum\nolimits_{i=0}^{s-1} w_i \times n_{\rho,i}\right)$$

where $\left|G_{ij}\right|$ is the number of moving objects contained in G_{ij}, $\rho \in N_\varepsilon(G_{ij})$ is a cell adjacent neighbor, s is the number of inner slots, $n_{\rho,i}$ is the number of moving objects in slot i index of neighboring cell ρ, and w_i is the ith decay weight.

We propose two discrete weight decay functions, a linear and an exponential one, illustrated by Fig. 3 (for $s = 4$ inner slots). The linear decay weights can be computed as $w_i = \frac{-i}{s} + 1$, where i is the given cell inner slot index, $e.g.$, considering $s = 4$ as inner slots, the slots weights are $w_0 = 1, w_1 = 0.75, w_2 = 0.50$, and $w_3 = 0.25$. The exponential decay weights can be computed as $w_i = k^i$, where k is a number between 0 and 1 such that $k^s \cong 0$. Based on this definition, k varies accordingly to the number of inner slots s. For example, considering that cells have $s = 4$ inner slots, k value is approximately 0.3162, $i.e.$, $0.3162^4 \approx 0$, while for grid cells that have $s = 10$ inner slots, k is approximately 0.6309, since $0.6309^{10} \approx 0$. To discover k, one can assume $k^s = 0.01$, then $\ln k^s = \ln 0.01$, which yields $k = e^{\frac{\ln 0.01}{s}}$. For example, as said, for $s = 4$, k is approximately 0.3162 and the slot weights are $w_0 = 1$, $w_1 \approx 0.3162$, $w_2 \approx 0.3162^2 \approx 0.099$ and $w_3 \approx 0.3162^3 \approx 0.0312$.

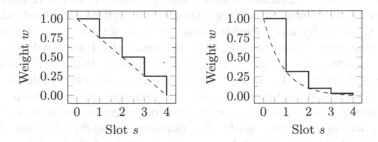

Fig. 3. Linear and exponential discrete weight for the density function (for $s = 4$ inner slots).

By applying the discrete weight function to the neighboring cells inner slots, the proposed heuristic can detect several answer loss clustering scenarios using a counting semantics. For example, consider the clustering scenario of Fig. 4(a) and parameters $s = 4$ and *minPts* = 4. Since the cell density is 2, the standard grid algorithm would not detect the cluster. Using the proposed heuristic, with a linear weight decay, the computed density will be $2 + (1 \times 1) + (4 \times 0.75) = 5 \geq minPts$, thus, the cluster would be detected. An exponential decay weight will also detect this cluster, since the computed density would be $2 + (1 \times 0.3162^0) + (4 \times 0.3162^1) = 4.26 \geq minPts$.

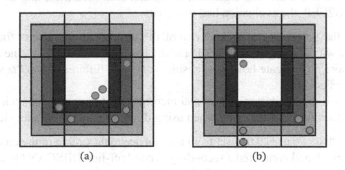

Fig. 4. Cell configurations scenarios. In (a) the scenario forms a cluster, while in (b) it does not.

On the other hand, as a collateral effect of considering moving objects of neighboring cells when calculating the cell density, the proposed heuristic would detect a non-existing cluster (a false positive) in some situations, as illustrated in the cell configuration of Fig. 4(b), for $s = 4$ and *minPts* = 4. In this scenario, the standard grid algorithm would correctly not detect the cluster, since the cell density is 1. However, the linear weight decay would wrongly detect the cluster, since the cell density in this case would be $1 + (1 \times 1) + (1 \times 0.75) + (2 \times 0.5) + (1 \times 0.25) = 4 \geq minPts$. Nevertheless, in this scenario, the exponential weight decay would correctly not detect such cluster, since the computed density would be $1 + \left(1 \times 0.3162^0\right) + \left(1 \times 0.3162^1\right) + \left(2 \times 0.3162^2\right) + \left(1 \times 0.3162^3\right) = 2.54 \leq minPts$.

To mitigate the heuristic collateral effect of detecting non-existing clusters, we propose to only apply the method when evaluating *transient cells*, that is, cells whose density are lower than *minPts*, but higher than a lower-bound *lowerPts* threshold, where *lowerPts* $\leq$ *minPts*. By using a lower-bound threshold, we can restrict the heuristic application to specific scenarios, *e.g.*, to almost dense cells. For example, consider a *lowerPts* = 2 threshold in the clustering scenario of Fig. 4(b). In this configuration, the linear weight heuristic will correctly not detect the cluster, since the cell density would be 2. In addition to mitigating false positive answers, transient cells also reduce the overall computational cost, since the heuristic will only apply to cells whose density are within the interval *lowerPts* $\leq$ *minPts*. For example, a *lowerPts* = 0 means that the heuristic is applied to every updated moving objects. Although the heuristic employs a counting semantic, for large data streams, this can increase the computation cost and delay the cluster discovery. However, using a high *lowerPts* threshold value may again raise the answer loss problem, since the heuristic would only apply to values close to *minPts*. For example, in Fig. 4(a), for *lowerPts* = 3, the cluster would not be detected, since the heuristic density would not be applied.

4   Evaluation

The proposed heuristic was evaluated using real-world data streams and DG2CEP [9], a grid-based on-line clustering algorithm for spatial data streams. Using the heuristic-enhanced DG2CEP, the evaluation had two goals:

- Measure the percentage and similarity of the on-line cluster results found, when compared with the original DG2CEP and the baseline DBSCAN off-line algorithm. Furthermore, investigate how these results vary with different *lowerPts* values that define transient cells.
- Investigate if the number of correct and incorrect clusters found and their similarity with the baseline algorithm vary when using different number of inner slots s.

To do so, first, we used a dataset from the bus fleet data stream of the city of Rio de Janeiro, in Brazil, and computed a second-by-second off-line DBSCAN algorithm over this dataset. This second-by-second off-line clustering output is used as the ground-truth result, which enables us to compare the on-line clustering result when using the proposed heuristic. By comparing the outputs, in addition to the cluster similarity (*e.g.*, precision), we can measure the number of detected and undetected (missed) clusters.

4.1   Implementation

Since we aim at discovering the spatial clusters in real-time, we implemented the proposed heuristic as an module in DG2CEP [9], an on-line grid-based spatial clustering algorithm which uses the counting semantic to cluster in real-time streams of spatial data. We opted to implement the heuristic in DG2CEP due to its on-line processing flow, rather than operating in batch as other grid-based algorithms do [8, 10]. To process on-line streams and provide real-time results, DG2CEP uses Complex Event Processing (CEP) concepts [15]. CEP provides a set of data stream concepts based on continuous queries and event-condition-action rules, such as filtering and enrichment. Each continuous operation is performed by a processing stage, named Event Processing Agent (EPA). DG2CEP defines a network of such stages (EPAs) to enable a continuous detection of the formation and dispersion of spatial clusters.

Both the heuristic-enhanced and the original DG2CEP algorithms were implemented in the Java programming language and made use of the popular open source Esper CEP engine [16]. The heuristic uses several CEP processing stages (EPAs), such as filter and projection, to identify and extract cells that are within the *lowerPts* transient threshold. In addition, when a location update is mapped to a different cell then a previous one, we recheck the cell density with the heuristic to verify if such cell are still dense *w.r.t. minPts*.

4.2   Input Data Stream

The real world dataset is based on the bus fleet of the city of Rio de Janeiro, Brazil. We crawled the dataset[1] from the data.rio open platform, which contains trajectory data for the city 11,324 buses, with an average location update interval of 60 s. It is delimited by the following latitude and longitude intervals: $[-23.0693, -22.7664]$ and $[-43.7849, -43.1429]$, and contains the data of a single day, July 24th, 2014. We choose this day since it was the day with the highest number of updates over the week period.

As mentioned, to evaluate the number of discovered clusters, correct and incorrect, and the results similarity of the proposed heuristic over an on-line data stream, we computed the second-by-second ground-truth cluster result using the off-line DBSCAN clustering algorithm. To do so, at every second, we execute a DBSCAN over the dataset, considering the location updates in the 60 s time window, and store the clustering result as ground-truth information. Since this is a costly task, we opted for restricting this analysis to a two hour rush period of the dataset, from 17:00–19:00, given the probability of having more clusters due to traffic jams of the rush-hour traffic load.

4.3   Measurement Metrics

Whenever the proposed heuristic discovers a cluster we take a snapshot of its moving objects to analyze at a later time. Using this information, we compare the discovered clusters with their counterparts in the ground-truth log. A cluster c is discovered if, in the ground-truth snapshot, there exists a cluster d such that the overlap between c and d is larger than 50 %, that is, the discovered cluster in the heuristic enhanced DG2CEP contains at least 50 % of the moving objects of the ground-truth result. If the heuristic

---

[1] The crawled dataset is available to be downloaded and reproduced at: http://www.lac.inf.puc-rio.br/answerloss/.

wrongly detects a cluster, *i.e.*, no similar cluster exists in the ground-truth log, then this detected cluster is marked as false positive (*FP*). All clusters not detected by the heuristic but present in the ground-truth log are marked as false negatives (*FN*). By comparing these metrics, the percentage of incorrectly detected clusters (*FP*) and missed clusters (*FN*), to the total number of clusters in the ground-truth log, we can measure the heuristic's effectiveness of handling the answer loss problem.

To understand the quality of detected clusters, after detecting a cluster we measure its similarity with its DBSCAN counterpart using the F_1 metric. F_1 is a accuracy metric that considers the precision and recall between the detected cluster and its ground-truth counterpart. Its values range from 0 to 1, such that the closer the value is to 1 the more accurate the results are. Precisely, when F_1 is 1 the cluster contain the same moving objects, while 0 means that they are totally different (no common objects). It is defined as $F_1 = 2 \times \dfrac{precision \times recall}{precision + recall}$, where $precision = \dfrac{tp}{tp + fp}$ and $recall = \dfrac{tp}{tp + fn}$, considering that *tp* is the number of pairs of moving objects in both clusters, *fp* as the number of pairs of moving objects that is in the detect cluster but not in the ground-truth result, and *fn* the number of pairs of moving objects that is in the ground-truth result but not in the detected cluster.

4.4 Parameters of the Experiment

We choose the parameters of the experiment based on a hypothetic application for detecting traffic jams or massive events caused by a concentration of buses in a given region. In this scenario, the spatial clusters are formed when a set of different buses that are close to each other exceeds a given size threshold. Such application can be useful to dynamically control the number of buses serving the citizens or, for example, to alert other drivers to avoid the specified region.

Based on the application scenario and the data stream frequency, we set the window size to be $\Delta = 60$ s, that is, we consider the location updates received within the last 60 s. Furthermore, we choose two grid sizes, $\varepsilon = 100$ m and $\varepsilon = 200$ m, for the respective density thresholds *minPts* = 20 and *minPts* = 40.

To measure the impact of the proposed heuristic, we considered *lowerPts* thresholds ranging from 90 % to 30 % of the *minPts* density threshold. For example, for *minPts* = 20, we evaluate the following *lowerPts* thresholds: 18, 16, 14, 12, 10, 8, and 6, while for *minPts* = 40 we considered the following *lowerPts* thresholds: 36, 32, 28, 26, 24, 20, 16, and 12. Finally, we choose the number of slots to be $s = 10$. Thus, for $\varepsilon = 100$ m, each inner slot width is approximately 10 m, while for $\varepsilon = 200$ m, it is 20 m. We choose these values considering that the GPS accuracy is approximately between 10 to 20 m. However, in this evaluation, we also investigated if the number of slots *s* impact the number and similarity of the detected clusters. For this test, we considered the following number of slots: 10, 50, and 100.

4.5   Experiment Setup

We executed all experiments in the Microsoft® Azure Cloud, using two virtual machines running Ubuntu GNU/Linux 14.04.3 64-bit and the OpenJDK 1.7.91 64-bit Java runtime. The first machine contains a DG2CEP deployment, while the second one contains a dataset playback tool. The virtual machines were interconnected through a Gigabit link/bus and had the following hardware configuration:

- 4 Cores Intel® Xeon CPU E5-2673 v3 @ 2.40 GHz
- 14 GiB Memory RAM

4.6   Result and Analysis

In this subsection, we present the evaluation results. Each experiment was run 10 times and the error bars in the graphs represent a 99 % confidence interval. Each test replays a two-hour dataset, 17:00–19:00, to the input stream of the on-line clustering algorithm.

Figure 5 illustrates the percentage of undetected clusters (false negative - *FN*) and incorrectly detected clusters (false positive - *FP*) of the total clusters found by the heuristic enhanced DG2CEP when compared with DBSCAN's ground-truth results, for parameters $\varepsilon = 100$, *minPts* = 20, and $s = 10$. It also illustrates how each of these metrics and the similarity (F_1) of the set of detected clusters vary when using different *lowerPts* thresholds, where the *lowerPts* threshold is represented as a percentage of the original *minPts* thresholds. Subfigure (a) and (b) of represents the values obtained when evaluating the heuristic with linear and exponential weights respectively.

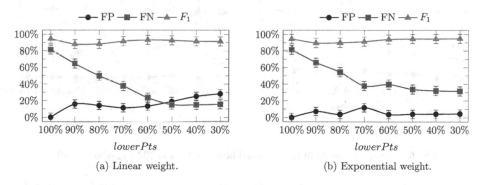

(a) Linear weight.   (b) Exponential weight.

Fig. 5.   Evaluation result of the proposed heuristic for $\varepsilon = 100$ and *minPts* = 20.

As expected, in both cases (linear and exponential), due to the *answer loss* problem, the graph shows that when *lowerPts* is equal or close to *minPts* (100 % and 90 %) the algorithm will fail to detect the majority of the clusters *(FN)*. However, since the threshold is closer to *minPts*, the original density threshold, there will be few collateral effects, hence, the low percentage of incorrect detected clusters *(FP)*.

According to the linear weight graph, Fig. 5(a), the *lowerPts* thresholds that yielded the best tradeoff results were 60 % and 50 % of *minPts*. These thresholds reduced the number of undetected clusters from 80 % to 23.57 % and 15.32 %, respectively, with a collateral effect of incorrect clusters of 13.51 % and 19.05 %, respectively. Interestingly, in the heuristic, using both weights (linear and exponential), the detected clusters of all thresholds presented a high precision and recall (F_1) of over 90 %, with its ground-truth counterpart.

The heuristic exponential weight graph, Fig. 5(b), presented better results as *lowerPts* decreases. This illustrates that, when using the exponential weight, the heuristic is more tolerant to collateral effects. For example, the number of incorrect clusters (false negative) results is 4.47 % for a *lowerPts* equal to 30 % of *minPts*. However, for this parameter, the heuristic reduced the number of cluster not detected due to the answer loss problem, from 80 % to 31.51 %, instead to 13.51 % when using linear weights.

Similarly to Fig. 5, Fig. 6 illustrates the number of *FN* (undetected clusters), *FP* (incorrectly detected) and the similarity of the clusters detected, for parameters $\varepsilon = 200$, *minPts* = 40, and $s = 10$. The graph results are similar to those obtained for $\varepsilon = 100$ and *minPts* = 20. The major difference is that, for both weight functions, the detected clusters similarity $\left(F_1\right)$ with their ground-truth counterpart decreases to an average of 80 %, while the clusters from the previous experiment had over 90 % of similarity, for all parameters configuration. The primary reasons for this reduction comes from having a bigger cell ε width. Indeed, remember that a grid cluster is a combination of dense cells and their borders cells, *i.e.*, in addition to the moving objects inside the dense cells, all border cells' moving objects are also included. Thus, a bigger cell width can potentially include a higher number of objects that are not part of the cluster.

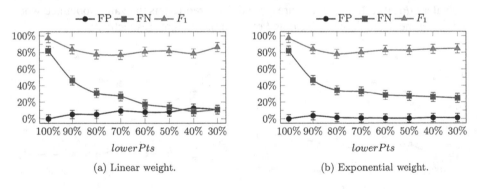

(a) Linear weight.   (b) Exponential weight.

Fig. 6. Evaluation result of the proposed heuristic for $\varepsilon = 200$ and *minPts* = 40.

Another difference between this experiment and the previous one is that, for the linear weight heuristic, Fig. 6(a), the graph presents a reduction of approximately half of the collateral effects, *i.e.*, the number of wrongly detected clusters. More precisely, the number of false positive results dropped from 13.51 % to 7.97 %, for a *lowerPts* threshold equal to 60 % of *minPts*. As the cell width increases, the slot size also increases, and it can lead to a better representation of the objects distribution.

With respect to the relationship between the number of cell divisions and the heuristic results, Figs. 7 and 8 shows how the number of undetected clusters (false negative) and incorrectly detected clusters (false positive) vary for different number of cell slots when using a linear and exponential weight respectively for parameters $\varepsilon = 100$ and

$minPts = 20$. Overall, the results presented a similar behavior. Intuitively, we expected that the heuristic would present a better result as the number of slots increased. However, this was not the case. It seems that the high number of cell slots combined with the GPS error failed to grasp the correct distribution of objects within the cell.

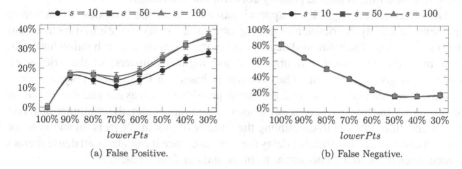

(a) False Positive.   (b) False Negative.

Fig. 7. Relationship between heuristic results and number of cell slots s, for linear weights

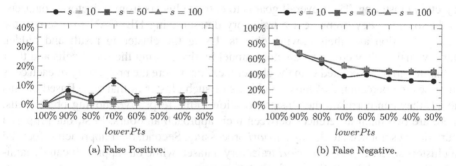

(a) False Positive.   (b) False Negative.

Fig. 8. Relationship between heuristic results and number of cell slots s, for exponential weights.

5   Related Work

In this section, we compare the proposed heuristic with other solutions that address the answer loss problem. In general, these approaches are aimed at off-line scenarios and extensively use spatial index and operators, contrarily to the counting semantic.

Ni *et al.* [13] presented two techniques, an exact and an approximate method that solve the *answer loss* problem in spatial dense queries. Both methods rely on a spatial index TPR-Tree, an R*-tree variant, to index the moving object trajectories. Furthermore, they extended the cell density concept to dense points, as if each point in the cell had its own *ε-Neighborhood* radius. The first method uses a filtering and refinement strategy. Cells that contain less than *minPts* moving objects, but when combined with their neighborhood cells surpass this threshold, are filtered to be further analyzed. The refinement step applies a detailed plane-sweep algorithm to count the number of moving objects in the cell's neighborhood. To do so, it executes a sequence of spatial-temporal range queries in the TPR-tree. Then, by combining the queries answers, they are able to discover the moving objects that are within the cell radius. Although interesting, the approach requires a sequence of range-queries and spatial index operations, during the plane-sweep step, which can be troublesome to guarantee in data streams scenarios that requires timely responses. Hence, this approach is better suited to off-line scenarios where response time is not the primary concern, but correctness.

Their second technique is an approximation method that provides a function to represent the density distribution of moving objects. Contrary to their first method, this function represents the entire grid density, rather than considering each individual cells. More precisely, the function returns all dense regions (clusters) of the grid, *w.r.t.* *minPts*, in a given timestamp. The function is based on Chebyshev polynomials, a recursive function that uses several geometric primitives such as *cos* and *arccos*. While using a single function to discover any dense cluster is appealing, the computational cost of such function is higher than counting the number of moving objects. In addition, the function can cause an overhead or delay the response since it calculates all dense regions at once, even those that do not suffer from the *answer loss* problem.

Jeung *et al.* [14] proposed an interesting supervised solution, which combines off-line processing and hidden Markov chains, for solving the *answer loss* problem in trajectory clustering. Their overall goal is to extract clusters from trajectories datasets. To do so, first, they preprocess a trajectory dataset using DBSCAN to discover the clusters location and their moving objects. Using the clustering result and hidden Markov chains, they create a trajectory model by discovering the set of cells and their probability to be associated with the cluster location, where the probability of each cells expresses the percentage of moving objects of such cluster in that cell. Based on this model, they can correlate the clusters location to the respective moving object cells. There are two main differences between their approach and ours. First, our proposed heuristic does not need to do any *a priori* processing. Second, their approach is focused on clusters solely from the *a priori* trajectory dataset, while our approach can dynamically discover and detect clusters from data streams.

6   Conclusion

This paper presented a density heuristic for on-line grid-based spatial clustering algorithms that addresses the *answer loss* problem, which occurs when clusters are not detected due to spatially and temporally close moving objects being mapped to adjacent cells. The heuristic provides a density function that is sensible to moving objects in adjacent cells while retaining the counting semantics of on-line grid-based approaches. To do so, we further divide each cell into horizontal and vertical slots. Then, we count the distribution of moving objects inside the slots of the adjacent cells when computing a cell density. Moreover, we proposed two discrete weight functions, a linear and an exponential, so that closer inner slots to the cell have higher weight than distant ones. To avoid collateral effects of wrongly detecting a cluster that does not exist, due to considering moving objects in adjacent cells, we proposed that the heuristic should be applied only when the evaluated cell has a *transient density*, *i.e.*, less than the required *minPts* threshold, but higher than a *lowerPts* threshold.

To investigate the relationship between the number of undetected clusters discovered and wrongly detected clusters, we evaluated the heuristic with several transient thresholds using real-world data streams and the on-line DG2CEP grid-cluster algorithm [9]. Based on the experiments, we concluded that, for linear weights, a *lowerPts* threshold between 60 % to 50 % of the original *minPts* threshold produced the best results. Specifically, for a *transient* threshold of 60 %, the number of undetected clusters reduced from 80 % to 23.57 %, with a collateral effect of incorrect clusters of 13.51 %. For exponential weights, we concluded that as *lowerPts* decreases *w.r.t. minPts*, and the number of undetected and incorrect clusters also decreased. Finally, although exponential weights are resilient to collateral effects, *e.g.*, 4.47 %, the number of undetected clusters discovered was lower, 31.51 %, instead of 13.51 % when using linear weights.

As future work, we intend to investigate a hybrid weight function, which combines the benefits of linear and resilience of exponential weights. In addition, we are interested in exploring our heuristic with different datasets and other grid-based algorithms.

Acknowledgment. This work was partly funded by CNPq under grants 557128/2009-9, 303332/2013-1, 442338/2014-7, by FAPERJ under grants E-26-170028/2008, E-26/201.337/2014 and E-01/209996/2015, and by Microsoft Research.

References

1. Zheng, Y., Capra, L., Wolfson, O., Yang, H.: Urban computing. ACM Trans. Intell. Syst. Technol. **5**, 1–55 (2014)
2. Dodge, S., Weibel, R., Lautenschütz, A.-K.: Towards a taxonomy of movement patterns. Inf. Vis. **7**, 240–252 (2008)
3. Amini, A., Wah, T., Saboohi, H.: On density-based data streams clustering algorithms: a survey. J. Comput. Sci. Technol. **29**, 116–141 (2014)
4. Ester, M., Kriegel, H., Sander, J., Xu, X.: A density-based algorithm for discovering clusters in large spatial databases with noise. In: Proceedings of the 2nd International Conference on Knowledge Discovery and Data Mining, pp. 226–231 (1996)
5. Ester, M., Kriegel, H.-P., Sander, J., Wimmer, M., Xu, X.: Incremental clustering for mining in a data warehousing environment. In: Proceedings of the 24th International Conference on Very Large Data Bases, San Francisco, CA, USA, pp. 323–333 (1998)
6. Han, J., Kamber, M., Pei, J.: Data Mining: Concepts and Techniques. Morgan Kaufmann Publishers Inc., San Francisco (2011)
7. Garofalakis, M., Gehrke, J., Rastogi, R.: Querying and mining data streams: you only get one look a tutorial. In: Proceedings of the 2002 ACM SIGMOD International Conference on Management of Data, p. 635. ACM, New York (2002)
8. He, Y., Tan, H., Luo, W., Mao, H., Ma, D., Feng, S., Fan, J.: MR-DBSCAN: an efficient parallel density-based clustering algorithm using mapreduce. In: 2011 IEEE 17th International Conference on Parallel and Distributed Systems, pp. 473–480 (2011)
9. Roriz Junior, M., Endler, M., da Silva e Silva, F.J.: An on-line algorithm for cluster detection of mobile nodes through complex event processing. Inf. Syst. (2016)
10. Chen, Y., Tu, L.: Density-based clustering for real-time stream data. In: Proceedings of the 13th ACM SIGKDD International Conference on Knowledge Discovery and Data Mining, USA, pp. 133–142 (2007)

11. Jensen, C.S., Lin, D., Ooi, B.C.: Continuous clustering of moving objects. IEEE Trans. Knowl. Data Eng. **19**, 1161–1174 (2007)
12. Jensen, C.S., Lin, D., Ooi, B.C., Zhang, R.: Effective density queries on continuously moving objects. In: Proceedings of the IEEE 22nd International Conference on Data Engineering (2006)
13. Ni, J., Ravishankar, C.V.: Pointwise-dense region queries in spatio-temporal databases. In: Proceedings of the IEEE 23rd International Conference on Data Engineering, pp. 1066–1075 (2007)
14. Jeung, H., Shen, H.T., Zhou, X.: Mining trajectory patterns using Hidden Markov Models. In: Song, I.-Y., Eder, J., Nguyen, T.M. (eds.) DaWaK 2007. LNCS, vol. 4654, pp. 470–480. Springer, Heidelberg (2007)
15. Etzion, O., Niblett, P.: Event Processing in Action. Manning Publications Co., USA (2010)
16. EsperTech: Esper - Complex Event Processing. http://www.espertech.com/esper/

An Exhaustive Covering Approach to Parameter-Free Mining of Non-redundant Discriminative Itemsets

Yoshitaka Kameya$^{(\boxtimes)}$

Department of Information Engineering, Meijo University, 1-501 Shiogama-guchi,
Tenpaku-ku, Nagoya 468-8502, Japan
ykameya@meijo-u.ac.jp

Abstract. Discriminative pattern mining is a promising extension of frequent pattern mining. This paper proposes an algorithm called ExCover, a shorthand for exhaustive covering, for finding non-redundant discriminative itemsets. ExCover outputs non-redundant patterns where each pattern covers best at least one positive transaction. With no control parameters limiting the search space, ExCover efficiently performs an exhaustive search for best-covering patterns using branch-and-bound pruning. During the search, candidate best-covering patterns are concurrently collected for each positive transaction. Formal discussions and experimental results exhibit that ExCover efficiently finds a more compact set of patterns in comparison with previous methods.

Keywords: Discriminative patterns · Sequential covering · Branch-and-bound search

1 Introduction

Discriminative pattern mining is a promising extension of frequent pattern mining, which has been studied under several different names [6,15]. In a typical setting of discriminative pattern mining, each transaction belongs to one of two or more pre-defined classes, and we are interested in a particular class c. For example, we may be studying mushrooms which are edible. Then, we attempt to find discriminative patterns that frequently occur in the transactions of class c (called *positive* transactions) and do not frequently occur in the transactions not of class c (called *negative* transactions). The patterns found are expected to characterize well c, the class of interest.

A main difficulty of discriminative pattern mining is that the quality score for a pattern is *not* anti-monotonic w.r.t. set-inclusion, and thus we cannot enjoy powerful pruning like the one with the minimum support threshold. In the literature, several authors have conducted branch-and-bound search (e.g. [19]). Another difficulty is redundancy among patterns. For example, if a pattern {A} has significantly high quality in characterizing a class c of interest, patterns including A, such as {A, B}, {A, C} and {A, B, C}, also tend to have high quality.

S. Madria and T. Hara (Eds.): DaWaK 2016, LNCS 9829, pp. 143–159, 2016.
DOI: 10.1007/978-3-319-43946-4_10

One remedy against redundancy is to use set-inclusion-based constraints among patterns. For example, closed pattern mining techniques [20,24] have also been exploited in discriminative pattern mining [8,22]. In the studies on subgroup discovery, the notion of irrelevant patterns has been introduced [8]. It is known that the productivity constraint [2] is efficiently tested in the search over suffix enumeration trees [1,14,16].

In this paper, we propose an algorithm called ExCover, which is a short-hand for *exhaustive covering*. ExCover outputs non-redundant patterns where each pattern x covers best at least one positive transaction t, namely, x is of the highest quality among the patterns that cover t. Hereafter we call this constraint the *best-covering* constraint. ExCover efficiently performs an exhaustive, depth-first search for best-covering patterns using branch-and-bound pruning techniques developed so far. During the search, candidate best-covering patterns are concurrently collected for each positive transaction.

The merit of ExCover is three-fold. First, ExCover outputs patterns that are non-redundant from the viewpoint of coverage over positive transactions. In previous work, HARMONY [25] employs the same strategy, and surely covering all positive transactions is counted as an advantage in its original paper. In fact, however, the best-covering constraint also works for removing redundancy — a formal result presented in this paper is that the best-covering constraint is tighter than productivity, the aforementioned constraint ensuring non-redundancy. The second merit of ExCover is that it finds patterns in an exhaustive manner. In practice, we occasionally use rule learners for knowledge discovery, rather than for black-box classification. Unfortunately, however, most of traditional learners rely on greedy search and some heuristics [7,26], and consequently, for some generated rules, the reason why such rules have been generated may not be clear to non-experts. We believe that the exhaustiveness of ExCover would enhance the explainability of the obtained results. Lastly, ExCover has no control parameters required to be tuned for limiting the search space. In frequent pattern mining, it is often said to be tedious or infeasible to tune the minimum support threshold for having a manageable amount of useful patterns. To alleviate this inconvenience, Han et al. [13] proposed a top-k mining method without tuning minimum support. Here k, the number of patterns to be output, is more user-centric than minimum support. ExCover follows this line of research and does not even require k since it automatically stops the search when every positive transaction has been covered best by some pattern already examined.

To illustrate, let us consider a dataset shown in Table 1(1). There are five positive transactions (belonging to class $+$) and five negative transactions (belonging to class $-$). Each transaction is associated with an identifier called TID. From this dataset, we attempt to find top-k discriminative patterns w.r.t. class $c = +$ with $k = 20$ and obtain the patterns in Table 1(2a). The quality of each pattern in the first field is measured by the F-score recorded in the second field. The third field contains the TIDs of positive transactions covered by the pattern in the first field. Table 1(2a) includes more than 20 patterns due to the tie score at the bottom. For example, for class $c = +$ and the top-ranked pattern

Table 1. Example transactions (1), and discriminative patterns (2a)–(2d) and (3).

(2b) Productive

Pattern	F-score	TIDs
{A, C}	0.750	2,3,4
{B}	0.727	1,2,4,5
{A}	0.667	1,2,3,4
{C}	0.600	2,3,4

(2a) Top-k

Pattern	F-score	TIDs
{A, C}	0.750	2,3,4
{B}	0.727	1,2,4,5
{A}	0.667	1,2,3,4
{A, B}	0.667	1,2,4
{A, D, E}	0.600	1,2,3
{A, E}	0.600	1,2,3
{C}	0.600	2,3,4
{A, B, C}	0.571	2,4
{A, C, D}	0.571	2,3
{A, C, D, E}	0.571	2,3
{A, C, E}	0.571	2,3
{A, D}	0.545	1,2,3
{A, B, D}	0.500	1,2
{A, B, D, E}	0.500	1,2
{A, B, E}	0.500	1,2
{B, C}	0.500	2,4
{B, D}	0.444	1,2
{B, D, E}	0.444	1,2
{B, E}	0.444	1,2
{C, D}	0.444	2,3
{C, D, E}	0.444	2,3
{C, E}	0.444	2,3

(2c) Closed

Pattern	F-score	TIDs
{A, C}	0.750	2,3,4
{B}	0.727	1,2,4,5
{A}	0.667	1,2,3,4
{A, B}	0.667	1,2,4
{A, D, E}	0.600	1,2,3
{A, B, C}	0.571	2,4
{A, C, D, E}	0.571	2,3
{A, B, D, E}	0.500	1,2
{A, B, C, D, E}	0.333	2

(1)

TID	Class	Transaction
1	+	{A, B, D, E}
2	+	{A, B, C, D, E}
3	+	{A, C, D, E}
4	+	{A, B, C}
5	+	{B}
6	−	{A, B, D, E}
7	−	{B, C, D, E}
8	−	{C, D, E}
9	−	{A, D, E}
10	−	{A, D}

(2d) Productive & Closed

Pattern	F-score	TIDs
{A, C}	0.750	2,3,4
{B}	0.727	1,2,4,5
{A}	0.667	1,2,3,4

(3) Best-covering (ExCover)

Pattern	F-score	TIDs
{A, C}	0.750	2,3,4
{B}	0.727	1,2,4,5

$x = \{A, C\}$, we have its (positive) support $p(x \mid c) = 0.6$ since three out of five positive transactions are covered by $\{A, C\}$. Similarly, we have x's confidence $p(c \mid x) = 1$ since all transactions covered by $\{A, C\}$ belong to class +. The F-score of $\{A, C\}$ is then obtained as the harmonic mean of support $p(x \mid c)$ and confidence $p(c \mid x)$ which amounts to $2 \times 0.6 \times 1/(0.6 + 1) = 0.75$. Remark here that the patterns in Table 1(2a) have been obtained under no constraint except the top-k constraint, and more patterns would be output with a larger k.

In contrast, the patterns in Table 1(2b)–(2d) are obtained under some additional, set-inclusion-based constraints. Specifically, Table 1(2b) only contains productive patterns, i.e. the patterns having no improvement in quality from their sub-patterns are excluded. For example, the patterns in Table 1(2a) containing item B together with some other items are all excluded in Table 1(2b), since their F-scores are lower than the F-score of pattern {B}. On the other hand, Table 1(2c) contains the patterns closed on the positive transactions. One may find that the patterns in Table 1(2a) that cover the same positive transactions are replaced with the largest pattern listed in Table 1(2c). For example, patterns

{A, D}, {A, E} and {A, D, E} in Table 1(2a) cover the same positive transactions 1, 2 and 3, whereas Table 1(2c) only contains the largest one {A, D, E}. Choosing productive patterns from the closed patterns in Table 1(2c) yields the patterns in Table 1(2d). Note that, even with a larger k, no other patterns will be output, and therefore additional constraints surely work for reducing the number of patterns to be output.

Furthermore, we wish to have fewer patterns that are sufficient to characterize the class of interest. Indeed, in the current example, ExCover only outputs two patterns listed in Table 1(3). Pattern {A} has been excluded here since each of transaction 1, 2, 3 and 4 covered by {A} is also covered by other patterns {A, C} and {B} of higher quality. We say that {A, C} covers best transactions 2, 3 and 4, and {B} covers best transactions 1 and 5. From the viewpoint of coverage over positive transactions, in the current example, performing top-k mining where $k = 1$ seems inappropriate, since we obviously lose the information from transactions 1 and 5. Having that said, however, we are not able to know it beforehand. So eliminating k in ExCover would reduce the user's effort.

The rest of the paper is outlined as follows. First, Sect. 2 gives several background notions and notations related to ExCover. We then describe the details of ExCover in Sect. 3. Section 4 presents some results of our experiments, and Sect. 5 discusses some related work. Finally Sect. 6 concludes the paper.

2   Background

2.1   Preliminaries

This paper shares several background notions and notations with [14]. We first consider a dataset $\mathcal{D} = \{t_1, t_2, \ldots, t_N\}$, a multiset of size N, where t_i ($1 \leq i \leq N$) is a set of items called a transaction. Each transaction belongs to one of predefined classes $\mathcal{C}$, and let c_i be the class of transaction t_i. The set of all items appearing in $\mathcal{D}$ is denoted by $\mathcal{X}$. A pattern $\boldsymbol{x}$ is a subset of $\mathcal{X}$. and we say that $\boldsymbol{x}$ covers a transaction t_i when $\boldsymbol{x} \subseteq t_i$. For convenience, we interchangeably denote a pattern as a vector $\boldsymbol{x} = (x_1, x_2, \ldots, x_n)$, as a set $\boldsymbol{x} = \{x_1, x_2, \ldots, x_n\}$, or as a conjunction $\boldsymbol{x} = (x_1 \wedge x_2 \wedge \ldots \wedge x_n)$.

We then define some subsets of a dataset $\mathcal{D}$: $\mathcal{D}_c = \{t_i \mid c_i = c, 1 \leq i \leq N\}$, $\mathcal{D}(\boldsymbol{x}) = \{t_i \mid \boldsymbol{x} \subseteq t_i, 1 \leq i \leq N\}$ and $\mathcal{D}_c(\boldsymbol{x}) = \{t_i \mid c_i = c, \boldsymbol{x} \subseteq t_i, 1 \leq i \leq N\}$, where $c \in \mathcal{C}$ is the class of interest. We use a symbol $\neg$ for negation, e.g. $\mathcal{D}_{\neg c} = \mathcal{D} \setminus \mathcal{D}_c$, $\mathcal{D}_c(\neg \boldsymbol{x}) = \mathcal{D}_c \setminus \mathcal{D}_c(\boldsymbol{x})$ and $\mathcal{D}_{\neg c}(\boldsymbol{x}) = \mathcal{D}(\boldsymbol{x}) \setminus \mathcal{D}_c(\boldsymbol{x})$. The transactions in $\mathcal{D}_c$ (resp. $\mathcal{D}_{\neg c}$) are called positive (resp. negative) transactions.

The probabilities treated in this paper are all empirical, i.e. they are computed from the dataset $\mathcal{D}$. Specifically, a joint probability $p(c, \boldsymbol{x})$ is obtained as $|\mathcal{D}_c(\boldsymbol{x})|/N$. Similarly we have $p(c, \neg \boldsymbol{x}) = |\mathcal{D}_c(\neg \boldsymbol{x})|/N$, $p(\neg c, \boldsymbol{x}) = |\mathcal{D}_{\neg c}(\boldsymbol{x})|/N$, and so on. Using joint probabilities, marginal probabilities and conditional probabilities are computed in a standard way, e.g. $p(\boldsymbol{x}) = p(c, \boldsymbol{x}) + p(\neg c, \boldsymbol{x})$, $p(c) = p(c, \boldsymbol{x}) + p(c, \neg \boldsymbol{x})$ or $p(c \mid \boldsymbol{x}) = p(c, \boldsymbol{x})/p(\boldsymbol{x})$. In the traditional terminology on pattern mining, we often call conditional probabilities $p(\boldsymbol{x} \mid c)$, $p(\boldsymbol{x} \mid \neg c)$

and $p(c \mid \boldsymbol{x})$ positive support, negative support and confidence, respectively. For brevity, 'support' means positive support unless explicitly noted.

2.2 Dual-Monotonicity

The quality of a pattern $\boldsymbol{x}$ for class c is written as $R_c(\boldsymbol{x})$, and most of popular quality functions are functions of positive support $p(\boldsymbol{x} \mid c)$ and negative support $p(\boldsymbol{x} \mid \neg c)$ [14]. As an instance of R_c, throughout the paper, we use *F-score* $\mathrm{F}_c(\boldsymbol{x}) = 2p(c \mid \boldsymbol{x})p(\boldsymbol{x} \mid c)/(p(c \mid \boldsymbol{x}) + p(\boldsymbol{x} \mid c))$, which can be simplified as Dice Index $2p(\boldsymbol{x}, c)/(p(\boldsymbol{x}) + p(c))$. Also F-score gives the same ranking over patterns as the one by Jaccard Index $p(\boldsymbol{x}, c)/(p(\boldsymbol{x}) + p(c) - p(\boldsymbol{x}, c))$ for a given class c of interest. Since we seek for the patterns characterizing a particular class c, we focus on the patterns $\boldsymbol{x}$ such that $p(\boldsymbol{x} \mid c) \geq p(\boldsymbol{x} \mid \neg c)$ or equivalently $p(c \mid \boldsymbol{x}) \geq p(c)$.[1] In the previous work, the *convexity* of quality scores has been exploited in branch-and-bound pruning [19], and recently, a relaxed condition called *dual-monotonicity* was introduced in [14]:

Definition 1. *Let R_c be a quality score for a class c. Then, R_c is dual-monotonic iff, for any pattern $\boldsymbol{x}$, $R_c(\boldsymbol{x})$ is monotonically increasing w.r.t. $p(\boldsymbol{x} \mid c)$ and monotonically decreasing w.r.t. $p(\boldsymbol{x} \mid \neg c)$ wherever $p(\boldsymbol{x} \mid c) \geq p(\boldsymbol{x} \mid \neg c)$.* □

ExCover works with any dual-monotonic quality score. Indeed, like the algorithm proposed in [14], dual-monotonicity plays a crucial role in various aspects of ExCover. Several well-known quality scores such as F-score, the Fisher score, information gain, Gini index, χ^2 and support difference are all dual-monotonic.

2.3 Branch-and-Bound Pruning in Top-k Mining

Suppose that we are performing a branch-and-bound search for top-k patterns under a dual-monotonic quality score R_c. Also consider an anti-monotonic upper bound $\overline{R}_c(\boldsymbol{x})$ of $R_c(\boldsymbol{x})$ of a pattern $\boldsymbol{x}$. Then, if it is found that $\overline{R}_c(\boldsymbol{x}) < R_c(\boldsymbol{z})$, where $\boldsymbol{z}$ is the pattern with the k-th greatest score at the moment, we can safely prune the subtree rooted by $\boldsymbol{x}$ in the enumeration tree. This pruning exploits the anti-monotonicity of $\overline{R}_c$ w.r.t. pattern-inclusion, which guarantees $R_c(\boldsymbol{x}') \leq \overline{R}_c(\boldsymbol{x}') \leq \overline{R}_c(\boldsymbol{x}) < R_c(\boldsymbol{z})$ for any super-pattern $\boldsymbol{x}'$ of $\boldsymbol{x}$. $R_c(\boldsymbol{x})$ and $\overline{R}_c(\boldsymbol{x})$ are computed from $p(\boldsymbol{x} \mid c)$ and $p(\boldsymbol{x} \mid \neg c)$, which in turn are computed from the statistics stored in the (compressed) databases such as FP-trees.

The next question is how to obtain such an anti-monotonic upper bound. Since $R_c(\boldsymbol{x})$ is dual-monotonic, by definition $R_c(\boldsymbol{x})$ is monotonically increasing (resp. decreasing) w.r.t. $p(\boldsymbol{x} \mid c)$ (resp. $p(\boldsymbol{x} \mid \neg c)$), and both $p(\boldsymbol{x} \mid c)$ and $p(\boldsymbol{x} \mid \neg c)$ are anti-monotonic w.r.t. pattern-inclusion. Thus, the most optimistic scenario when extending $\boldsymbol{x}$ into $\boldsymbol{x}'$ is that $p(\boldsymbol{x}' \mid c)$ remains $p(\boldsymbol{x} \mid c)$ and $p(\boldsymbol{x}' \mid \neg c)$ turns to be zero. One general way for obtaining an upper bound $\overline{R}_c(\boldsymbol{x})$ is then to

---

[1] In the previous example with the target class $c = +$, a pattern $\boldsymbol{x} = \{D\}$ is excluded, since $p(\boldsymbol{x} \mid c) = 3/5 = 0.6$ and $p(\boldsymbol{x} \mid \neg c) = 5/5 = 1$.

Item x	$p(x \mid +)$	$p(x \mid -)$	$F_+(x)$
A	0.8	0.6	0.667
B	0.8	0.4	0.727
C	0.6	0.4	0.600
D	0.6	1.0	0.462
E	0.6	0.8	0.500

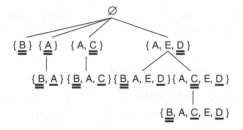

Fig. 1. F-scores of items (left) and the enumeration tree with SPC extension (right).

substitute $p(\boldsymbol{x} \mid \neg c) := 0$ into the definition of $R_c(\boldsymbol{x})$.[2] The upper bound $\overline{R}_c(\boldsymbol{x})$ where $p(\boldsymbol{x} \mid \neg c)$ is constant at zero is always anti-monotonic w.r.t. pattern-inclusion thanks to the dual-monotonicity of R_c. For example, the upper bound of F-score is obtained as $\overline{F}_c(\boldsymbol{x}) = 2p(\boldsymbol{x} \mid c)/(1+p(\boldsymbol{x} \mid c))$. The operations described here are applicable to any dual-monotonic quality score, but we should remark that some quality score such as confidence $p(c \mid \boldsymbol{x})$, its upper bound obtained by the way above goes into infinity and pruning does not work in a practical sense.

2.4   The Closedness Constraint

As stated in the introduction, one popular technique for redundancy elimination is to use the closedness constraint. For that, we first introduce a closure operator Γ such that $\Gamma(\boldsymbol{x}, \mathcal{D}) = \bigcap_{t \in \mathcal{D}(\boldsymbol{x})} t$, where $\mathcal{D}$ is the transactions and $\boldsymbol{x}$ is some pattern. Here $\Gamma(\boldsymbol{x}, \mathcal{D})$ is called a *closure* of $\boldsymbol{x}$ w.r.t. $\mathcal{D}$. A closed pattern is then a pattern $\boldsymbol{x}$ such that $\boldsymbol{x} = \Gamma(\boldsymbol{x}, \mathcal{D})$. Each closed pattern $\boldsymbol{x}$ is the largest pattern in an equivalence class $[\boldsymbol{x}] = \{\boldsymbol{x}' \mid \mathcal{D}(\boldsymbol{x}) = \mathcal{D}(\boldsymbol{x}')\} = \{\boldsymbol{x}' \mid \boldsymbol{x} = \Gamma(\boldsymbol{x}', \mathcal{D})\}$ and seen as a representative of $[\boldsymbol{x}]$. Since the size of $[\boldsymbol{x}]$ can be exponential, focusing only on closed patterns often leads to a significant reduction of the search space.

Closed patterns are also beneficial from the viewpoint of quality, especially when the closure operator is applied only to the positive transactions $\mathcal{D}_c$. Such patterns are often said to be *closed on the positives*. Let c be a class of interest, $\mathcal{D}_c$ be positive transaction, and $\boldsymbol{x}$ be some pattern. Also let $\boldsymbol{x}^* = \Gamma_c(\boldsymbol{x})$, where $\Gamma_c(\boldsymbol{x})$ is an abbreviation of $\Gamma(\boldsymbol{x}, \mathcal{D}_c)$. We further note that $\mathcal{D}_c(\boldsymbol{x}^*) = \mathcal{D}_c(\boldsymbol{x})$ since $\boldsymbol{x}^*$ and $\boldsymbol{x}$ are in the same equivalence class $[\boldsymbol{x}]$, and $\mathcal{D}_{\neg c}(\boldsymbol{x}^*) \subseteq \mathcal{D}_{\neg c}(\boldsymbol{x})$ since $\boldsymbol{x}^*$ is the largest pattern in $[\boldsymbol{x}]$. Then, under a dual-monotonic quality score R_c, we have $R_c(\boldsymbol{x}^*) \geq R_c(\boldsymbol{x})$ since $p(\boldsymbol{x}^* \mid c) = p(\boldsymbol{x} \mid c)$ and $p(\boldsymbol{x}^* \mid \neg c) \leq p(\boldsymbol{x} \mid \neg c)$ [8,14,22]. This means that the quality of a patterns closed on the positives is no lower than the qualities of the patterns in the same equivalence class.

To enumerate patterns closed on the positives without duplicate visits to a pattern, we perform a variant of prefix-preserving closure (PPC) extension used in LCM [24], called *suffix-preserving closure* (SPC) *extension* [14]. Here we will explain SPC extension by illustration. First, consider again the dataset in Table 1(1), where we have a total order $B \prec A \prec C \prec E \prec D$ over items as the

---

[2] Equivalent substitutions are also possible: $p(c \mid \boldsymbol{x}) := 1$, $p(\neg \boldsymbol{x} \mid \neg c) := 1$, and so on.

descending order of F-score in Fig. 1 (left). Here the quality $R_c(x)$ of an item x is defined as $R_c(\{x\})$. Then, Fig. 1 (right) is the enumeration tree obtained by exhaustive applications of SPC extension. In enumeration, we introduce a new pattern by adding a *core item* to a pattern already visited. In Fig. 1 (right), the core items are underlined, and among them, the core items lastly added are doubly underlined. Other items, called *accompanying items* here, are those taken along into a pattern by the closure operation. In SPC extension, a pattern having *no* accompanying item which is a successor w.r.t. $\prec$ of the core item lastly added are considered to preserve the suffix of the original pattern. Then, a pattern *not* preserving the suffix is immediately pruned. At each branch in the enumeration tree, the core items to be added are chosen from unadded predecessors of the core item lastly added, in the ascending order w.r.t. $\prec$.

For example, given an empty pattern $\emptyset$, we apply an SPC extension by item D to the positive transactions in Table 1(1) and obtain $\Gamma_c(\{D\} \cup \emptyset) = \{A, E, D\}$. In this case, D is the last core item in $\{A, E, D\}$, while A and E are taken along into the pattern by the closure operation. For a new pattern $\{A, E, D\}$, we further add B and C which have not been added yet, following the ascending order w.r.t. $\prec$. One may see that adding E into an empty pattern also yields the same pattern $\Gamma_c(\{E\} \cup \emptyset) = \{A, E, D\}$, where an accompanying item D is a successor of the core item E lastly added. The pattern $\{A, E, D\}$ obtained in this way is not suffix-preserving and so is immediately pruned. We finally obtain the enumeration tree in Fig. 1 (right) which has nine non-root nodes which correspond to the patterns closed on the positives shown in Table 1(2c).

In practice, the choice of the total order $\prec$ is important. Suppose we have $x \prec x'$ iff $R_c(x) \geq R_c(x')$ like the example above. Then, in a depth-first search, we visit patterns including high quality items earlier, and hence there would be more chances for pruning described in Sects. 2.3 and 2.5.

2.5   The Productivity Constraint

We explain another constraint called productivity, whose original version is defined with confidence [2]. This constraint has also been used in associative classification [17] and explanatory analysis of Bayesian networks [28]. Productivity is defined as follows:

Definition 2. *Let c be a class of interest. Then, for a pair of patterns x and x', x is* weaker *than x' iff $x \supset x'$ and $R_c(x) \leq R_c(x')$. A pattern x is* productive *iff x is not weaker than any sub-pattern of x.* $\square$

It is mentioned in [14] that, under a dual-monotonic quality score, the productivity constraint above is tighter than the irrelevancy constraint [8,10] among the patterns closed on the positives. Also, it is desirable to test easily the productivity constraint. In a depth-first search, the following property is useful [14]:

Proposition 1. *When a pattern x is visited in a depth-first search with SPC extension, all of x's sub-patterns have already been visited.* $\square$

Algorithm 1. SEQCOVER

1: $L :=$ an empty set
2: **while** $\mathcal{D}_c \neq \emptyset$ **do**
3:     Induce the best rule $x \Rightarrow c$ from $\mathcal{D}_c$ and $\mathcal{D}_{\neg c}$
4:     $L := L \cup \{x \Rightarrow c\}$
5:     Remove all positive examples covered by x from $\mathcal{D}_c$
6: **end while**
7: Output the rules in L

In Fig. 1 (right), $\{B, A, E, D\}$ is visited after all its sub-patterns $\{B\}$, $\{A\}$, $\{B, A\}$ and $\{A, E, D\}$ being visited in a depth-first search with SPC extension. So, when visiting $\{B, A, E, D\}$, we can easily compare the quality of $\{B, A, E, D\}$ with the qualities of $\{B\}$, $\{A\}$, $\{B, A\}$ and $\{A, E, D\}$ to test whether $\{B, A, E, D\}$ is productive. This is not the case with PPC extension in the original LCM.

Moreover, we are able to conduct an aggressive pruning based on an extended notion of weakness which is defined as follows:

Definition 3. *Let c be a class of interest and (x, x') be a pair of patterns. Then, x is prunably weaker than x' iff $x \supset x'$ and $\overline{R}_c(x) \leq R_c(x')$.* □

If a pattern x is prunably weaker than a pattern x' in the current top-k candidates, any super-pattern of x is also weaker than x', and thus we can safely prune the subtree rooted by x. Proposition 1 is also useful for this pruning.

2.6   Sequential Covering

Sequential covering, also known as separate-and-conquer, is a traditional search strategy in rule learning [7,26]. Also in the literature of discriminative pattern mining, several methods such as DDPMine [4] take this strategy. Algorithm 1 shows a typical workflow of sequential covering. Sequential covering takes as input the class c of interest and the dataset $\mathcal{D}_c \cup \mathcal{D}_{\neg c}$ and outputs a set of rules for class c. In sequential covering, we iteratively build a new rule (Line 3) and remove all positive examples covered by the new rule (Line 5). The iteration continues until there remain no positive examples to be covered (Line 2). DDPMine performs branch-and-bound search in building new rules.

The removal of positive examples surely reduces the overlap of coverage among generated rules, but there seem to be two problems. First, such a removal prevents us from *declarative* understanding of the generated rules. In other words, the meanings of the generated rules reflect a *procedural* behavior of Algorithm 1, which may not be clear to non-experts. For example, the statistics used in exploring the first new rule are different from those used in exploring the second new rule. Second, as Domingos [5] pointed out, available positive examples dwindle as the iteration continues, and therefore the rules generated at later iterations can be less accurate in a statistical sense.

Domingos proposed a greedy algorithm in which each rule is learned from the entire dataset, and referred to his own strategy by *conquering-without-separating* [5]. Later, Rijnbeek et al. [21] proposed an algorithm that directly finds a rule condition in disjunctive normal form (DNF), instead of finding conjunctive rule conditions one by one. A main drawback of this method is its computational cost, and therefore some extra control parameters limiting the size of the DNF condition are often required. ExCover, which will be explained next, also takes the conquering-without-separating approach in an exhaustive but light-weight manner without control parameters limiting the search space.

3 The Proposed Method

3.1 The Best-Covering Constraint

From now on, we explain the details of ExCover. As illustrated in Sect. 1, ExCover seeks for best-covering patterns that are closed on the positives. First, let us formally define the patterns output by ExCover:

Definition 4. *Let c be a class of interest and $t \in \mathcal{D}_c$ be a positive transaction. then, a pattern x is said to* cover t best *when x covers t (i.e. $x \subseteq t$) and satisfies the following two conditions for any other pattern x' that also covers t:*

1. *$R_c(x) \geq R_c(x')$ if x' is not a subset of x*
2. *$R_c(x) > R_c(x')$ if x' is a subset of x*

A pattern x is said to be best-covering *if there is at least one positive transaction in $\mathcal{D}_c$ which is covered best by x.*   □

The definition of the "covers-best" relation above says that a pattern x is said to cover best a positive transaction t when x is of the highest quality among the patterns covering t. One technical point here is that the requirements on the quality score of x slightly differ depending on whether the pattern x' in comparison is a sub-pattern of x or not. Similarly to the "weaker-than" relation in Definition 2, no pattern covering t best is allowed to have the same quality as those of its sub-patterns. These slightly different requirements make it easy to prove a key property that the best-covering constraint is tighter than the productivity constraint (Definition 2):

Proposition 2. *A pattern x is productive if x is best-covering.*   □

Proof. We prove this by contraposition. Suppose that x is not productive. That is, there exists a pattern x' such that x is weaker than x', i.e. $x' \subset x$ and $R_c(x') \geq R_c(x)$. For this x', any transaction t covered by x is also covered by x' since $x' \subset x \subseteq t$. Then, for any positive transaction t covered by x, the second condition of the "covers-best" relation is violated, and therefore x is not best-covering.   □

Based on the definition of the best-covering constraint, we can immediately introduce the following pruning condition:

Algorithm 2. EXCOVER

1: L := an array indexed by $t \in \mathcal{D}_c$
2: Initialize $L[t]$ as an empty set for each $t \in \mathcal{D}_c$
3: $\boldsymbol{x}$:= an empty pattern
4: T := an initial database constructed from $\mathcal{D}$
5: Call GROW$(\boldsymbol{x}, T)$
6: Output the patterns $\bigcup_{t \in \mathcal{D}_c} L[t]$

Proposition 3. *Let c be a class of interest and $\boldsymbol{x}$ be a pattern. Suppose that, for every positive transaction t covered by $\boldsymbol{x}$, there exists some other pattern $\boldsymbol{x}'$ covering t such that $\overline{R}_c(\boldsymbol{x}) < R_c(\boldsymbol{x}')$. Then, $\boldsymbol{x}$ and its super-patterns are not best-covering.*   □

Proof. Let $\boldsymbol{u}$ be $\boldsymbol{x}$ or its super-pattern. Note that $R_c(\boldsymbol{u}) \leq \overline{R}_c(\boldsymbol{x})$ always holds and any transaction covered by $\boldsymbol{u}$ is also covered by $\boldsymbol{x}$. Then, for any positive transaction t covered by $\boldsymbol{u}$, $R_c(\boldsymbol{u}) \leq \overline{R}_c(\boldsymbol{x}) < R_c(\boldsymbol{x}')$ holds for $\boldsymbol{x}'$ in the proposition. From this fact, $\boldsymbol{u}$ cannot cover t best since either of the two conditions in Definition 4 is unsatisfied.   □

In addition, the "covers-best" relation in Definition 4 is restated in a different form which reduces the computational cost required for set-inclusion check:

Proposition 4. *Let c be a class of interest and $t \in \mathcal{D}_c$ be a positive transaction. Then, a pattern $\boldsymbol{x}$ covers t best iff $\boldsymbol{x}$ covers t and satisfies one of the following conditions for any other pattern $\boldsymbol{x}'$ covering t:*

1. $R_c(\boldsymbol{x}) > R_c(\boldsymbol{x}')$, *and*
2. $R_c(\boldsymbol{x}) = R_c(\boldsymbol{x}')$ *and $\boldsymbol{x}$ is not a superset of $\boldsymbol{x}'$.*   □

Proof. Let us introduce three Boolean variables A, B and C which respectively indicate "$\boldsymbol{x}'$ is a subset of $\boldsymbol{x}$," "$R_c(\boldsymbol{x}) > R_c(\boldsymbol{x}')$" and "$R_c(\boldsymbol{x}) = R_c(\boldsymbol{x}')$." The "covers-best" relation in Definition 4 is then written as $(\neg A \Rightarrow B \vee C) \wedge (A \Rightarrow B)$. This condition is simplified as $B \vee (C \wedge \neg A)$, which coincides with the proposition.   □

3.2   Algorithm Description

Based on the notions and properties explained so far, ExCover efficiently performs a branch-and-bound search for best-covering patterns closed on the positives. An underlying strategy of ExCover is to conduct a top-1 mining (top-k mining where $k = 1$) concurrently for each positive transaction t. To be specific, we show the main routine of ExCover in Algorithm 2. Like sequential covering, ExCover takes as input the class c of interest and the dataset $\mathcal{D}_c \cup \mathcal{D}_{\neg c}$. We first introduce a global array variable L referring to the *candidate table* which stores candidate patterns (Lines 1–2). Formally, for each positive transaction

---

Algorithm 3. GROW(x, T)

Require: x: the current pattern, T: conditional database corresponding to x
1:   $B := \{x \in \mathcal{X} \mid x \notin x$ and x is a predecessor of the core item lastly added into $x\}$
2:   **for** each $x \in B$ enumerated in the ascending order w.r.t. $\prec$ **do**
3:      $x' := \{x\} \cup x$ and compute $\overline{R}_c(x')$ from T
4:      **if** $\bigcup_{t \in \mathcal{D}_c(x')} L[t] \neq \emptyset$ and $\overline{R}_c(x') < \min_{t \in \mathcal{D}_c(x'), z \in L[t]} R_c(z)$ **then continue**
5:      $x^* := \Gamma_c(x')$
6:      **if** x^* does not preserve x's suffix **then continue**
7:      Construct T^* corresponding to x^* from T
8:      Compute $R_c(x^*)$ from T^* together with $p(x^* \mid c)$ and $p(x^* \mid \neg c)$
9:      Call ADD(x^*) if $p(x^* \mid c) \geq p(x^* \mid \neg c)$
10:     Call GROW(x^*, T^*)
11: **end for**

---

t, $L[t]$ is a candidate set of t's best-covering patterns, i.e. those covering t and having the same highest quality score. Then, we call the GROW procedure with an empty pattern and an initial FP-tree-like database for finding best-covering patterns closed on the positives in a depth-first manner (Lines 3–5). After the call, ExCover outputs all patterns stored in L, removing duplicates (Line 6).[3]

The GROW procedure, presented in Algorithm 3, recursively visits patterns in a depth-first order over the enumeration tree like the one illustrated in Fig. 1 (right). When visiting a pattern x, as a part of SPC extension, we collect the core items to be added from unadded predecessors of the core item lastly added (Line 1) and pick them up one by one in the ascending order w.r.t. $\prec$ (Line 2). Then, for each collected item x, we first obtain a temporary pattern x' by adding x into the current pattern x and then compute its upper bound of quality from the conditional database corresponding to x (Line 3). In the pruning condition in Line 4, the former part $\bigcup_{t \in \mathcal{D}_c(x')} L[t] \neq \emptyset$ checks whether there exists a candidate pattern stored in L covering a positive transaction covered by x'.[4] If there does not exist, x' is free from being pruned regardless of its quality. The latter part, derived from Proposition 3, then checks the quality of the temporary pattern x'. If the pruning condition is not satisfied, we generate a new pattern x^* closed on the positives from the temporary pattern x' (Line 5). We then we check the validity of x^* w.r.t. SPC extension (Line 6), and invalid ones are pruned immediately. For a valid pattern x^*, we construct a new conditional database T^* (Line 7), evaluate the quality of x^* (Line 8), add x^* to the candidate table L (Line 9), and further visit x^* (Line 10). Since we only need the patterns characterizing class c, we exclude x^* such that $p(x^* \mid c)$ is lower than $p(x^* \mid \neg c)$, but in any case we continue to visit the super-patterns of x^*.

The ADD procedure, presented in Algorithm 4, adds a pattern satisfying the best-covering constraint into the candidate table L. The procedure is derived directly from Proposition 4. Since $L[t]$ contains candidate patterns with the

---

[3] In other words, ExCover outputs all patterns having the same best score. We only exclude apparently redundant patterns to avoid the loss of crucial information.
[4] Remind that $\mathcal{D}_c(x)$ denotes the set of positive transactions covered by a pattern x.

Algorithm 4. ADD(x)

Require: x: a pattern to be added
1: **for each** $t \in \mathcal{D}_c(x)$ **do**
2:    Let r be the quality $R_c(z)$ of an arbitrary pattern z in $L[t]$
3:    **if** $R_c(x) > r$ **then**
4:      $L[t] := \{x\}$
5:    **else if** $R_c(x) = r$ and x is not a superset of any pattern z in $L[t]$ **then**
6:      $L[t] := L[t] \cup \{x\}$
7:    **end if**
8: **end for**

---

same highest quality score, to obtain the highest score for a positive transaction t, in Line 2, it is sufficient to pick up z arbitrarily from $L[t]$ as a representative.

Lastly, we add three remarks. First, the property of SPC extension on the visiting order in Proposition 1 effectively works in Line 4 of the GROW procedure. That is, since it is guaranteed that all sub-patterns of x' have already been visited, each $L[t]$ is not empty for $t \in \mathcal{D}_c(x')$ at least when x' has two or more core items, and hence the latter part of the pruning condition is not skipped. Second, for reducing the computational effort, we delay the closure operation Γ_c, which is costly in general, until Line 5. Instead, we refer to $\overline{R}_c(x')$ rather than $\overline{R}_c(x^*)$ in the pruning condition in Line 4. This operation is justified by the nature of the patterns closed on the positives and the way of obtaining the upper bound $\overline{R}_c$, i.e. we have $\overline{R}_c(\Gamma_c(x)) = \overline{R}_c(x)$ for any pattern x. The third remark is that, we frequently refer to $\mathcal{D}_c(x)$ for a given pattern x. So for quick reference to $\mathcal{D}_c(x)$, a data structure for conditional databases in a vertical layout [29] is crucial from a practical point of view. In the implementation used in our experiments, we introduced a simple extension of FP-trees [12] in which each node contains a list of TIDs for each class in addition to the counts.

4   Experimental Results

We conducted two experiments. The first experiment aims at confirming whether we can obtain a more compact set of discriminative itemsets for the mushroom dataset, which is available from the UCI Machine Learning Repository (http://archive.ics.uci.edu/ml/datasets/Mushroom). The second one compares the search space using the datasets available from http://dtai.cs.kuleuven.be/CP4IM/datasets/.

In the first experiment, we transform the original dataset into a transaction dataset, treating each attribute-value pair as an item. We then obtained the patterns shown in Table 2 (above) by a previous method [14] for mining top-k productive patterns closed on the positives and the patterns shown in Table 2 (below) by ExCover. $k = 30$ was specified in the previous method but only eight patterns were generated. Among the patterns obtained by the previous method, the top-2 patterns cover 4,112 out of 4,208 positive transactions. Although the third-ranked pattern has high quality but the positive transactions covered by it

Table 2. Discriminative patterns for the *edible* class in the mushroom dataset, obtained by a previous method [14] (above) and obtained by ExCover (below).

Rank	Pattern	F-score
1	{odor=n, veil-type=p}	0.881
2	{gill-size=b, stalk-surface-above-ring=s, veil-type=p}	0.866
3	{gill-size=b, stalk-surface-below-ring=s, veil-type=p}	0.837
4	{gill-size=b, veil-type=p}	0.798
5	{stalk-surface-above-ring=s, veil-type=p}	0.776
6	{ring-type=p, veil-type=p}	0.771
7	{stalk-surface-below-ring=s, veil-type=p}	0.744
8	{veil-type=p}	0.682
Rank	Pattern	F-score
1	{odor=n, veil-type=p}	0.881
2	{gill-size=b, stalk-surface-above-ring=s, veil-type=p}	0.866
3	{stalk-surface-above-ring=s, veil-type=p}	0.776

Table 3. Statistics on the datasets used in the second experiment.

Dataset	#Transactions	#Items	Dataset	#Transactions	#Items
anneal	812	93	lymph	148	68
audiology	216	148	mushroom	8,124	119
australian-credit	653	125	primary-tumor	336	31
german-credit	1,000	112	soybean	630	50
heart-cleveland	296	95	splice-1	3,190	287
hepatitis	137	68	tic-tac-toe	958	27
hypothyroid	3,247	88	vote	435	48
kr-vs-kp	3,196	73	zoo-1	101	36

are also covered by the top-2 patterns. The remaining 96 positive transactions, on the other hand, are covered the fifth-ranked pattern. From this result, we can say that the patterns like the third-ranked one are redundant and specifying $k < 5$ implies losing information from 96 positive transactions. In contrast, thanks to the best-covering constraint, redundant patterns including the third-ranked one in Table 2 (above) are automatically excluded from the output of ExCover.

The statistics on the datasets used in the second experiment is summarized in Table 3. With these datasets, we compare the search space, i.e. the number of visited patterns in the enumeration tree, between the previous method above and ExCover.[5] For the previous method, we made three runs varying $k = 10, 100$

---

[5] More formally, a visited pattern is a pattern closed on the positives produced at Line 5 in the GROW procedure.

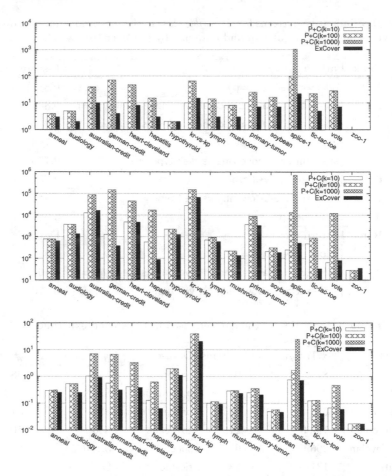

Fig. 2. The number of output patterns (top) and visited patterns (middle), and the average run time (bottom, in seconds). The y-axis is logarithmic and only one pattern were found in zoo-1. The error bars for average run time are narrow and omitted.

and 1,000. First, we show the number of output patterns in Fig. 2 (top). In the graphs, P+C(k=10) corresponds to the case with the previous method under $k = 10$ (P+C stands for productivity and closedness), and so on. The results show that ExCover outputs a more compact set of patterns, as Proposition 2 formally suggests. For most of the datasets, there are more than ten productive patterns, and there are fewer (sometimes by an order of magnitude) best-covering patterns. A more drastic difference was observed in the number of visited patterns shown in Fig. 2 (middle). The search space of the previous method varies with k and hence a suitable k is sometimes unclear in advance, whereas ExCover totally explores the search space of moderate size. Finally, Fig. 2 (bottom) shows the run time averaged over 30 runs. Our implementation is written in Java and we used Intel Core i7 3.6 GHz. The result clearly shows that ExCover finishes in practical time, i.e. within one second in most datasets.

5   Related Work

Associative classification [23] is a task for building classifiers from class association rules (CARs) and is closely related to discriminative pattern mining. CBA [18] is the first associative classifier, in which we first extract CARs using two user-specified control parameters: minimum support and minimum confidence. To filter out redundant CARs, CMAR [17] further introduces the productivity constraint (based on confidence) and χ^2 testing. The closest method to ExCover should be HARMONY [25] since it seeks for CARs which have the highest confidence for at least one positive transaction. However, as said earlier, its original paper does not mention redundancy among CARs. In addition, the pruning mechanism in HARMONY is fairly complicated and heavily relies on the minimum support threshold given by the user. One possible reason for this is that a finite upper bound of confidence is not easy to obtain. On the other hand, using a dual-monotonic quality score whose upper bound is finite, ExCover performs simple branch-and-bound pruning with no user-specified control parameters. To achieve parameter-freeness, as done by the fitcare algorithm [3], automatic tuning of control parameters looks promising, but it may weaken the explainability unless the objective function for tuning is clear to the user.

In addition, a generic problem-solving framework, which includes pattern set mining, based on constraint programming was recently developed [11]. We are rather focusing on exploring a desirable combination of comprehensible constraints, and developing a specific algorithm that works with less resources. For example, ExCover chooses a depth-first search over a suffix enumeration tree that brings less memory consumption and more chances for pruning. Lastly, Xin et al. proposed a method for mining non-redundant frequent patterns by minimizing the overlaps among the patterns' coverage over transactions [27]. The best-covering constraint balances well the relevance to the class of interest with the degrees of overlaps.

6   Concluding Remarks

This paper proposed an algorithm called ExCover that finds best-covering patterns closed on the positives in an exhaustive manner. Formal discussions and experimental results exhibit that ExCover efficiently finds a more compact set of patterns in comparison with previous methods. Although ExCover was motivated by knowledge discovery situations, it is interesting to build a practical classifier using best-covering patterns. For that, an extension for handling continuous attributes seems indispensable. In addition, the idea behind ExCover seems not limited to closed itemsets. For example, it would be useful to extend ExCover for discriminative sequential pattern mining [9].

References

1. Aggarwal, C.C.: Data Mining: The Textbook. Springer, Switzerland (2015)
2. Bayardo, R., Agrawal, R., Gunopulos, D.: Constraint-based rule mining in large, dense databases. Data Min. Knowl. Discov. **4**, 217–240 (2000)
3. Cerf, L., Gay, D., Selmaoui, N., Boulicaut, J.-F.: A parameter-free associative classification method. In: Song, I.-Y., Eder, J., Nguyen, T.M. (eds.) DaWaK 2008. LNCS, vol. 5182, pp. 293–304. Springer, Heidelberg (2008)
4. Cheng, H., Yan, X., Han, J., Yu, P.S.: Direct discriminative pattern mining for effective classification. In: Proceedings of ICDE 2008, pp. 169–178 (2008)
5. Domingos, P.: The RISE system: conquering without separating. In: Proceedings of ICTAI 1994, pp. 704–707 (1994)
6. Dong, G., Bailey, J. (eds.): Contrast Data Mining: Concepts, Algorithms, and Applications. CRC Press, Boca Raton (2012)
7. Fürnkranz, J., Gamberger, D., Lavrač, N.: Foundations of Rule Learning. Springer, Heidelberg (2012)
8. Garriga, G.C., Kralj, P., Lavrač, N.: Closed sets for labeled data. J. Mach. Learn. Res. **9**, 559–580 (2008)
9. Grosskreutz, H., Lang, B., Trabold, D.: A relevance criterion for sequential patterns. In: Blockeel, H., Kersting, K., Nijssen, S., Železný, F. (eds.) ECML PKDD 2013, Part I. LNCS, vol. 8188, pp. 369–384. Springer, Heidelberg (2013)
10. Grosskreutz, H., Paurat, D.: Fast and memory-efficient discovery of the top-k relevant subgroups in a reduced candidate space. In: Gunopulos, D., Hofmann, T., Malerba, D., Vazirgiannis, M. (eds.) ECML PKDD 2011, Part I. LNCS, vol. 6911, pp. 533–548. Springer, Heidelberg (2011)
11. Guns, T., Nijssen, S., De Raedt, L.: k-Pattern set mining under constraints. IEEE Trans. Knowl. Data Eng. **25**(2), 402–418 (2013)
12. Han, J., Pei, J., Yin, Y.: Mining frequent patterns without candidate generation. In: Proceedings of SIGMOD 2000, pp. 1–12 (2000)
13. Han, J., Wang, J., Lu, Y., Tzvetkov, P.: Mining top-k frequent closed patterns without minimum support. In: Proceedings of ICDM 2002, pp. 211–218 (2002)
14. Kameya, Y., Asaoka, H.: Depth-first traversal over a mirrored space for non-redundant discriminative itemsets. In: Bellatreche, L., Mohania, M.K. (eds.) DaWaK 2013. LNCS, vol. 8057, pp. 196–208. Springer, Heidelberg (2013)
15. Novak, P.K., Lavrač, N., Webb, G.I.: Supervised descriptive rule discovery: a unifying survey of contrast set, emerging pattern and subgroup mining. J. Mach. Learn. Res. **10**, 377–403 (2009)
16. Li, J., Li, H., Wong, L., Pei, J., Dong, G.: Minimum description length principle: generators are preferable to closed patterns. In: Proceedings of AAAI 2006, pp. 409–414 (2006)
17. Li, W., Han, J., Pei, J.: CMAR: accurate and efficient classification based on multiple class-association rules. In: Proceedings of ICDM 2001, pp. 369–376 (2001)
18. Liu, B., Hsu, W., Ma, Y.: Integrating classification and association rule mining. In: Proceedings of KDD 1998, pp. 80–86 (1998)
19. Morishita, S., Sese, J.: Traversing itemset lattices with statistical metric pruning. In: Proceedings of PODS 2000, pp. 226–236 (2000)
20. Pasquier, N., Bastide, Y., Taouil, R., Lakhal, L.: Discovering frequent closed itemsets for association rules. In: Beeri, C., Bruneman, P. (eds.) ICDT 1999. LNCS, vol. 1540, pp. 398–416. Springer, Heidelberg (1998)

21. Rijnbeek, P.R., Kors, J.A.: Finding a short and accurate decision rule in disjunctive normal form by exaustive search. Mach. Learn. **80**, 33–62 (2010)
22. Soulet, A., Crémilleux, B., Rioult, F.: Condensed representation of emerging patterns. In: Dai, H., Srikant, R., Zhang, C. (eds.) PAKDD 2004. LNCS (LNAI), vol. 3056, pp. 127–132. Springer, Heidelberg (2004)
23. Thabtah, F.: A review of associative classification mining. Knowl. Eng. Rev. **22**(1), 37–65 (2007)
24. Uno, T., Asai, T., Uchida, Y., Arimura, H.: An efficient algorithm for enumerating closed patterns in transaction databases. In: Proceedings of DS 2004, pp. 16–31 (2004)
25. Wang, J., Karypis, G.: HARMONY: efficiently mining the best rules for classification. In: Proceedings of SDM 2005, pp. 205–216 (2005)
26. Witten, I.H., Frank, E.: Data Mining: Practical Machine Learning Tools and Techniques, 2nd edn. Morgan Kaufmann, San Diego (2005)
27. Xin, D., Han, J., Yan, X., Cheng, H.: On compressing frequent patterns. Data Knowl. Eng. **60**(1), 5–29 (2007)
28. Yuan, C., Lim, H., Lu, T.C.: Most relevant explanation in Bayesian networks. J Artif. Intell. Res. **42**, 309–352 (2011)
29. Zaki, M.J.: Scalable algorithms for association mining. IEEE Trans. Knowl. Data Eng. **12**(3), 372–390 (2000)

Applications of Big Data Mining II

A Maximum Dimension Partitioning Approach for Efficiently Finding All Similar Pairs

Jia-Ling Koh$^{(\boxtimes)}$ and Shao-Chun Peng

Department of Information Science and Computer Engineering,
National Taiwan Normal University, Taipei 106, Taiwan, ROC
jlkoh@ntnu.edu.tw

Abstract. For solving the All Pair Similarity Search (APSS) problem efficiently, this paper provides a maximum dimension partitioning approach to effectively filter non-similar pairs in an early stage. At first, for each data point, the dimension with the maximum value is used to decide the corresponding segment of data partition. An adjusting method is designed to balance the number of elements in each data segment. The similar pairs consist of inter-segment similar pairs and intra-segment similar pairs, where most effort of computing APSS comes from the computation of finding inter-segment similar pairs. For speeding up the computation, a pilot-vector is used to represent each segment for estimating the upper bound of similarity between each segment pair. Only the segment pairs, whose upper bounds of similarity are larger than the given similarity threshold, need to generate the inter-segment data pairs as candidates. Moreover, based on the proposed partitioning method, we designed a MapReduce framework to solve the APSS problem in parallel. The performance evaluation results show the proposed method provides better pruning effectiveness on non-similar data pairs than the related works. Moreover, the proposed partition-based method can properly fit into the MapReduce programming scheme to effectively reduce the response time of solving the APSS problem.

1 Introduction

In real-world applications of data mining, a crucial problem is to perform similarity search, such as collaborative filtering for similarity-based recommendations, near duplicate document detection, and coalitions of click fraudster identification. Given a function $Sim(x, y)$ and a similarity threshold t, a similarity search aims to find all objects in a dataset with a similarity value of at least t compared to a query object. The All Pair Similarity Search (APSS) problem performs a similarity search for each object in a dataset to find all similar pairs in the dataset.

A data object in an application is generally numerically represented by a high dimensional vector, where each dimension is a feature extracted from the object. Suppose that the dataset consists of n objects and each object has a m dimensional feature vector. The time complexity of a brute force algorithm for APSS is $O(mn^2)$, which is infeasible in practice. Accordingly, there have been many works studied how to improve the performance efficiency for solving APSS [1–3, 5, 9, 12, 14]. In order to

© Springer International Publishing Switzerland 2016
S. Madria and T. Hara (Eds.): DaWaK 2016, LNCS 9829, pp. 163–178, 2016.
DOI: 10.1007/978-3-319-43946-4_11

reduce the computation cost of solving APSS, it is necessary to effectively reduce the search space of the problem. In other words, it requires some pruning strategies to reduce the number of generated data pairs which need similarity computation. The cost of a pruning strategy is count into the total cost for solving the problem. Therefore, the pruning strategy should be both effective and efficient. Moreover, to develop a parallelized approach for reducing response time is the recent direction for solving the issue of huge amount of data [10, 11, 15].

For solving the All Pair Similarity Search (APSS) problem efficiently, this paper provides a maximum dimension partitioning approach to effectively filter non-similar pairs in an early stage. At first, for each data point, the dimension with the maximum value is used to decide the corresponding segment of data partition. An adjusting method is designed to balance the number of elements in each data segment. The similar pairs consist of inter-segment similar pairs and intra-segment similar pairs, where most effort of computing APSS comes from the computation of finding inter-segment similar pairs. For speeding up the computation, a pilot-vector is used to represent each segment for estimating the upper bound of similarity between each segment pair. Only the segment pairs, whose upper bounds of similarity are larger than the given similarity threshold, need to generate the inter-segment data pairs as candidates. Moreover, the prefix filtering strategy is used to improve the efficiency of computing similarity of both segment pairs and intra-segment data pairs. Based on the partitioning method, we designed a MapReduce framework to solve the problem in parallel. The performance evaluation results show the proposed method provides better pruning effectiveness on non-similar data pairs than the related works. Moreover, the proposed partition-based method can properly fit into the MapReduce programming scheme to effectively reduce the response time of solving the APSS problem.

This paper is organized as follows. The problem definition and related work are introduced in Sect. 2. In Sect. 3, the details of the proposed partitioning method and the pruning strategy are introduced. The MapReduce extension is proposed in Sect. 4. The performance evaluation on the proposed methods and related works is reported in Sect. 5. Finally, in Sect. 6, we conclude this paper.

2   Preliminaries and Related Work

2.1   Problem Definition

Let $D = \{d_1, d_2, d_3, \ldots, d_n\}$ denote a set of data, where each data d_i is represented by a m dimensional vector $d_i = <d_i[1], d_i[2], d_i[3], \ldots d_i[m]>$. It is assumed that each vector is normalized. Accordingly, the similarity score between two data d_i and d_j is computed by the cosine-similarity function as follows:

$$Sim(d_i, d_j) = \sum_{k=1}^{m} d_i[f_k] \times d_j[f_k].$$

Given a threshold t, two vectors d_i and d_j form a *similar pair* if their similarity score is larger than or equal to the threshold value t, i.e. $Sim(d_i, d_j) \geq t$.

The All Pair Similarity Search (APSS) problem is to find all (d_i, d_j) pairs, where d_i, $d_j \in D$ and $Sim(d_i, d_j) \geq t$.

2.2 Related Work

When a dataset consists of high-dimensional data vectors, it is costly to perform similarity computation for data pairs. Accordingly, many strategies were designed to approximately estimate the similarity of a pair of data with the assistance of an inverted list index on the dataset [2, 4, 5]. In order to save the computation time, when the partially computed similarity value of a data pair can decide that the pair is dissimilar, it is not necessary to compute all dimensions of the feature vectors for getting the exact similarity of the pair.

An inverted list is a data structure of indexing, for each feature dimension f, which constructs a linked list to store the object identifiers with a non-zero value on the dimension. It is the most popular data structure used in document retrieval systems for finding documents containing a query keyword. The studies in [5, 6] considered that it is not necessarily to build a complete inverted index over the vector input. In [5], for a data object x, its cumulative estimated similarity on features with other objects is computed. The suffix feature values of x start being indexed only when the cumulative estimated similarity with other objects has reached the similarity threshold t. Besides, the prefix filtering principle was proposed to reduce the generated candidate pool size. Bayardo et al. [5] showed that, by using this method, the indexed feature values of the object x are enough to identify any object y that is potentially similar to x during the similarity search. [2] applied the Cauchy-Schwarz inequality to provide l^2-norm filtering. The l^2-norm was used to improve the estimation bound of similarity between feature vectors for getting a smaller inverted index size and pruning more candidates of APSS. However, to compute l^2-norm is costly for a high-dimensional dataset. [4] extended the prefix-filtering strategy in [5] for solving the Incremental All Pair Similarity Search (IAPSS) problem, which performs APSS multiple times over the same dataset by varying the similarity threshold. The previously mentioned works focused on reducing candidates of APSS, but obtaining better pruning effect in candidate generation may cause much processing effort. Accordingly, [12] aimed to decrease the computational cost of candidate generation by reducing the number of indexed objects. [16] also stated that the length of inverted index has significant effect on processing efficiency, i.e. prefix filtering does not always achieve high performance. Therefore, a cost model was proposed to decide the length of its prefix index for each object.

The key idea of Locality-sensitive hashing (LSH) methods is to hash the points using several hash functions so as to ensure that, for the vectors hashed into one bucket, the probabilities of similar pairs are much higher than non-similar pairs [9, 13]. Since the Locality-sensitive hashing (LSH) approach groups similar vectors into one bucket with approximation, it has a trade-off between precision and recall. Besides, redundant computation occurs when multiple hash functions are used. The study [4] provided an exact algorithm to perform set-similarity join. The algorithm first generates signatures for input sets, where all pairs of sets whose signatures have overlap are candidates to find the similarity set pairs. Its experiments showed that, when inverted indexing and computation filtering methods are applied, the exact algorithm performs competitive to LSH.

In recent years, many studies solved the APSS by parallel processing for reducing the response time, mostly by providing a MapReduce framework [8, 10, 11, 15]. [8] implemented a basic algorithm for APSS, which consists of two separate MapReduce jobs: (1) construct the inverted list, (2) compute the similarities of data pairs according to the inverted list. Furthermore, three variations are provided to prune the non-similar pairs early from the computation for reducing its cost. [11] proposed the V-SMART-Join framework, which consists of two phases of MapReduce processing: the first phase generates the candidate pairs and the second phase computes the similarity between all candidate pairs according to the constructed inverted index. The MapReduce framework proposed in [15] introduces a partitioning method where a data pre-processing phase first selects random points as centroids and the data points are assigned to the nearest centroid for computing centroid statistics. In the Similarity computation phase, the original dataset as well as the centroid statistics are read to construct the independent work sets, i.e. the set pairs for generating candidate pairs. Besides, an optional repartitioning step is used to enhance the load-balancing of the partitioning before computing similar pairs from the work sets.

[1] considered that most parallel methods of APSS using an inverted index to perform computation filtering suffer from excessive I/O and communication overhead of the intermediate partial results. Accordingly, a partition-based approach was proposed to solve APSS in parallel. This approach statically groups data vectors into partitions such that the dissimilar partitions can be revealed in an early stage to avoid unnecessary data loading and comparison. There are two steps to perform partitioning in [1]. The first step is to sort the data vectors according to their l^1-norm values in a non-decreasing order. The ordered list of data vectors is divided evenly into consecutive groups. In the second step, for the i-th group G_i, its vectors are further divided into i disjoint subgroups $G_{i,1}$; $G_{i,2}$; …; and $G_{i,i}$ such that the vectors in $G_{i,j}$, where $j \le i$, must be dissimilar with G_k for each $k \le j$. For each data vector d in G_i, $maxw(d)$ denote the maximum value in d. Besides, $Leader(G_j)$ denote the vector in G_j with the maximum l^1-norm length. It is induced that $Sim(G_j, d) \le maxw(d) \times Leader(G_j)$. Therefore, $maxw(d) \times Leader(G_j)$ is an estimation for the upper bound of similarity between d and the vectors in G_j. If $maxw(d) \times Leader(G_j)$ is less than the similarity threshold t, it implies that d is dissimilar with all the vectors in G_j such that d is assigned to G_{ij}. From the generated groups, except for the group pairs whose vectors are sure dissimilar to each other, the other group pairs need to generate inter-group data pairs as candidates of APSS. The intra-group data pairs generated in each group are also candidate of APSS. Furthermore, $Sim(d_i, d_j) \le \min(maxw(d_i) \times \|d_j\|_1, maxw(d_j) \times \|d_i\|_1)$ is used estimate the upper bound of similarity for pruning the dissimilar data pairs in the candidates. Finally, the remained data pairs need exactly compute their similarities to find similar pairs. [14] applies $\|d_i\|_r \times \|d_j\|_s$ to estimate the upper bound of similarity between d_i and d_j to identify more dissimilar vectors. Besides, the work aimed to solve the problem of load balance for [1] and tried to reduce the size gap among partitions. The partition-based approach is a good strategy for pruning non-similar pairs in an early stage. However, the weakness of the PSS (Partition-based Similarity Search)

method in [1, 14] occurs when applying to high dimensional vectors. As the dimension of vectors grows, the maximum l^1-norm length of the normalized vectors will grow as well. Accordingly, in most cases, the estimated upper bound of similarity in [1] is much higher than the exact similarity value. It causes the effectiveness of pruning dissimilar group pairs not well.

To summarize the above related works, this paper would combine the benefits of 2-level pruning strategies for speeding up the processing of APSS. We proposed a new static data partitioning in data pre-processing to reduce similarity computation of inter-segment data pairs, which improve the weakness of the PSS method [1]. Our proposed approach also applied the inverted index to perform dynamic filtering but requires less communication cost in a MapReduce framework.

3   A Partition-Based Approach for Solving APSS

In the maximum partitioning approach, the processing is divided into 4 main tasks for computation: (1) data partitioning, (2) find intra-segment similar data pairs, (3) generate candidate segment pairs, (4) prune inter-segment dissimilar data pairs and find inter-segment similar pairs. The strategies provided for processing each task are described in the following subsections.

3.1   Partitioning Method

According to the definition of cosine similarity, it is more likely that two vectors are similar if the maximum values in their vectors are located on the same dimension. Therefore, in our partitioning strategy, the vectors with the same maximum dimension are initially assigned to the same segment. There are three steps in the proposed partitioning method: (1) dimension reordering, (2) maximum-dimension partitioning, and (3) pilot vector computation.

Step 1: Dimension reordering

For each data dimension f_i, the number of data objects in D whose f_i has a non-zero value is counted, which is denoted as non-zero-count(f_i). The dimensions of the data objects in D are re-ordered by their non-zero-count values in descending order.

For example, in the sample dataset shown in Table 1, the non-zero-count values of dimensions f_1, f_2, f_3, and f_4 are 4, 3, 4, and 2, respectively. Therefore, the dimensions are reordered as f_1, f_3, f_2, and f_4.

Step 2: Maximum-dimension partitioning

For each data object d_i, find its dimension with the maximum value, denoted as $maxf(d_i)$. The data objects with the same $maxf$ value are assigned to the same segment. The partitioning result of the sample dataset D is shown in Table 2.

Step 3: Pilot vector computation

For each segment S_i, its pilot vector $S_i.pilot$ is defined as follows:

$$S_i.pilot[f_j] = \max(d_k[f_j]|\forall d_k \in S_i) \text{ for } j = 1, \ldots, m.$$

According to the partitioning result shown in Table 2, the pilot vectors of the three segments are shown as Table 3.

The number of the initial generated segments by the maximum-dimension partitioning method depends on the distribution of data. Accordingly, we provide the following adjusting methods to control the number of the generated segments.

(1) Increase the number of segments:

The segment with the maximum number of data objects, say S_i, is selected to be divided into two segments. The data objects in S_i are sorted in decreasing order according to the l^1-norm length of their vectors. Then S_i is divided into two segments with equal size at the middle of the sorted objects. If there is more than one segment with the maximum number of data objects, let *max_1-norm(S)* and *min_1-norm* (S) denote the maximum l^1-norm length and the minimum l^1-norm length among the vectors in such a segment S, respectively. The segment S_j with a larger difference between *max_1-norm(S_j)* and *min_1-norm(S_j)* has a sparser data distribution. Therefore, in this case, the segment containing vectors with the largest range on their l^1-norm length is selected to be split.

Table 1. A sample dataset D.

	f_1	f_2	f_3	f_4
d_1	0	0	1	0
d_2	0	1	0	0
d_3	0.45	0.9	0	0
d_4	0.65	0	0.76	0
d_5	0.6	0	0.6	0.52
d_6	0.5	0.5	0.5	0.5

Table 2. The partitioning result on the sample dataset D.

	f_1	f_3	f_2	f_4	*maxf*	l^1-norm	Assigned segment
d_6	0.5	0.5	0.5	0.5	f_1	2	S_1
d_5	0.6	0.6	0	0.52	f_1	1.72	S_1
d_3	0.45	0	0.9	0	f_2	1.35	S_2
d_2	0	0	1	0	f_2	1	S_2
d_4	0.65	0.76	0	0	f_3	1.41	S_3
d_1	0	1	0	0	f_3	1	S_3

Performing the above processing one time will increase 1 of the number of segments. The pilot vectors of the resultant two segments have to be recomputed. However, the resultant two segments remain their *maxf* values unchanged, which are the same with the *maxf* value of the segment before splitting.

Table 3. The pilot vectors of the segmentations for the sample dataset D.

	f_1	f_3	f_2	f_4
$S_1.pilot$	0.6	0.6	0.5	0.52
$S_2.pilot$	0.45	0	1	0
$S_3.pilot$	0.65	1	0	0

Table 4. The postfix-based inverted index of the segment S_1.

	Postfix-based inverted list
f_1	Null
f_3	Null
f_2	$(d_6, 0.5)$
f_4	$(d_5, 0.52) \rightarrow (d_6, 0.5)$

(2) Decrease the number of segments

Two segments are selected to be merged into one segment. At first, the smallest data segment S_i is selected. Then another segment, denoted as S_c, should be selected to be merged with S_i. Suppose that S_i and S_c are selected to be merged to get the resultant segment denoted as S_{merge}. $S_{merge}.pilot$ will be no less than both $S_i.pilot$ and $S_c.pilot$ on each dimension. A larger value difference on $S_{merge}.pilot$ with respect to $S_i.pilot$ and $S_c.pilot$ implies that the vectors in S_{merge} will have wider range of feature values. It will cause the similarity estimation between segments less effective. Accordingly, the segment whose pilot vector has the smallest distance with $S_i.pilot$ is a better candidate to merge with S_i.

In order to reduce the computation cost when selecting the segment S_c, only the data segment, whose size is less than the average size of all segments, are used as candidates. Besides, the differences between $S_i.pilot$ and $S_c.pilot$ on the features $maxf$ (S_i) and $maxf(S_c)$ are used to estimate the dissimilarity of two segments S_i and S_c as follows:

$$dis\_similar(S_i, S_c) =$$
$$\max(|S_i.pilot[maxf(S_i)] - S_c.pilot[maxf(S_i)]|, |S_i.pilot[maxf(S_c)] - S_c.pilot[maxf(S_c)]|).$$

Accordingly, the candidate segment has the least dissimilarity with S_i is selected to be merged with S_i to get a new segment S_{merge}.

The pilot vector of the resultant segment S_{merge} is updated as follows:

$$S_{merge}.pilot[f_k] = max(S_i.pilot[f_k], S_c.pilot[f_k]) \text{ for } k = 1, \ldots, m.$$

Performing the above processing one time will decrease 1 of the number of segments.

We call the above partitioning method the max-d partitioning (MD) method. In order to prevent generating unbalanced sizes of segments, we proposed an alternative method to set an upper bound constraint on the segments for the MD method, which is called the max-d partitioning with balance constraint (BMD). We will compare the performance of the partitioning method in the experiments.

3.2   Find Intra-segment Similar Pairs

For each data segment S_i, we have to find all the intra-segment similar pairs in S_i. In order to dynamically reduce the cost of similarity computation, an inverted list index is constructed for the data objects in each segment and the prefix filtering strategy [5] is applied.

In our implementation, the boundary of constructing the inverted list for the postfix vector of a data d_j is determined according to the maximum similarity estimated between d_j and $S_i.pilot$. For example, the segment S_1 in the sample dataset has S_1. $pilot$ = <0.6, 0.6, 0.5, 0.3>. The data d_5 in S_1 has feature vector = <0.6, 0.6, 0, 0.52>. Let the similarity threshold t set to be 0.8. The partial inner product results between d_5 and $S_1.pilot$ on the first 3 dimensions, i.e. 0.72, is less than t and on the 4 dimensions is larger than t. Therefore, the prefix vector of d_5 consists of its first 3 dimensions: <0.6, 0.6, 0> and the postfix vector of d_5 consists of the last dimension: <0.3>. Accordingly, the constructed postfix-based inverted index of the segment S_1 is shown in Table 4.

Due to space limit, please refer [5] about the detailed constructing process of inverted list and the prefix filtering strategy. From the sample dataset, the discovered intra-segment similar pairs are (d_5, d_6) and (d_2, d_3).

3.3   Dissimilar Segment Pairs Pruning Strategy

A *dissimilar segment pair* (S_i, S_j) means that for each data d in segment S_i and each data d' in segment S_j, $Sim(d, d') < t$. To discover dissimilar segment pairs in early stage can eliminate the similarity computations on the data pairs across the dissimilar segment pairs. Therefore, we estimate the upper bound of similarity between two segments to prune some dissimilar segment pairs.

For each data segments S and S', the similarity between $S.pilot$ and $S'.pilot$ is computed. According to the definition of the pilot vector of a segment, for each d in S and each d' in S', $Sim(d, d') \leq Sim(S.pilot, d') \leq Sim(S.pilot, S'.pilot)$. Therefore, if $Sim(S.pilot, S'.pilot) < t$, it implies $Sim(d, d') < t$ for each d in S and each d' in S'. Accordingly, (S, S') is a dissimilar segment pair and is removed from the candidate segment pairs.

For example, according to the partitioning result shown in Table 2 and the pilot vectors of the 3 segments shown in Table 3, the upper bounds of the similarity of the segment pairs are computed as follows:

$$Sim(S_1.pilot, S_2.pilot) = 0.6 * 0.45 + 0.5 * 1 = 0.77$$
$$Sim(S_1.pilot, S_3.pilot) = 0.6 * 0.65 + 0.6 * 1 = 0.99$$
$$Sim(S_2.pilot, S_3.pilot) = 0.45 * 0.65 = 0.2925$$

Suppose that the similarity threshold $t = 0.8$, (S_1, S_2) and (S_2, S_3) are discovered to be dissimilar segment pairs and pruned. Only the segment pair (S_1, S_3) is remained as a candidate segment pair to generate inter-segment data pairs.

To speed up the computation, we also construct an inverted list index for the pilot vectors of segments. The strategy described in 3.2 is used to find similar pilot vector pairs such that the corresponding segment pairs are candidate segment pairs.

3.4   Inter-segment Data Pairs Pruning Strategy

Let (S_k, S_l) denote a segment pair where $Sim(S_k.pilot, S_l.pilot) \geq t$, such that the segment pair is remained after performing the pruning strategy introduced in the previous subsection. It is possible to find a data object $d \in S_k$ and a data object $d' \in S_l$ such that (d, d') is a similar pair. To prevent from performing similarity computations for all inter-segment data pairs between S_k and S_l, an efficient pruning strategy is proposed to prune part of the dissimilar objects pairs as follows.

Let $max\_v(S_k)$ denote the maximum feature value contained in the data vectors in segment S_k. $max\_v(S_k)$ is obtained by computing $\{v | \exists i(m \geq i \geq 1), v = S_k.pilot[f_i] \wedge \forall m \geq j \geq 1 \wedge j \neq i, v \geq S_k.pilot[f_j]\}$. For a data object d_x in S_l, if $\|d_x\|_1 < t/max\_v(S_k)$, it implies that $Sim(d_x, d_y) < t$ for all d_y in S_k. Therefore, when generating inter-segment data pairs between S_k and S_l, it is not necessary to generate data pair consisting d_x.

For example, after performing the pruning strategy described in Sect. 3.3 on the sample dataset, the inter-segment data pairs are generated between segments S_1 and S_3. From Table 3, we obtain $max\_v(S_1) = 0.6$ and $max\_v(S_3) = 1$. Accordingly, any data object in S_3 with l^1-norm length less than $0.8/0.6 = 1.3$ is dissimilar with all data objects in S_1. It implies that the data object d_1 can be pruned. Only the data pairs (d_4, d_6) and (d_4, d_5) are generated for further computing the similarities. Finally, the discoverer inter-segment similar pair is (d_4, d_5).

4   A MapReduce Framework for Solving APSS

In this section, we will introduce the proposed MapReduce framework for performing the 4 tasks of the maximum-dimension partitioning approach. At first, the partitioning is performed centralized. After that, the other three tasks can be performed in parallel as described in the following three sub-sections.

4.1   Parallel Processing for Finding Intra-segment Similar Pairs

For each segment, the task of finding the intra-segment similar pairs is performed independently by 2 stages of MapReduce tasks. The first stage is to construct the inverted list index for the data objects in the segment, and the second stage is to perform similarity computation for the intra-segment data pairs.

(1) Index construction

<Mapper>:

Input: the set of data objects in the database, where each object has its object id d_i, the assigned segment id S_k, and the list of non-zero feature and value on its vector.

The mappers output the <key: value> pair for each non-zero feature f_j on a vector of object d_i. The feature id f_j is the key. Besides, the feature value, object id d_i, and the segment id S_k are combined to be the value.

For example, the object d_3 is assigned to segment S_2, which has non-zero features $f_1 = 0.45$ and $f_2 = 0.9$. Two <key: value> pairs are generated: $<f_1: (0.45, d_3, S_2)>$ and $< (f_2: (0.9, d_3, S_2)>$.

<Reducer>:

The reducers combine the <key: value> pairs with the same key value f_i to generate the inverted list of the feature f_i.

By collecting the results of the reducers, the inverted list of the dataset in a segment is constructed.

(2) Find intra-segment similar pairs

<Mapper>:

Input: the inverted list of each feature.

Let the inverted list of a feature f_i be denoted as $<(d_{i1}, d_{i1}.f_i, d_{i1}.seg\_id), (d_{i2}, d_{i2}.f_i, d_{i2}.seg\_id), \ldots, (d_{im}, d_{im}.f_i, d_{im}.seg\_id)>$. For each d_{ik} and d_{il} in the inverted list, if $d_{ik}.seg\_id$ is equal to $d_{il}.seg\_id$, a key value pair: $<(d_{ik},d_{il}): d_{ik}.f_i \times d_{il}.f_i>$ is generated.

<Reducer>:

The reducers combine the <key: value> pairs with the same key value (d_{ik}, d_{il}) to compute the sum of the $d_{ik}.f_i \times d_{il}.f_i$ on various feature to get $Sim(d_{ik}, d_{il})$. If $Sim(d_{ik}, d_{il}) \geq t$, (d_{ik}, d_{il}) is output as a similar pair with its similarity value.

The similar pairs with similarity value no less than t are intra-segment similar pairs.

4.2   Parallel Processing for Pruning Dissimilar Segment Pairs

This task is to generate the candidate segment pairs (S_i, S_j) which are possible to find inter-segment similar pairs. The MapReduce processing described in 4.1 is used to find the candidate segment pairs, where the pilot vector of each segment is considered a data object. Besides, all their corresponding segment ids are set to be the same.

4.3   Parallel Processing for Finding Inter-segment Similar Pairs

After finding the candidate segment pairs (S_i, S_j) which are possible to find inter-segment similar pairs, the strategy introduced in 3.4 is used to prune some inter-segment dissimilar object pairs centralized. Then the following MapReduce approach is used to find the inter-segment similar pairs.

<Mapper>:

Input: the candidate object pairs (d_i, d_j) and the individual non-zero feature values of the two objects. $<f_{i1} = d_i.f_{i1}, f_{i2} = d_i.f_{i2}, ..., f_{in} = d_i.f_{in}, f_{j1} = d_j.f_{j1}, f_{j2} = d_j.f_{j2}, ..., f_{jn} = d_j.f_{jm}>$.

Output: Generate the key value pairs $<(d_i, d_j, f_{ik}): d_i.f_{ik} >$ for $k = 1 ... n$, and $<(d_i, d_j, f_{jk}): d_j.f_{jk}>$ for $k = 1 ... m$.

<Combiner>:

The key value pairs $<(d_i, d_j, f): d_i.f>$ with the same key value are collected, which gets the feature value for both objects in the data pair (d_i, d_j). Accordingly, the values of the same key value are multiplied if both d_i and d_j have non-zero values on f. In addition, the key value pair $<(d_i, d_j): d_i.f \times d_j.f>$ is output.

<Reducer>:

The key value pairs $<(d_i, d_j): d_i.f \times d_j.f>$ with the same (d_i, d_j) pair are collected. Therefore, the summation of the multiplication result on various feature values is computed to get $Sim(d_i, d_j)$. If $Sim(d_{ik}, d_{il}) \geq t$, (d_{ik}, d_{il}) is output as a similar pair with its similarity value.

The similar pairs with similarity value no less than t are inter-segment similar pairs.

5   Performance Study

5.1   Experimental Environment

The dataset used in the experiments is a real dataset, which consists of the collected documents posted in Yahoo! Answer. The vector of each document is represented by the TF-IDF of the bag of words in the document.

We performed the experiments on the following two types of hardware:

(1) A client running Windows7 on 2 cores (2.4 GHz), 4 GB RAM.
(2) The Hadoop parallelized environment consists of 1 master and 2 slave nodes, where each node running Ubuntu Linux on 2 core (3.4 GHz) and 16G memory.

In addition to our approach, the following related works are also implemented for comparison.

(1) Prefix-filtering approach (PF): an inverted-list index structure is constructed for prefix-filtering when computing similarity as the strategy used in [2].
(2) The PSS algorithm (PSS): a partition-based method proposed in [1], where the initial partitioning is based on the descending order of the l^1-norm length of data.

The proposed approach and related works are implemented by Java language.

5.2 Performance Evaluation on the Partitioning Methods

In the first part of experiments, we would like to compare the pruning effectiveness of the partitioning methods. The following three partitioning methods are compared:

(1) the maximum dimension partitioning method (MD),
(2) the maximum dimension partitioning method with balance constraint (BMD), and
(3) the l^1-norm length partitioning (1N) method, where even size partitioning is performed based on the descending order of the l^1-norm length of data. This method is the initial partitioning result of the PSS algorithm [1].

In this part of experiments, a set consisting of 7800 data vectors, where each vector has 11800 dimensional features, was used. In the dataset, there are 299 similarity pairs when the similarity threshold t is set to be 0.8.

[Exp. 1-1] Compare the percentage of similar data pairs discovered from intra-segment data pairs.

The goal of this experiment is to observe the effectiveness that the partitioning methods can group similar pairs in the same segment. Therefore, the percentages of similar pairs discovered from intra-segment data pairs are computed for the MD, BMD, and 1N methods, respectively. According to the results shown in Fig. 1(a), by using the MD method, about 2/3 similar pairs can be discovered from the intra-segment data pairs when the number of segments is set from 2000 to 3000. It indicates that the MD method can effectively group most similar pairs within segments. Besides, the BMD method keeps more similar pairs within the same segments than the 1N method.

[Exp. 1-2] Compare the percentage of pruned dissimilar segment pairs among all segment pairs.

The goal of this experiment is to observe the effectiveness that the partitioning methods combined with the pruning strategy by computing similarity of pilot vectors to prune dissimilar segment pairs. According to the number of the remained candidate segment pairs, we can compute the percentage of pruned segment pairs among all possible segment pairs as shown in Fig. 1(b). It shows that the BMD method can prune 99.7 % segment pairs, which is better than both the MD and 1N methods when the number of segments is less than 3200. It indicates that fewer inter-segment data pairs are generated by using the BMD method.

The pruning effect of the MD method is not as good as the other twos is because that the MD method may generate some large segments whose pilot vectors tend to similar with the pilot vectors of other segments.

[Exp. 1-3] Compare the percentage of pruned data pairs by the partitioning methods combined with the pruning strategies.

In this experiment, in addition to the MD, BMD, and 1N method, we also observe the pruning effectiveness of the PSS method [1]. As the result shown in Fig. 1(c), the pruning effectiveness of the MD, BMD, and 1N are consistent with the results shown in Fig. 1(b) because most data pairs are generated from the inter-segment data pairs. It shows the pruning effectiveness of using pilot vector is much better than the strategy proposed in PSS for high dimensional data.

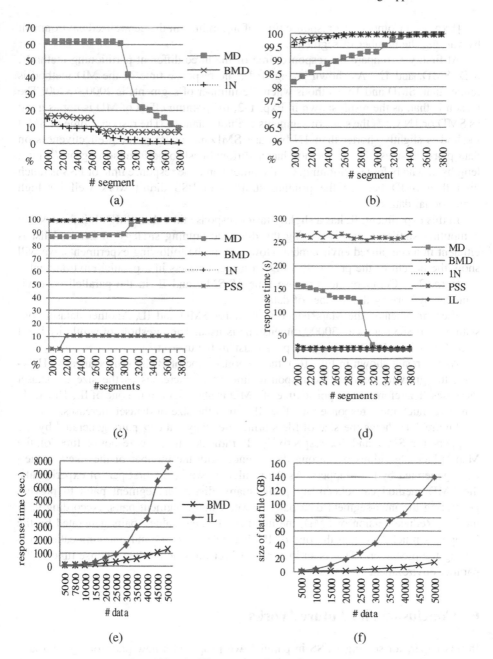

Fig. 1. The results of experiments

In the second part of experiments, we compared the response time for the proposed methods and the related works.

[Exp. 2-1] Compare the response time of algorithms in the centralized environment by varying the number of segments.

At first, we compare the response time of the three different partitioning methods: MD, SMD, and 1N. As shown in Fig. 1(d), the response time of the MD method is longer than SMD and 1N methods when the number of segments is $2000 \sim 3200$. The reason is that, as the result shown in Exp. 1-2, the pruning effect of MD is not as good as SMD or 1N when the sizes of segments are not balanced. The response time of SMD method is slightly shorter than 1N because SMD has better pruning effectiveness on data pairs. Besides, the 1N method has additional cost to perform sorting on l^1-norm lengths of data before partitioning. On the other hand, the response time of PSS is much high than SMD because the pruning strategy of PSS didn't work well for high dimensional dataset.

In this experiment, IL has a slightly faster.response time than SMD. It indicates that computing similarity pairs by using the dynamic pruning strategy of inverted list is efficient in a centralized environment. However, in the following experiments, we will show the benefit of the proposed SMD on larger datasets in a parallel environment.

[Exp. 2-2] Compare the response time of SMD and IL in the parallelized environment by varying the number of data.

We implemented the MapReduce version for SMD and IL. Another dataset consisting of 49885 data with 30000 dimensions is used. The threshold of similarity is set to be 0.8, where 5047 similar data pairs exist in the dataset.

As the result shown in Fig. 1(e), the response time of BMD grows linearly. However, the growing trend of the response time of IL increases as the size of dataset increases. Therefore, the response time of SMD is about half of the one of IL. The SMD can save much more response time than IL when the size of dataset increases.

Figure 1(f) shows the size of file storing the <key, value> pairs generated by the Mappers for SMD and IL, respectively. It indicates that the response time of the MapReduce algorithms has strong dependence with the number of the <key, value> pairs generated by the Mappers. As the results shown in the first part of experiments, the SMD method can effectively prune many dissimilar segment pairs to prevent generating the inter-segment data pairs between these segment pairs. Accordingly, in the MapReduce version of SMD, the number of <key, value> pairs generated by the Mapper is much less than the one of IL. It causes less I/O and communication cost among the processors. That is an important effect influencing the response time in the parallel environment.

6   Conclusion and Future Works

In this paper, for solving APSS in parallel, we proposed a new partitioning approach based on the maximum dimension of data vectors. Moreover, a pilot-vector is designed to represent each segment for estimating the upper bound of similarity between each segment pair. The proposed pruning strategy on segment pairs effectively reduces the number of candidate data pairs in an early stage. Besides, the prefix filtering strategy is used to improve the efficiency of computing similarity of both segment pairs and intra-segment data pairs. The results of experiment show the proposed BMD approach improves the weakness of the PSS method performed on high dimensional data sets. Moreover, we implemented a MapReduce framework for the proposed partitioning method. It reveals the benefit of candidate pruning achieved by BMD to effectively reduce the response time. Furthermore, Instead of using a global inverted index to perform computation filtering, this approach prevents the problem of excessive I/O and communication overhead of the intermediate partial results.in a MapReduce framework.

The proposed partitioning method can combine with a load assignment policy to achieve further load balance in a parallel environment. Another future work is to extend the proposed partitioning method to support incremental all pairs similarity search efficiently.

References

1. Alabduljalil, M., Tang, X., Yang, T.: Optimizing parallel algorithms for all pairs similarity search. In: Proceedings of the 6th ACM International Conference on Web Search and Data Mining (WSDM) (2013)
2. Anastasiu, D.C., Karypis, G.: L2AP: fast cosine similarity search with prefix L-2 norm bounds. In: Proceedings of the 30th IEEE International Conference on Data Engineering (ICDE) (2014)
3. Arasu, A., Ganti, V., Kaushik, R.: Efficient exact set-similarity joins. In: Proceedings of the 32nd International Conference on Very Large Data Bases (VLDB) (2006)
4. Awekar, A., Samatova1, N.F., Breimyer, P.: Incremental all pairs similarity search for varying similarity thresholds with reduced I/O overhead. In: Proceedings the 3rd SNA-KDD Workshop (2009)
5. Bayardo, R.J., Ma, Y., Srikant, R.: Scaling up all pairs similarity search. In: Proceedings of the 16th International Conference on World Wide Web (WWW) (2007)
6. Chaudhuri, S., Ganti, V., Kaushik, R.: A primitive operator for similarity joins in data cleaning. In: Proceedings of the 24th IEEE International Conference on Data Engineering (ICDE) (2006)
7. Dean, J., Ghemawat, S.: MapReduce: simplified data processing on large clusters. In Proceedings of OSDI (2004)
8. Francisci, G.D., Lucchese, C., Baraglia, R.: Scaling out all pairs similarity search with MapReduce. In: Proceedings of 8th Workshop on Large-Scale Distributed Systems for Information Retrieval (LSDS-IR) (2010)
9. Gionis, A., Indyky, P.: Similarity search in high dimensions via hashing. In: Proceedings of 25th International Conference on Very Large Data Bases (VLDB) (1999)
10. Lin, J.: Brute force and indexed approaches to pairwise document similarity comparisons with MapReduce. In: Proceedings the 32nd International ACM SIGIR Conference on Research and Development in Information Retrieval (SIGIR) (2009)
11. Metwally, A., Faloutsos, C.: V-SMART-join: a scalable MapReduce framework for all-pair similarity joins of multisets and vectors. Proc. VLDB Endowment 5(8), 704–715 (2012)
12. Ribeiro, L.A., Härder, T.: Efficient set similarity joins using min-prefixes. In: Grundspenkis, J., Morzy, T., Vossen, G. (eds.) ADBIS 2009. LNCS, vol. 5739, pp. 88–102. Springer, Heidelberg (2009)
13. Satuluri, V.: Bayesian locality sensitive hashing for fast similarity search. Proc. VLDB Endowment 5(5), 430–441 (2012)

14. Tang, X., Alabduljalil, M., Jin, X., Yang, T.: Load balancing for partition-based similarity search. In: Proceedings of the 37th International ACM SIGIR Conference on Research and Development in Information Retrieval (SIGIR) (2014)
15. Wang, Y., Metwally, A., Parthasarathy, S.: Scalable all-pairs similarity search in metric spaces. In: Proceedings of the 19th ACM SIGKDD International Conference on Knowledge Discovery and Data Mining (KDD) (2013)
16. Wang, J., Li, G., Feng, J.: Can we beat the prefix filtering: an adaptive framework for similarity join and search. In: Proceedings of the 2012 ACM SIGMOD International Conference on Management of Data (SIGMOD) (2012)

Power of Bosom Friends, POI Recommendation by Learning Preference of Close Friends and Similar Users

Mu-Yao Fang and Bi-Ru Dai[✉]

Department of Computer Science and Information Engineering,
National Taiwan University of Science and Technology, Taipei, Taiwan, ROC
M10315072@mail.ntust.edu.tw, brdai@csie.ntust.edu.tw

Abstract. With the emergence of social networks, mining interesting information from the social media datasets becomes a popular research direction. Previous
researches on social networks, such as POI (point of interest) recommendation,
usually ignore the social tie strength between users. If we can further consider the
closeness between friends in the analysis, it is possible to improve the results.
Therefore, in this paper, we focus on analyzing the social tie strength between
users in the location-based social network. The proposed method analyzes the
movement of users and the interaction between them by the spatial-temporal data.
Furthermore, the social relationship structure is also taken into consideration for
the calculation of the social tie strength. Finally, the location list for POI recommendation will be constructed accordingly. Experimental results show that the
proposed method significantly outperforms the competitor on both precision and
recall.

Keywords: Social networks · Location-based social network · Social tie strength ·
POI recommendation · Data mining · Spatial-temporal data

1 Introduction

With the emergence of GPS and portable devices, it is easier to share our locations and
experiences after visiting some places. More and more social networks provide the
check-in feature, that is, people can upload pictures, tag friends and leave some
comments for locations or venues where they have been. Thus, large amounts of data
are available for studying user mobility and further extending to location prediction,
POI (point of interest) recommendation, and so on [1–3, 6, 7, 9–12]. Spatial and temporal
data are important for modeling the movements of users. For example, people tend to
visit locations near their home or work place [1, 11], and the user mobility is influenced
by time and date [1, 13]. In addition to spatial and temporal data, the power of social
relationship cannot be ignored either [2, 4, 9, 11].

People are usually influenced more by close friends. Previous works on location-based
social network (LBSN) mainly focus on the binary social relationship [1, 2, 6, 11], in other
words, being friend or not. As a result, the social tie strength between users is usually
ignored. In our opinion, this assumption does not conform to the real life situations. Generally speaking, people tend to go along with close friends. However, it is much easier to

© Springer International Publishing Switzerland 2016
S. Madria and T. Hara (Eds.): DaWaK 2016, LNCS 9829, pp. 179–192, 2016.
DOI: 10.1007/978-3-319-43946-4_12

become friends in the social network, even though two users have never met before or have not talked to each other over three times. Therefore, if we observed from the social network dataset that two persons, who are friends in the social network, have the same movement records, we cannot identify whether it belongs to a coincidence or not via the binary relationship directly. In addition, previous works often consider only the number of common friends or times of visiting same locations. However, the number of common friends is not able to represent the tie strength between users completely [4, 8]. Besides, including all friends in the social network for analyzing will bring in lots of noises because all acquaintances are taken into consideration with the same importance as close friends.

In view of this, we further consider the social tie strength by the user movement data and the social relationship structure in our method. The proposed method, named **Social Tie Strength-Based POI Recommendation** (SSR), includes the concepts of location diversity, co-occurrence behavior, and personal check-in diversity to discover the closeness between users from the user movement data. Then, the social relationship is combined together to estimate the social tie strength. Finally, the location list for POI recommendation will be constructed accordingly. Experimental results show that the proposed method significantly outperforms the competitor on both the precision and the recall.

In the rest sections of this paper, we summarize related works in Sect. 2. The proposed method is introduced in Sect. 3, and the design of location scores for recommendation is presented in Sect. 4. In Sect. 5, we provide experiment results to evaluate our method. Finally, we make a conclusion for this paper in Sect. 6.

2   Related Works

We aim to estimate the social tie strength and exploit it in POI recommendation on LBSN. Thus, in this section, we will describe some related works of POI recommendation and social tie strength estimation on LBSN. In the end of this section, we summarize our work with the previous POI recommendation works in Table 1. $\sqrt{}$ indicates that the feature is addressed by the work.

Table 1. Summary of related POI recommendation works according to the utilized features, including social tie strength (SS), social features (S), geographical features (e.g. distance) (G), user preference similarity (U), time awareness (e.g. date, periodic) (T), and others.

Title	SS	S	G	U	T	Others
SSR	$\sqrt{}$	$\sqrt{}$		$\sqrt{}$		
[10]				$\sqrt{}$		Opinion
[11]		$\sqrt{}$	$\sqrt{}$	$\sqrt{}$		
[12]		$\sqrt{}$	$\sqrt{}$	$\sqrt{}$		
[13]					$\sqrt{}$	
[20]					$\sqrt{}$	
[21]						Topic, Profile, POI property
[22]						Topic, POI property
[23]			$\sqrt{}$			

2.1 POI Recommendation System

Mining user check-in data and make recommendation for users has become a popular issue, after we began able to gathering large amounts of user movement data. Main approaches of recommendation system are collaborative filtering and content based filtering [15, 18]. Collaborative filtering base on collecting and analyzing data of user activities, preferences and surmise their interest by computing the similarity between them and other users [10–12, 15, 18, 19]. Content based filtering base on the preference of users and description of items. User profile indicates that what user would like and the content based filtering will find items which have similar description with that [15, 18, 19]. In addition, there are some kinds of recommendation in LBSN which combined other features like time awareness [13, 20], content awareness [21, 22], geographical feature [23] and so on. These works usually contain similarity calculation between users or locations.

People are always affected by others and most of related work had considered the interaction and relationship between users, but in our view, some of them are more influential. If we treat all friends as have same influence to us, we may underestimate the power of social relation on LBSN.

2.2 Social Tie Strength Estimation

Occupying some location in the social graph can means a lot, and connections between users lead to different tie strength to users. [7] develops a unsupervised model utilizing interaction activities data and similarity between users to estimate social tie strength. [24] utilize the homophily, structural theory and triadic closure, etc. to find out social role or status of users on social networks. There are some works using other data to infer social role or tie strength. [3] utilizes check-in data to analyze the user mobility and attribute of locations. If a pair of users usually travel together and always visit places which are unpopular, then they probably are good friends.

We usually travel together to somewhere with friends. So we can investigate the closeness between target user and his/her friends by their behaviors. In the other hand, estimate the social tie strength by their social relationship can help us to find out more close friends for target users, then we can generate a more completely recommendation list for him/her.

3 Estimating the Social Tie Strength

In this paper, we aim to find out the social tie strength between users and suggest some points of interest for users accordingly. The proposed method, named **Social Tie Strength-Based POI Recommendation** (SSR), analyzes the movement of users and the interaction between them by the spatial-temporal data, e.g. the check-in data. Furthermore, the social relationship structure is also taken into consideration. The social tie strength will be discovered in Sect. 3 and be applied in Sect. 4 to construct the location list for POI recommendation. The overview diagram is shown in Fig. 1, and Table 2 summarizes the symbols used in this paper.

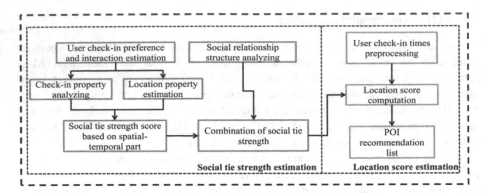

Fig. 1. Framework structure diagram

Table 2. The summary of notations

Symbols	Description
s_{ij}^l	Co-occurrence numbers of user i and j at location l
p_{ij}^l	Probability of user i and j co-occurrence at l
c_{ul}	Check-in records of user u at location l
H_{ij}	Renyi entropy of co-occurrence of user i and j
H^l	Location entropy of l
D_{ij}^c	Co-occurrence diversity of user i and j
D_i^p	Personal check-in records diversity of user i
D_{ij}^f	Check-in consistency between user i and j
K_{ij}	Katz of user i and j
$S_{u,l}$	Saturation of user check-in at location l
$R_{i,l}^p$	Score of location l by users similar to user I
$R_{i,l}^S$	Score of location l by close friends of user i
w_{ij}	Combined score of check-in consistent between user i and j
f_{ij}	Combined score of social strength between user i and j
R_{il}^F	Combined score of location l to user i

We will describe our method in detail in the following subsections. In Sect. 3.1 we investigate user behaviors on LBSN to estimate the social strength. In Sect. 3.2, a social strength estimation feature is integrated into our method to generate a unified social tie strength score which can reflect the closeness between friends.

3.1 Check-in Behavior and Location Property

We can investigate the social strength between a pair of people by their interaction. Some previous methods find influential friends of target users by counting their common friends or the number of identical places they visited. However, in our view, co-occurrence is more important than co-location and the property of visited places are influential too. Note that co-occurrence considers check-ins at the same place and the same time, but co-location considers only check-ins at the same location.

Although co-occurrence exists among many users, who have check-in records at the same location and at the same time, only a few of them are actually friends and visited these places together. As illustrated in Fig. 2, we can observe that co-location records between the target user and user A are relatively more in both of popular and unpopular venues. Hence, it makes sense that they are close friends because these co-occurrence records seem to be the case that they had talked about somewhere and made an appointment to visit it together. On the other hand, although the target user and user B also usually check-in at the same time and the same locations, the closeness between them is not as high as that between the target user and user A because venues of their co-location are popular, which means that there are usually lots of people visiting these places. If we only consider the common places they visited, in the case of Fig. 2, the importance of user A and user B will be similar. Consequently, the deeper influence from close friends, such as user A, will be ignored. In order to reflect the significance of close friends, we design a personal check-in diversity based on the co-occurrence diversity [3] to find out who always accompanies with the target user and determine whether their check-ins are consistent or not. In this way, the closeness between a friend and the target user can be estimated. In addition, to reduce the influence of visiting the same place caused by coincidence, the concept of weighted frequency [3] is included in our model to capture the property of locations.

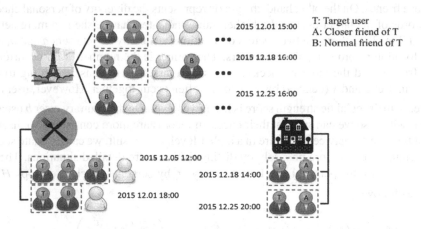

Fig. 2. Graph representation of check-in behaviors. Note that people in the dotted rectangle are actually friends and visit these places together.

Check-in Property Analyses. As we mentioned above, if two users have more co-occurrence records among various types of places, they are more likely to be closer friends. In this point of view, the diversity of co-occurrence records between the target user and his/her friends will be a useful indicator. We will utilize the concept of entropy to measure the check-in diversity between two users. As [3] mentioned, to measure the social tie strength,

Renyi entropy [17] is more influential and has a better performance on distinguishing cases of coincidence. The Renyi entropy H_{ij} of order d, where $0 \leq d < 1$, is defined as:

$$H_{ij} = -\frac{1}{1-d} log\left(\Sigma_l(p_{ij}^l)^d\right), \quad \text{where } p_{ij}^l = \frac{S_{ij}^l}{S_{ij}}. \tag{1}$$

s_{ij}^l is the number of co-occurrence times between user i and user j at location l, and s_{ij} denotes the total number of co-occurrence between them. $p_{i,j}^l$ represents the probability of co-occurrence between users i and j at location l. According to [3], the co-occurrence diversity D_{ij}^c can be obtained from the entropy H_{ij} according to the following formulation:

$$D_{ij}^c = exp(H_{ij}) = exp\left[-\frac{1}{1-d} log\left(\Sigma_l(p_{ij}^l)^d\right)\right]. \tag{2}$$

Co-occurrence diversity D_{ij}^c can help us to find out some close friends of the target user. A higher value of D_{ij}^c represents that users i and j usually travel around together, so that they are more likely to be close friends. However, in our opinion, further taking the personal check-in diversity of the target user into consideration is able to discover closer friends more precisely. As illustrated in Fig. 3, there are two pairs of users, target users a and b and their friends x and y respectively, belong to two types of situations. In Fig. 3(A), the x-axis represents the diversity of co-occurrence of user i and j, which is the value D_{ij}^c introduced above. The higher value represents that they are more likely to be close friends. On the other hand, the y-axis represents the diversity of personal check-in records of the target user, and a higher value indicates that he/she is a more active user. In Fig. 3(B), there are two vectors of locations visited by target users a and b, and each location record is a pair of numbers. The former is the number of co-occurrence with friends, and the latter is check-in times of a target user. Thus, according to the diagram, user a and x (denoted by $u_{a,x}$) do not often occur together. However, user a is active, so their social tie strength score is in a lower level. On the contrary, user b usually appear with y, so we can say that their check-in records are more consistent, so that the social tie strength between them are in a higher level. As a result, we can estimate social tie strength of users more precisely by distinguishing more detailed situations. Therefore, we defined the personal check-in diversity D_i^p by personal check-in entropy H_i of user i as follows:

$$D_i^p = exp(H_i) = exp\left(-\Sigma_l \, p_{i,l} \, log \, p_{i,l}\right) = exp\left(-\Sigma_l \frac{c_{i,l}}{R_i} log \frac{c_{i,l}}{R_i}\right), \tag{3}$$

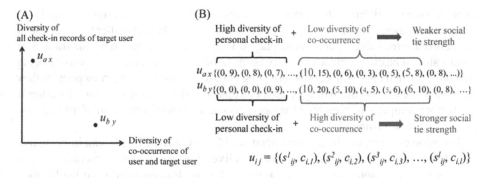

Fig. 3. The illustration of different types of check-in behavior. (A) Two pairs of users with different situations, where x-axis represents the diversity of co-occurrence and y-axis represents the personal check-in diversity. (B) Personal check-in records of the target user and co-occurrence records with friends. User a and b are target users, and different types of check-in behavior lead to different social tie strength. (Color figure online)

where $c_{i,l}$ denotes the number of check-in records at location l from user i, and R_i is the total number of check-ins of user i. Note that D_i^p is the value of y-axis in Fig. 3(A). Based on our observation, for the case of $u_{b\,y}$, they are considered closer friends because the value of D_{by}^c is high, and the personal check-in diversity D_b^p is low, which indicate that user b does not usually visit somewhere alone but is always accompanied by user y. In other words, their check-in records are more consistent. Thus, we design a unified score of check-in diversity D_{ij}^f being inversely proportional to D_i^p and proportional to D_{ij}^c. In this way it can estimates the social tie strength by the user movement more precisely. Finally, we combine the co-occurrence diversity and the personal check-in diversity as follows:

$$D_{ij}^f = \frac{D_{ij}^c}{D_i^p}. \tag{4}$$

D_{ij}^f represents the consistency of check-in records between a pair of users. In this design, the personal check-in behavior and the co-occurrence with a friend are integrated together to express the consistency of check-in behavior between users. This will further help us to estimate the social tie strength more precisely.

Location Property Estimation. In addition to co-occurrence of users, properties of locations are also influential to the movement of users. Next, we will discuss the location properties. For a popular place, the co-occurrence in the check-in data is often caused by a coincidence because such place is usually visited by a crowd of unrelated people. Like the situation of the target user and user B in Fig. 2, although they are friends, they still have the chance to visit a place independently, especially for popular places. Therefore, analyzing how popular a location is can help us to identify the possibility of being a coincidence for co-occurrence records. As shown in Fig. 3(B), numbers in warmer colors (closer to red)/larger fonts in vectors represent that the co-occurrence records are at popular locations. In the contrary, colder colors (closer to blue)/smaller fonts represent co-occurrence records at more private places. Therefore, according to the diagram, user b and y should be closer friends not only because of their co-occurrence diversity being high but also their check-in records spreading over locations no matter how popular they are. Consequently, we apply the location entropy to estimate how popular a location is and further compute the weighted frequency to show the social strength of users by the property of their visited places.

Location entropy, originally mentioned in [16] and utilized by [3], can indicate that how diverse of user check-ins at locations. Given the user check-in records $c_{u,l}$ at a location l and c_l denotes all check-ins at there, the check-in probability of the location is $P_{u,l} = c_{u,l}/c_l$. The corresponding location entropy is:

$$H_l = -\Sigma_u P_{u,l} \, log \, P_{u,l}, \text{ where } P_{u,l} \neq 0. \tag{5}$$

According to the location entropy, the co-occurrence at a private location, which has smaller entropy, can be regarded as closer social interaction. Then we introduce weighted frequency $F_{i,j}$ inspired by tf-idf [3] to give users a weight by their visiting records, where the lower location entropy leads to the higher social relation between users. $F_{i,j}$ is defined as:

$$F_{ij} = \Sigma_l c_{ij,l} exp(-H_l). \tag{6}$$

$F_{i,j}$ considers the property of locations for users who visited them. If a pair of users usually check-in at private locations, their social tie strength score is higher. It allows us to deal with coincidence more precisely. Next we will combine two features of social strength scores on spatial temporal data to a unified score.

Social Tie Strength Score by Spatial Temporal Part. The weighted frequency and the check-in diversity are normalized and combined as the following equation, where α is a parameter ranging within 0 to 1.

$$w_{ij} = \alpha D_{ij}^f + (1 - \alpha)F_{ij}. \tag{7}$$

w_{ij} estimates the social tie strength score by the user movement data. However, some close friends cannot be found by only utilizing the check-in data because they are probably not active. Therefore, in the next step, we will further integrate the social relationship data into our framework.

3.2 Combining Social Tie Strength Score with Social Relation Feature

Social Relationship Structure Estimation. If the movement of users is not obvious to discover the strength between them and leads to a short recommendation list, then we can investigate the social strength by their social connection to help us to find out more locations or more precisely.

Katz [3, 4, 7] has a better performance than other methods such as Jaccard's coefficient, Adamic/Adar, etc. [5, 14]. It considers the social relation structure to measure the tie strength. Given an undirected social connection graph of user i, a connection between i and others means that they are friends. Katz of user i and j is defined as follows:

$$K_{ij} = \Sigma_l \varepsilon^l \times |path_{ij,l}|. \tag{8}$$

$path_{ij,l}$ is all paths from node i to node j with length of l, and ε is the attenuation factor. If the density of the connection between user i and j is high, they are closer. As we mentioned above, because of the data sparseness, sometimes we cannot discover all closer friends from check-in data completely. Thus, we apply Katz for finding more close friends to enhance the accuracy.

Combination of Social Tie Strength Score. Now we have two social strength scores from spatial-temporal and social relationship data respectively. In the next step, we combine them to a unified social tie strength score after normalization to further measure the interest of locations.

$$f_{ij} = \beta \, w_{ij} + (1 - \beta) \, K_{ij}. \tag{9}$$

Here β is a parameter of score. We utilize this score to measure the interest of locations, but in our observation, there exist more social relation data than check-in data. That is, users may never check-in but still have social relation data, and it is harder to find information of his preference of movement. Therefore, β is usually larger than 0.5 to confirm to the real situation, and we set the value of β as 0.8 in our experiments.

This is the unified score of the social tie strength to every pair of user. We think that the tie strength can help us to know more about the preference of movement, So we will describe how to utilize the social tie strength score for rating locations in detail in the next section.

4   POI Recommendation

After knowing who are close friends and their check-in preference, we can further predict interests of a target user and make recommendation lists for him/her. As a result, we design an estimation method to measure the interest of locations for a target user.

In Sect. 3, we finally have a unified score to measure the tie strength between users. In the next step, we compute the location score for a target user. In our design, locations which are more often visited by closer friends of the target user will gain more scores. However, In LBSN, check-in times of a user is not limited, users can visit anywhere or just check-in at one place for thousands of times. If there is a location which is checked in by only one user u_j for lots of times, for example, there is his/her private house, then other users would be influenced by this because the check-in numbers is too high. Thus, we define a saturation function $S_{u,l}$ to make the influence of this kind of noise lower:

$$S_{u,l} = \frac{1}{1 + exp(c_{u,l} \times \lambda)}. \tag{10}$$

$c_{u,l}$ is check-in times of a user at location l, λ is a parameter to control the influence of users who had checked-in at somewhere so many times. As a result, locations visited by many people are relatively more important because of the saturation function. By considering $S_{u,l}$, Then, $R_{i,l}^S$, which is the interest score of location l by close friends of a target user, is defined as:

$$R_{i,l}^S = \Sigma_{u_j} S_{u,l} \times f_{ij}. \tag{11}$$

Thus, locations visited by closer friends obtain higher scores and the influence of noises is limited. In addition to the social tie strength, similarity between users is considerable because the recommendation list by the social strength is short for some users, for example, who prefer to visit places alone or have few friends.

If the preference between the target user and another user A is similar, we can recommend a POI which is visited lots of times by user A and is never visited by the target user before. In our method, we adopt the cosine similarity to compute the similarity of user check-in records, denoted by sim_{ij}. We then define a location score $R_{i,l}^P$ of location l to user i by the preference of check-ins between the target user and others. As we mentioned above, some users might visit only one location but for lots of times, so we also apply $S_{u,l}$ to compute the $R_{i,l}^P$.

$$R_{i,l}^P = \Sigma_{uj} S_{u,l} \times sim_{ij}. \tag{12}$$

Users who are similar to each other might not be close friends in real world. Sometimes friends of a user cannot provide enough movement data for us to estimate his/her interest, so we apply $R_{i,l}^P$ to find out more locations for recommendation. So far we have two location scores for users and a parameter α will be provided for fusion.

$$R_{i,l}^F = \alpha \, R_{i,l}^S + (1 - \alpha) \, R_{i,l}^P. \tag{13}$$

$R_{i,l}^F$ represents the score of the interest of user i to location l. As more close friends of a target user love to visit a place, he/she is more probably to visit there in the future. We further consider the preference of users who are similar to the target user to make the recommendation list more completely. So that this estimation is more confirm to the real situation.

5   Evaluation

In this section, we conduct some experiments to measure the performance of the proposed method. We will describe the data set in Sect. 5.1, the settings of evaluation metrics in Sect. 5.2, compared approach in Sect. 5.3, and we provide evaluation results of in Sect. 5.4.

5.1 Data Set

Our data sets are crawled from LBSN – Gowalla and Brightkite, originally from [1]. It contains 6,442,890 check-ins between Feb. 2009 and Oct. 2010 in Gowalla and 4,491,143 check-ins over the period of Apr. 2008 – Oct. 2010. Each of check-ins contains user ID, time, longitude, latitude and location ID. We preprocess the data by removing users who checked in fewer than 5 times and locations which are visited less than 5 users, and choose a part of the data set after preprocessing randomly to reduce noises and shorten the execution time. Finally, Gowalla contains 836,801 check-ins and there are 856,402 check-ins in Brightkite. These two data sets also contain social relationship data and they are undirected. There are 196,591 nodes and 950,327 edges in Gowalla and 58,228 nodes and 214,078 edges in Brightkite. After preprocessing, they contain 547,235 and 96,229 edges respectively.

5.2 Evaluation Metrics

Our method computes interest scores for locations and generates the top-N recommendation list ordered by the score. We utilize precision and recall, two well known metrics which are also utilized in [11, 13, 20], etc., to evaluate our method. Precision is the fraction of recommended locations that are actually visited by the target user, and recall is the fraction of visited locations that are recommended. Precision@N and Recall@N are defined as:

$$Precision@N = \frac{|U_{visited} \cap Rec_N|}{N}, \tag{14}$$

$$Recall@N = \frac{|U_{visited} \cap Rec_N|}{|U_{visited}|}, \tag{15}$$

where $U_{visited}$ represents locations visited by the target user in the testing data, and Rec_N is the recommendation list for users. We compute the average precision and recall of all testing users in experiments.

5.3 Compared Approach

We compare the performance of the proposed SSR to that of USG [11]. USG combined 3 features which are user preference (check-in similarity), social relationship and geographical influence to estimate ratings of locations for recommendation. USG considers the influence of friends, but only focuses on the ratio of common neighbors who visited the same places.

5.4 Evaluation Results

Experiment results of precision@N and recall@N are summarized in Tables 3 and 4 respectivily (N = 5, 10, 20). It is clear that our method-SSR, is better than the compared approach USG on both precision and recall. In our point of view, we consider the social tie strength by analyzing co-occurrence instead of only co-location, and we handle the case of extreme numbers of check-in by only a few of users. So that our method can find more correlated users of the target user and generate recommendation lists which are not influenced by noises. These two data sets have different properties, although they have similar numbers of check-in records. In Brightkite, there are fewer users in our data set but they checked-in more times in average. On the other hand, in Gowalla, check-in times of users are less in average but there are more users in this dataset. Further, check-in records in Brightkite are sparser than gowalla, that is, there are lots of locations that are visited by very few users (even only one user), so it is more difficult to find out the interaction between users by co-occurrence in Brightkite. In future works, we will enhance this part by investigating other features to find out other preference of users.

Table 3. Precision@N

Data set	Methods	@5	@10	@20
Brightkite	SSR	**0.0276**	**0.0217**	**0.0164**
	USG	0.0028	0.0019	0.0015
Gowalla	SSR	**0.0682**	**0.0529**	**0.0358**
	USG	0.0157	0.0142	0.0106

Table 4. Recall@N

Data set	Methods	@5	@10	@20
Brightkite	SSR	**0.0032**	**0.0045**	**0.0065**
	USG	0.0003	0.0005	0.0006
Gowalla	SSR	**0.0057**	**0.0085**	**0.0119**
	USG	0.0019	0.0031	0.0049

6   Conclusion

This paper tried to find out whether the social tie strength is influential to user movement analysis on LBSN. We think that the movement of a user is affected by close friends but this feature was not utilized by previous works. Therefore, we estimated the social tie strength by both of user movement and social relation data. Then, we reduced the influence of noises in the data set and fused the user similarity feature to a unified location score to generate recommendation lists. Finally we conducted experiments to evaluate our method and results outperformed the compared approach. Thus, in our point of view, the social tie strength between users is also an important feature of user movement analysis, and co-occurrence is more meaningful than co-location. In addition, because the check-in number is unlimited, there are some extremely situation needed to be handled. In future works, we will further consider geographical and periodic check-in features to find out more venues which will attract users. Furthermore, we will try to evaluate the influence of the social tie strength in other kinds of works on LBSN.

References

1. Cho, E., Myers, S.A., Leskovec, J.: Friendship and mobility: user movement in location-based social networks. In: Proceedings of the 17th ACM SIGKDD International Conference on Knowledge Discovery and Data Mining, pp. 1082–1090. ACM (2011)
2. Sadilek, A., Kautz, H.A., Bigham, J.P.: Finding your friends and following them to where you are. In: Proceedings of the fifth ACM International Conference on Web Search and Data Mining, pp. 723–732. ACM (2012)
3. Pham, H., Shahabi, C., Liu, Y.: Ebm: an entropy-based model to infer social strength from spatiotemporal data. In: Proceedings of the 2013 ACM SIGMOD International Conference on Management of Data, pp. 265–276. ACM (2013)
4. Wang, D., Pedreschi, D., Song, C., Giannotti, F., Barabási, A.L.: Human mobility, social ties, and link prediction. In: Proceedings of the 17th ACM SIGKDD International Conference on Knowledge Discovery and Data Mining, pp. 1100–1108. ACM (2011)
5. Liben-Nowell, D., Kleinberg, J.: The link-prediction problem for social networks. J. Am. Soc. Inf. Sci. Technol. 58(7), 1019–1031 (2007)
6. Scellato, S., Noulas, A., Mascolo, C.: Exploiting place features in link prediction on location-based social networks. In: Proceedings of the 17th ACM SIGKDD International Conference on Knowledge Discovery and Data Mining, pp. 1046–1054. ACM (2011)
7. Xiang, R., Neville, J., Rogati, M.: Modeling relationship strength in online social networks. In: Proceedings of the 19th International Conference on World Wide Web, pp. 981–990. ACM (2010)
8. Backstrom, L., Kleinberg, J.: Romantic partnerships and the dispersion of social ties: a network analysis of relationship status on facebook. In: Proceedings of the 17th ACM Conference on Computer Supported Cooperative Work & Social Computing, pp. 831–841. ACM (2014)
9. Wang, Y., Yuan, N.J., Lian, D., Xu, L., Xie, X., Chen, E., Rui, Y.: Regularity and conformity: location prediction using heterogeneous mobility data. In: Proceedings of the 21th ACM SIGKDD International Conference on Knowledge Discovery and Data Mining, pp. 1275–1287. ACM (2015)
10. Bao, J., Zheng, Y., Mokbel, M.F.: Location-based and preference-aware recommendation using sparse geo-social networking data. In: Proceedings of the 20th International Conference on Advances in Geographic Information Systems, pp. 199–208. ACM (2012)
11. Ye, M., Yin, P., Lee, W.C., Lee, D.L.: Exploiting geographical influence for collaborative point-of-interest recommendation. In: Proceedings of the 34th International ACM SIGIR Conference on Research and Development in Information Retrieval, pp. 325–334. ACM (2011)
12. Ference, G., Ye, M., Lee, W.C.: Location recommendation for out-of-town users in location-based social networks. In: Proceedings of the 22nd ACM International Conference on Information & Knowledge Management, pp. 721–726. ACM (2013)
13. Yuan, Q., Cong, C., Ma, Z., Sun, A., Magnenat-Thalmann, N.: Time-aware point-of-interest recommendation. In: Proceedings of the 36th International ACM SIGIR Conference on Research and Development in Information Retrieval, pp. 363–372. ACM (2013)
14. Adamic, L.A., Adar, E.: Friends and neighbors on the web. Soc. Netw. 25(3), 211–230 (2003)
15. Bao, J., Zheng, Y., Wilkie, D., Mokbel, M.: Recommendations in location-based social networks: a survey. GeoInformatica 19(3), 525–565 (2015)
16. Cranshaw, J., Toch, E., Hong, J., Kittur, A., Sadeh, N.: Bridging the gap between physical location and online social networks. In: Proceedings of the 12th ACM International Conference on Ubiquitous Computing, pp. 119–128. ACM (2010)

17. Renyi, A.: On measures of entropy and information. In: Fourth Berkeley Symposium on Mathematical Statistics and Probability. vol. 1, pp. 547–561 (1960)
18. Tang, J., Hu, X., Liu, H.: Social recommendation: a review. Soc. Netw. Anal. Min. 3(4), 1113–1133 (2013)
19. Pazzani, M.J., Billsus, D.: Content-based recommendation systems. In: Brusilovsky, P., Kobsa, A., Nejdl, W. (eds.) Adaptive Web 2007. LNCS, vol. 4321, pp. 325–341. Springer, Heidelberg (2007)
20. Gao, H., Tang, J, Hu, X., Liu, H.: Exploring temporal effects for location recommendation on location-based social networks. In: Proceedings of the 7th ACM Conference on Recommender Systems, pp. 93–100. ACM (2013)
21. Yin, H., Sun, Y., Cui, B., Hu, Z., Chen, L.: Lcars: a location-content-aware recommender system. In: Proceedings of the 19th ACM SIGKDD International Conference on Knowledge Discovery and Data Mining, pp. 221–229. ACM (2013)
22. Gao, H., Tang, J., Hu, X., Liu, H.: Content-aware point of interest recommendation on location-based social networks. In: AAAI, pp. 1721–1727 (2015)
23. Liu, B., Fu, Y., Yao, Z., Xiong, H.: Learning geographical preferences for point-of-interest recommendation. In: Proceedings of the 19th ACM SIGKDD International Conference on Knowledge Discovery and Data Mining, pp. 1043–1051. ACM (2013)
24. Zhao, Y., Wang, G., Yu, P.S., Liu, S., Zhang, S.: Inferring social roles and statuses in social networks. In: Proceedings of the 19th ACM SIGKDD International Conference on Knowledge Discovery and Data Mining, pp. 695–703. ACM (2013)

Online Anomaly Energy Consumption Detection Using Lambda Architecture

Xiufeng Liu[1]([✉]), Nadeem Iftikhar[2], Per Sieverts Nielsen[1], and Alfred Heller[1]

[1] Technical University of Denmark, Kongens Lyngby, Denmark
{xiuli,pernn}@dtu.dk, alfh@byg.dtu.dk
[2] University College of Northern Denmark, Aalborg, Denmark
naif@ucn.dk

Abstract. With the widely use of smart meters in the energy sector, anomaly detection becomes a crucial mean to study the unusual consumption behaviors of customers, and to discover unexpected events of using energy promptly. Detecting consumption anomalies is, essentially, a real-time big data analytics problem, which does data mining on a large amount of parallel data streams from smart meters. In this paper, we propose a supervised learning and statistical-based anomaly detection method, and implement a *Lambda* system using the in-memory distributed computing framework, *Spark* and its extension *Spark Streaming*. The system supports not only iterative refreshing the detection models from scalable data sets, but also real-time anomaly detection on scalable live data streams. This paper empirically evaluates the system and the detection algorithm, and the results show the effectiveness and the scalability of the lambda detection system.

Keywords: Anomaly detection · Real-time · Lambda architecture · Data mining

1 Introduction

Anomaly detection, also known as outlier detection, is the process of discovering patterns in a given data set that do not conform to expected behavior [2]. Anomaly detection is to find the events that happen relatively infrequently, which has been extensively used in a wide variety of applications, including fraud detection for credit cards, insurance, health care, intrusion detection for cyber-security, fault detection in safety critical systems, and many others [2]. In this paper, we will show how anomaly detection can be applied to analyze live energy consumption, aiming at identifying unusual behaviors for consumers (e.g., forgetting to turn off stoves after cooking), or detecting extraordinary events for utilities (e.g., energy leakage and theft). Since abnormal consumption may also be resulted from user activities, such as using inefficient appliances, or over-lighting and working overtime in office buildings, anomalous feedback can warn energy consumers to minimize usage and help them identify inefficient

© Springer International Publishing Switzerland 2016
S. Madria and T. Hara (Eds.): DaWaK 2016, LNCS 9829, pp. 193–209, 2016.
DOI: 10.1007/978-3-319-43946-4_13

appliances or over-lighting. Furthermore, anomaly detection can be used by utilities to establish the baseline of providing accurate demand-response programs to their customers [35]. Abnormal consumption detection is related to finding patterns in data where the statistical and data mining techniques are intensively used, e.g., [8,9,18,35], and it can perform close to or better than domain experts.

Energy consumption time series are recorded by smart meters at the regular interval of an hour or fewer [21,24]. Smart meters read the detailed energy consumption in a real-time or near real-time fashion, which provides the opportunity to monitor timely unusual events or consumption behaviors [22,23]. However, the enabling detection technologies combining smart meters typically using data mining technologies, which require large amounts of training data sets, and significantly complex systems. In a typical application of data mining to anomaly detection, the detection models are produced off-line because the learning algorithms have to process tremendous amounts of data [17]. The produced models are naturally used by off-line anomaly detections, i.e., analyzing consumption data after being loaded into an energy management system (EMS). However, we argue that effective anomaly detection should happen real-time in order to minimize the compromises to the use of energy. The efficiency of updating the detection model and the accuracy of the detection results are the important consideration for constructing a real-time anomaly detection system.

In this paper, we propose a statistical anomaly detection algorithm to detect the anomalous daily electricity consumption. The anomaly detection is based on consumption patterns, which are usually quite similar for a customer, such as in weekdays, or in weekends/holidays. We define the anomaly as the difference from the expected consumption as illustrated in Fig. 1. The proposed detection method is not limited to daily patterns, but can be easily adapted to the periodicity of the underlying data set. It is also important to note that our methods are applicable not only to electricity consumption, but other energy types of consumption, such as gas, water, and heat. This is caused by the general nature of time series data, and the generality of our detection methods presented in this paper.

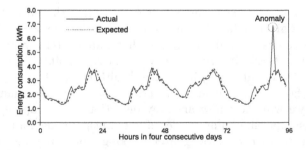

Fig. 1. Daily consumption pattern and anomalies

To detect anomalies in time and obtain a better accuracy, we make use of the so-called *Lambda* architecture [27], that can detect anomalies near real-time, and can efficiently update detection models regularly according to a user-specified time interval. A lambda architecture enables real-time updates through a three-layer structure, including speed layer (or real-time layer), batch layer and serving layer. It is a generic system architecture for obtaining near real-time capability, and its three layers use different technologies to process data. It is well-suited for constructing an anomaly detection system that requires real-time anomaly detection and efficient model refreshment (we will detail it in the next section). To support big data capability, we choose Spark Streaming as the speed layer technology for detecting anomalies on a large amount of data streams, Spark as the batch layer technology for computing anomaly detection models, and PostgreSQL as the serving layer for saving the models and detected anomalies; and sending feedbacks to customers. The proposed system can be integrated with smart meters to detect anomalies directly. We make the following contributions in this paper: (1) we propose the statistical-based anomaly detection algorithm based on customers' history consumption patterns; (2) we propose making use of the lambda architecture for the efficiency of the model updating and real-time anomaly detection; (3) we implement the system with a lambda architecture using hybrid technologies; (4) we evaluate our system in a cluster environment using realistic data sets, and show the efficiency and effectiveness of using the lambda architecture in a real-time anomaly detection system.

The rest of this paper is organized as follows. Section 2 discusses the anomaly detection algorithm used in the paper. Section 3 describes the implementation of the lambda detection system. Section 4 evaluates the system. Section 5 surveys the related works. Section 6 concludes the paper and provides the direction for the future works.

2   Preliminaries

2.1   Anomaly Detection Model

The used anomaly detection model is a combination of a short-term energy consumption prediction algorithm, called *periodic auto-regression with eXogenous variables (PARX)* [3], and *Gaussian statistical distribution*. We now first describe the PARX algorithm, which is used to predict the daily consumption. Generally speaking, residential electricity consumption is highly correlated to temperature. For example, in winter, electricity consumption increases since the temperature decreases because of the heating needs. In summer, electricity consumption increases when the temperature is higher because of cooling loads. The daily consumption pattern of a customer typically demonstrates the periodic characteristics, due to the living habit of the customer, e.g., the morning peak appears between 7 and 8 o'clock if a customer usually gets up at 7 o'clock; and evening peak appears between 17 and 20 o'clock (due to cooking and washing) if the customer gets home at 5 o'clock after work.

The PARX model, thus, uses a daily period, taking 24 hours of the day as the seasons, i.e., $t = 0...23$, and uses the previous p days' consumptions at the hour at t for auto-regression. The PARX model at the s-th season and at the n-th period is formulated as

$$Y_{s,n} = \sum_{i=1}^{p} \alpha_{s,i} Y_{s,n-i} + \beta_{s,1} XT1 + \beta_{s,2} XT2 + \beta_{s,3} XT3 + \varepsilon_s, \quad s \in t \quad (1)$$

where Y is the data point in the consumption time-series; p is the number of order in the auto-regression; $XT1, XT2$ and $XT3$ are the exogenous variables accounting for the weather temperature, defined in the equations of (2); α and β are the coefficients; and ε is the value of the white noise.

$$XT1 = \begin{cases} T - 20 & \text{if } T > 20 \\ 0 & \text{otherwise} \end{cases} \quad XT2 = \begin{cases} 16 - T & \text{if } T < 16 \\ 0 & \text{otherwise} \end{cases} \quad XT3 = \begin{cases} 5 - T & \text{if } T < 5 \\ 0 & \text{otherwise} \end{cases} \quad (2)$$

The variables represent the cooling (temperature above $20°$), heating (temperature below $16°$), and overheating (temperature below $5°$), respectively. The anomaly detection algorithm is of using unique variate Gaussian distribution, described in the following. Given the training data set, $X = \{x_1, x_2, ..., x_n\}$ whose data points obey the normal distribution with the mean μ and the variance δ^2, the detection function is defined as

$$p(x; \mu, \delta) = \frac{1}{\delta \sqrt{2\pi}} e^{-\frac{(x-\mu)^2}{2\delta^2}} \quad (3)$$

where $\mu = \frac{1}{n} \sum_{i=1}^{n} x_i$ and $\delta^2 = \frac{1}{n} \sum_{i=1}^{n} (x_i - \mu)^2$. For a new data point, x, this function computes its probability density. If the probability is less than a user-defined threshold, i.e., $p(x) < \varepsilon$, it is classified as an anomaly, otherwise, it is a normal data point. In our model training process, we compute the L1 distance between the actual and predict consumptions, i.e., $||Y_t - \hat{Y}_t||$, where Y_t is the actual hourly consumption at the time t, and $\hat{Y}_t$ is the predict hourly consumption at the time t. The predict hourly consumption, $\hat{Y}_t$, is computed using the PARX model in Eq. 1. We find that the L1 distances observe to a lognormal distribution (see Sect. 4.2). Therefore, the x in the normal distribution will be the log value of the distance, i.e., $ln||Y_t - \hat{Y}_t||$.

2.2   Lambda Architecture

We now introduce the lambda architecture used in our anomaly detection system. As mentioned in Sect. 1, the lambda architecture consists of three layers, including speed layer, batch layer and serving layer, illustrated in Fig. 2. The speed layer directly ingests data streams from data sources, processes them, and continuously updates the results into the real-time views in the database in the serving layer. The speed layer does not keep any history records, and typically uses main memory based technologies to analyze the incoming data. In contrast, the batch layer runs iteratively and starts from the beginning of the data set once a batch job has finished. When a batch job starts, all the available data in the batch layer storage will be reprocessed. Therefore, the data arriving after the job starts will be processed by the next job. Since all the data are analyzed in each iteration, each of the new result views will replace its predecessor. As the batch layer does not rely on incremental processing, it is robust to any system failures, which the batch job simply processes all the available data sets in each iteration. The speed and batch usually use different technologies because of their distinct requirements regarding read and write operations. Any query against the data is answered through the serving layer, i.e., the query processor queries both the views from the speed and the batch layers, and merges them.

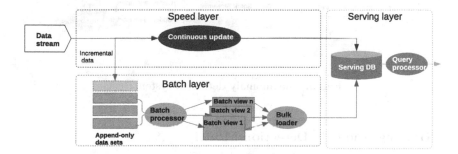

Fig. 2. Lambda architecture

The lambda architecture itself is only a paradigm. The technologies with which the different layers are implemented are independent of the general idea. The speed layer only deals with new data and compensates for the high latency updates of the batch layer. It can typically leverage stream processing systems, such as Storm, S4, and Spark Streaming, etc. The batch layer needs to be horizontally scalable and supports random reads, where the technologies like Hadoop with Cascading, Scalding, Pig, and Hive, are suitable. The serving layer requires a system with the ability to perform fast random reads and writes. The system can be a high-performance RDBMS (e.g., PostgreSQL), an in-memory data store (e.g., Redis, or Memcache), or a high scalable NoSQL system (e.g., HBase, Cassandra, ElephantDB, MongoDB, or DynamoDB).

3 Implementation

3.1 System Overview

We now describe the implementation of the anomaly detection system. We choose *Spark Streaming, Spark*, and *PostgreSQL* as the speed layer, batch layer and serving layer technology, respectively (see Fig. 3). The system employs Spark to compute the models for anomaly detection, which reads the data from the

Hadoop distributed file system (HDFS) in the batch layer. The batch job runs at a regular time interval, computes, and updates the detection models to the table in PostgreSQL database. Spark Streaming is used to process real-time data streams, e.g., directly reads the readings from smart meters, and detects abnormal consumption with the detection algorithm. The detection algorithm always uses the latest models getting from the PostgreSQL database. Spark Stream writes the detected anomalies back to the PostgreSQL database, which will be used for the notification of customers.

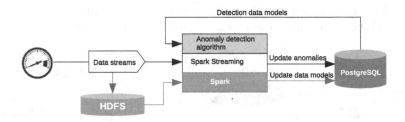

Fig. 3. The anomaly detection system

3.2   Training Anomaly Detection Models

We employ Spark to train the detection models by running regular batch jobs. All the consumption data from smart meters are written to the append-only HDFS. In each iteration of the batch jobs, Spark uses all the available data in HDFS to compute the detection models. The use of Spark and HDFS supports the computation of the models based on scalable data sets. Since they both are the distributed computing technology, the computation can be finished within a certain time limit, meaning that the detection algorithm can always use the latest models. Figure 4 illustrates the training process of generating PARX and Gaussian models using energy consumption and weather temperature time series at the season from 0 to 23. That is, for each season we create a new time series, e.g., for $s = 0$, the time series is created using the readings at 0 o'clock of all days. Then the Eq. 1 is used to compute the PARX model (or parameters), and to compute the Gaussian model, i.e., $N(\mu, \delta^2)$. Therefore, there are 24 PARX and 24 Gaussian models in total.

Algorithm 1 gives more details about the implementation. This algorithm computes the anomaly detection models with the given training time series collection $\mathcal{TS}$, weather temperature time series ts', and auto-regression order p. Each time series in $\mathcal{TS}$ represents the hourly energy consumption of a customer. To compute the detection models for each season s, we first need to create a new consumption time series and a new temperature time series (see line 7), then use the two new time series to compute the PARX model (see line 8). According to our analysis in Sect. 4.2, the $L1$ distances between predict consumption and actual consumption at season s for all days observes to a log-normal distribution. Therefore, we compute Gaussian statistical model based on the $L1$ distance

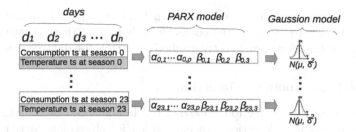

Fig. 4. Process of training detection models

log values (see lines 12–18). The total number of PARX models for all the time series is $||\mathcal{TS}|| \times 24$, which is same as the number of the Gaussian models. In the end, all the models are updated to the PostgreSQL database that will be used for the online anomaly detection in the speed layer.

Algorithm 1. Training of the anomaly detection models

1: **function** TRAIN(TimeSeriesCollection $\mathcal{TS}$, TemperatureTimeSeries ts' Order p)
2:     $\mathcal{M} \leftarrow \{\}$ ▷ Initialize the collection of PARX parameters
3:     $\mathcal{N} \leftarrow \{\}$ ▷ Initialize the collection of the statical model parameters
4:     **for all** $ts \in \mathcal{TS}$ **do**
5:         $id \leftarrow$ Get the unique identity of ts
6:         **for all** $s \in 0...23$ **do**
7:            $ts^c, ts^t \leftarrow$ Construct a new consumption time series using ts, and a new temperature time series ts^t using ts' at the season of s
8:            $\alpha_1, ..., \alpha_p, \beta_1, \beta_2, \beta_3 \leftarrow$ Compute PARX model using ts^c and ts^t
9:            Insert $(id, s, \alpha_1, ..., \alpha_p, \beta_1, \beta_2, \beta_3)$ into $\mathcal{M}$
10:          $\mathcal{L} \leftarrow \{\}$
11:          $\mathcal{D} \leftarrow$ Get the days of ts
12:          **for all** $d \in \mathcal{D}$ **do**
13:            $\hat{v} \leftarrow$ Compute the predict reading of the season s using PARX
14:            $v \leftarrow$ Get the actual hourly reading from ts of the day d
15:            $l \leftarrow$ Compute the ln value of L_1 distance of the day d, $ln(||\hat{v} - v||)$
16:            Add l into $\mathcal{L}$
17:          $\mu, \beta \leftarrow$ Compute the mean and standard deviation using the normal distribution statistical model on $\mathcal{L}$
18:         Insert (id, s, μ, δ) into N
         return $\mathcal{M}, \mathcal{N}$

The implementation is a Spark program. The consumption time series, as well as temperature time series, are read into the distributed memory as *resilient distributed datasets (RDDs)*, which are fault-tolerant, immutable and partitioned parallel data structures that can be operated in parallel, e.g., by using the operators, including map, reduce, groupByKey, filter, collect, etc. [33]. To generate the new time series, we use the *groupByKey* operator to aggregate the consumption series by the composite key of meter ID and season (or hours); while use only the season as the key to the temperature time series. Then, we merge the generated time series by the join operator on the key of the season. The PARX, in fact, can be regarded as a multi-linear regression model, which simply takes the auto-regressors and the exogenous variables as the independent variables. We, then, apply the multiple linear regression function from the Spark machine learning library, MLib [28], to compute the coefficients. For all of these operations, the transformation functions are directly applied on RDDs for data processing.

3.3 Real-Time Anomaly Detection

The real-time anomaly detection is carried out in the speed layer. Algorithm 2 describes the anomaly detection process, which is self-explanatory. First, the detection algorithm reads meter readings from all the incoming data streams, and reads the weather temperature and detection models from the PostgreSQL database each hour. For each data stream, the algorithm predicts the reading using the PARX algorithm, with the pre-computed parameters, the previous p day's readings at the current hour, i.e., the season s, and weather temperature (see lines 4–7). Then, the algorithm calculates the log value of the $L1$ distance between the predict and the actual readings, then uses it compute the probability using the Gaussian model (see lines 8–10). In the end, the algorithm decides whether the current reading is an anomaly or not based on the computed probability value, i.e., if its value is below the user-defined threshold, ε. If the current reading is classified as an anomaly, it will be written into the database for the customer notification (see lines 11–12).

Algorithm 2. Real-time anomaly detection

1: **function** DETECT(CurrentReadingCollection $\mathcal{V}$, Temperature t, PredictModel $\mathcal{M}$, StaticalModel $\mathcal{N}$, Threshold ε)
2:     $\mathcal{R} \leftarrow \{\}$ ▷ Initialize the detection results
3:     **for all** $v \in \mathcal{V}$ **do**
4:         $id \leftarrow$ Get the unique identity of v
5:         $s \leftarrow$ Get the season of v
6:         $\alpha_1, ..., \alpha_p, \beta_1, \beta_2, \beta_3 \leftarrow$ Get the parameters from $\mathcal{M}$ by id
7:         $\hat{v} \leftarrow$ Compute the predict reading at s using PARX with the parameters, the p days' readings at s, and temperature t
8:         $x \leftarrow ln||\hat{v} - v||$ ▷ Compute the ln value of L_1 distance at the season s
9:         $\mu, \delta \leftarrow$ Get the statical model parameters from $\mathcal{N}$ by id and s
10:         $p \leftarrow$ Compute the probability using the normal distribution function, $\frac{1}{\delta\sqrt{2\pi}}e^{-\frac{(x-\mu)^2}{2\delta^2}}$
11:         **if** $p < \varepsilon$ **then**
12:             Add $(id, s, p, v, \hat{v})$ into $\mathcal{R}$
         return $\mathcal{R}$

We implement the algorithm to process the real-time data on Spark Streaming. Spark Streaming allows for continuous processing via short interval batches, and its basic data abstraction is called *discretized streams (D-Streams)*, a continuous stream of data [34]. The data are received in each interval batch, *hourly* in our case, and operations will run upon the data for doing transformations, such as filter unnecessary attribute values, extracting the hour from the timestamp, etc. (see Fig. 5). When using the PARX for prediction, we fetch the previous p days' readings of the current hour for auto-regression. For example, in Fig. 5 we set the order, $p = 3$, therefore, the window size is set to 72 hours (i.e., 3 days) to keep the past three days' readings at the particular hour within the same window (e.g., the RDDs colored by green). This is done by using the window function, `reduceByKeyAndWindow(func, windowLength, slideInterval)`, to aggregate the data with specified key, window length and slide interval (e.g., meter ID and season as the composite key in this case, $windowLength = 72$ hours and $slideInterval = 1$ hour). In the underlying, Spark uses the data *checkpointing* mechanism to keep the past RDDs in HDFS. At the beginning of each interval batch, the data models are read from the PostgreSQL database in the serving layer, and broadcast to all DStreams. Therefore, the detection program always uses the latest models to detect anomalies.

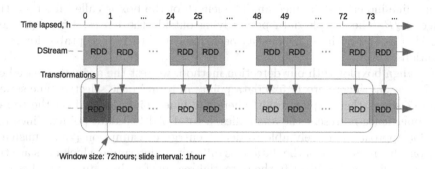

Fig. 5. Slide windows in the real-time anomaly detection (Color figure online)

4 Evaluation

4.1 Experimental Settings

In this section, we will evaluate the effectiveness and the scalability of our anomaly detection system. We conduct the experiments in a cluster with 17 servers. Five servers are used for running the speed layer, while twelve servers are used for the batch layer. We also exploit one of the servers in the speed layer as the serving layer for managing the detection models and sending anomaly detection messages. All the servers have the identical settings, configured with an Intel(R) Core(TM) i7-4770 processor (3.40 GHz, 4 Cores, hyper-threading is enabled, two hyper-threads per core), 16 GB RAM, and a Seagate Hard driver (1TB, 6 GB/s, 32 MB Cache and 7200 RPM), running Ubuntu 12.04 LTS with 64 bit Linux 3.11.0 kernel. The serving layer uses PostgreSQL 9.4 database with the settings "shared buffers = 4096 MB, temp buffers = 512 MB, work mem = 1024 MB, checkpoint segments = 64" and default values for the other configuration parameters.

We have a real-world residential electricity consumption data set (27,300 time series), which will be used to evaluate anomaly detection accuracy. The time-series has a two-year length and hourly resolution. To evaluate the scalability, we use the synthetic data set generated by our data generator seeded by the real-world data. The size of data tested in the cluster environment is scaled up to one terabyte, corresponding to over twenty million time series.

4.2 Anomaly Detection Accuracy

We start by evaluating the accuracy of our anomaly detection system using a randomly-selected time series from the real-world data set.

To provide a basis for comparison, we perform the anomaly detection using a standard boxplot analysis as well. Boxplot is a quick graphic approach for examining data sets, and has been used for decades. A boxplot uses five parameters to describe a numeric data set, including lower fence, lower quartile, media, upper quartile and upper fence (see Fig. 6). According to Fig. 6, a boxplot is constructed by drawing a rectangle between the upper and lower quartiles with a solid line indicating the median. The length of the box is called interquartile range, IQR. The sample data points lying outside the fences, $1.5 * IQR$, are classified as the outliers, which has been indicated to be acceptable for most situations [10].

To align boxplot with our detection method, we test the anomalies based on the 24 seasons. There are 17,520 data points in total in the selected time series. Figure 7 shows the boxplot result where the blue points located on the top of the upper fence represent the anomalies, a total of 1,260 data points. Since the boxplot approach is merely able to detect energy consumption lying unusually far from the main body of the data, it is difficult to determine which ones are the true anomalies, and to identify the potential reasons for these anomalies because there are too many false positives.

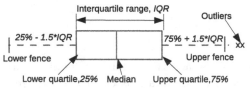

Fig. 6. Box plot

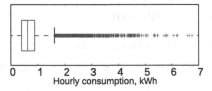

Fig. 7. Anomaly detection using box plot (Color figure online)

We now use the proposed detection algorithm to analyze the same time series. Figure 8 depicts the distribution of the $L1$ distances of a season by the histogram. As shown, the distribution has the shape of a log-normal distribution. We have checked the $L1$ distance distributions for all the 24 seasons, and found that they all share a similar shape. This is the reason that we choose log-normal distribution in our statistical-based anomaly detection. Besides, we test the anomalies by treating all the days the same, and differently, i.e., discriminating the days into workdays, weekend & holidays. Moreover, we increase the threshold value, ε, from 0.05 to 0.15, and do the evaluation. The results in Fig. 9 demonstrate that the detection identifies more anomalies for treating all the days the same than differently. The reason is that during weekends and holidays, people tend to stay at home more time, thus use more energy. The consumptions are more likely higher than the weekdays. For the threshold parameter, its value is for

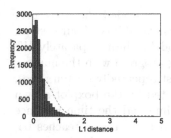
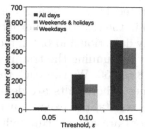
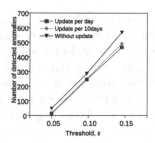

Fig. 8. Log-normal distributions across the L1 distances

Fig. 9. Anomaly detection using PARX and statistical method

Fig. 10. Impact of detection model update frequency

classifying a usual or unusual reading. According to the results, if the value increases, the number of detected anomalies changes significantly. For the real-world deployment of this system, the threshold value can be set by the residents to decide when to receive anomaly alerting messages.

We now evaluate the impact of the model update frequency on the detection accuracy. We use half-year's time series as the initial data set to train the detection models. We design the following three scenarios for the model update: (1) update per day; (2) update per 10 days behind the detection; and (3) without update. We measure the detected anomalies of the three scenarios by treating all days the same. According to the results in Fig. 10, the frequent updates of the models help to decrease the detected anomalies. It is due to the improvement of the prediction accuracy of the PARX model. Thus, less large $L1$ distances are identified as the anomalies. However, although the update frequency does help to determine the real anomalies, the results do not show a big difference if the models are updated within a certain short-time interval, e.g., the results of the scenario (1) and (2).

In the end, we compare our approach with the boxplot, and the result shows that the number of the anomalies reported by the PARX prediction and statistical method can be decreased notably. This increases the chance to determine accurately real anomalies for an energy consumption time series.

4.3 System Scalability

We now evaluate the scalability of our anomaly detection system. As our system can scale-out and to efficiently cope with large amounts of data, we vary both the number of executors and the volume of the input data in the following experiments.

Scale-Out Experiment. Parallel processing is a key feature of the proposed system. To evaluate the scalability of our implementation, we conduct the experiment by varying the number of executors in Spark. In this experiment, we use a fix-sized synthetic data set with eight million time series of a one-year length

(275 GB), which were generated by our data generator seeded the real-world data. Since we are interested in the real-time and batch capability of our system, we test the real-time anomaly detection and batch model training separately. We first test the batch capability by running the training program with the number of executors increased from 8 to 256. We run each test repeatedly for ten times, and record its execution times. The results are depicted by the boxplot shown in Fig. 11. According to the results, the execution time and the time variance decrease when more executors are added. But, when the number reaches 64, the increasing parallelism does not speed up the batch processing further, which is due to the overhead of the Spark master when managing a large number of executors. We conduct the real-time anomaly detection on Spark Streaming, and likewise, we scale the number of executors from 8 to 256. Figure 12 shows the results, which indicate that the variance of execution time is larger than the batch model training. It might be due to the variability of real-time batch executions on Spark Streaming when doing the anomaly detection for each hour.

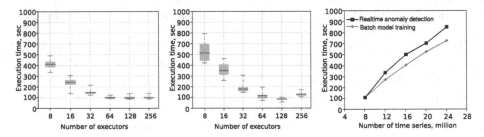

Fig. 11. Batch model training   **Fig. 12.** Real-time detection   **Fig. 13.** Size-up experiment

Increased Data Load Experiment. To evaluate the scalability of the algorithms over large volumes of data, we compare different workloads. According to the above experiments, the optimal number of parallel executors for model training and anomaly detection are 64 and 128, respectively. We use the optimal executor number (the memory of executor is configured to 4 GB) in our experiments, but vary the number of time series from 8 to 24 million (corresponding to the size from 275 GB to 825 GB). The processing times of varying data workloads are displayed in Fig. 13. We observe that both of the training and detection processes can scale near linearly with the quantity of the time series. The time of detection, in this case, is the total execution time of handling all the time series of a one-year length, e.g., it takes less than two minutes to finish eight million time series with the optimized settings. The average time of each real-time batch only takes a few seconds (recall that a batch in Spark Streaming processes the data of each hour). In the real-world deployment, the detection program can be set to run every hour to inspect hourly smart meter readings. According to the results, the anomaly detection system has a very good scalability which can meet the fast-growth of smart meter data.

Unlike the anomaly detection, the training process uses the full set of the data to generate new models each time. Anomaly detection, instead, is performed for each hour where Spark Streaming runs periodical batch (or impulse) to process the data, which needs more time in overall. The training and detection programs can be deployed either in different clusters or the same cluster. If deployed in the same cluster, it is necessary to allocate computing resources in a reasonable way. For example, since the batch job takes a much longer time, it can be scheduled to run immediately after the anomaly detection job. A scheduling system is, thus, necessary, and this will leave to our future work.

5   Related Works

Anomaly Energy Consumption Detection. Anomaly detection is an important aspect in energy consumption time series management. Chandola et al. present a survey of different anomaly detection techniques in various application domains including energy [6]. Statistical and data mining are the commonly used techniques for discovering abnormal consumption behaviors [14]. Statistical methods are based on modeling data using distributions, and see if the data under test observes to the distributions. Accordingly, the approaches presented in this paper combine PARX and log-normal distribution function to detect anomalies in energy consumption time series. Jakkula and Cook use statistics and clustering to identify outliers in power datasets collected from smart environments [13], but they have not considered the impact of the exogenous variables, e.g., weather temperature, on the electricity consumption. Linear regression can extract time series features when the dependent variables are well-defined [25]. The early experience of identifying outliers in linear regression is through setting a threshold limit, but this yields many false positives for large data sets [16]. Adnan et al. combine linear regression with clustering techniques for getting better results [1]. Zhang et al. [35] further use piecewise linear regression to fit the relation between energy consumption and weather temperature. The results obtained are more favorable than entropy and clustering methods. But, their approach does not take the changes of consumption pattern into account. Brown et al. use K-nearest neighborhood (KNN) in fast kernel regression to predict electricity consumption [4], which requires large datasets. The resulting models are static, thus it is not preferable for online anomaly detection and the situation when consumption pattern is changed. Nadai et al. combine ARIMA and adaptive artificial neural network (ANN) to detect anomaly consumption [9] using a relatively small data set that is from a few buildings. In comparison, we propose the prediction and statistical anomaly models and combine with the lambda architecture for supporting regular model refreshment, and real-time anomaly detections. Besides, the proposed approach can handle scalable data sets, and consumption pattern changing owing to its use of the PARX model.

Batch and Stream Processing on Big Data. Batch and realtime/stream processings have attracted much research effort in recent year, with the popularity of Internet of Things (IoT). Liu et al. make a survey of the existing stream processing systems, and discuss the potential technologies used for lambda architecture [20]. Cheng et al. propose a smart city data platform that supports both batch and real-time data processings [7], and they suggest that anomaly detection should be implemented as the chief component of any platform for processing sensor data. Different to proposing the generic lambda architecture [27], Preuveneers et al. [29] and Gao [11] present the big data architectures for processing domain-specific big data, including health care, context-aware user authentication and social media. Schneider et al. study batch data and streaming data anomaly detection, respectively [30]. The used detection model, however, is static, and the use case is different to ours which employs the batch job to update the models only while use the real-time job to detect the anomalies online in data streams.

The Use of Lambda Architecture. Lambda architecture has attracted a growing interest due to its mix capabilities to process both real-time and batch data. Sequeira et al. use lambda architecture in an industrial EMS solution with cloud computing capabilities [31]. Kroß et al. develop on-demand stream processing within the lambda architecture to optimize computing resource usage in a cluster [15]. Martnez-Prieto et al. adapts the lambda architecture in semantic data processing [26]; Liu et al. applies it to smart grid complex event processing (CEP) [19]; Villari et al. proposes AllJoyn Lambda, the platform for managing embedded devices of smart homes [32]; and Hasani et al. use it for real-time big data analytics [12]. Besides, the works [5,20] both give an extensive review of the technologies of the lambda architecture. Although there are various use cases of the lambda architecture, we focus on its use in the particular use case, anomalous energy consumption detection. More specifically, we use it in the model update and the real-time anomaly detection, which is significant to the large deployment of smart meters and sensors of IoT today.

6   Conclusions and Future Work

Analyzing and detecting anomalies is an important task for live energy consumption data while the improvement of detection accuracy and scalability is challenging. In this paper, we applied the novel lambda architecture technique to an anomaly detection system in order to support batch updates of the detection models, and real-time detection. We have proposed the detection algorithm for finding the anomalies based on one's history consumption pattern via the supervised learning and statistical algorithms. Furthermore, the system supports personalized alerting service by setting a threshold value for suspicious energy consumption. We have evaluated the accuracy of the anomaly detection algorithm on a real-world data set, and the scalability of the system on a large synthetic data set. The results have validated the effectiveness and the efficiency of the proposed system with a lambda architecture.

For the future work, we will implement a scheduling system that can coordinate the running of the batch and real-time jobs within the same cluster. We intend to explore the ways to detect a greater range of anomalies, such as missing values, negative energy consumption and device errors. Besides, we plan to support additional types of data, such as gas, heating and water data, and to implement the corresponding detection algorithm.

Acknowledgements. This research was supported by the CITIES project (NO. 1035-00027B) funded by Innovation Fund Denmark.

References

1. Adnan, R., Setan, H., Mohamad, M.N.: Multiple outliers detection procedures in linear regression. Matematika **19**, 29–45 (2003)
2. Akyildiz, I.F., Su, W., Sankarasubramaniam, Y., Cayirci, E.: Wireless sensor networks: a survey. J. Comput. Netw. **38**(4), 393–422 (2002)
3. Ardakanian, O., Koochakzadeh, N., Singh, R.P., Golab, L., Keshav, S.: Computing electricity consumption profiles from household smart meter data. In: EDBT/ICDT Workshops, vol. 14, pp. 140–147 (2014)
4. Brown, M., Barrington-Leigh, C., Brown, Z.: Kernel regression for real-time building energy analysis. J. Build. Perform. Simul. **5**(4), 263–276 (2011)
5. Casado, R., Younas, M.: Emerging trends and technologies in big data processing. Concurrency Comput. Pract. Exp. **27**(8), 2078–2091 (2015)
6. Chandola, V., Banerjee, A., Kumar, V.: Anomaly detection: a survey. ACM Comput. Surv. **41**(3), 15 (2009)
7. Cheng, B., Longo, S., Cirillo, F., Bauer, M., Kovacs, E.: Building a big data platform for smart cities: experience and lessons from santander. In: IEEE International Congress on Big Data, pp. 592–599. IEEE Press, New York (2015)
8. Chou, J.S., Telaga, A.S.: Real-time detection of anomalous power consumption. Renew. Sustain. Energ. Rev. **33**, 400–411 (2014)
9. De Nadai, M., van Someren, M.: Short-term anomaly detection in gas consumption through arima and artificial neural network forecast. In: IEEE Workshop on Environmental, Energy and Structural Monitoring Systems, pp. 250–255. IEEE Press, New York (2015)
10. Frigge, M., Hoaglin, D.C., Iglewicz, B.: Some implementations of the boxplot. Am. Stat. **43**(1), 50–54 (1989)
11. Gao, X.: Scalable Architecture for Integrated Batch and Streaming Analysis of Big Data. Doctoral dissertation, Indiana University (2015)
12. Hasani, Z., Kon-Popovska, M., Velinov, G.: Lambda architecture for real time big data analytic. In: ICT Innovations (2014)
13. Jakkula, V., Cook, D.: Outlier detection in smart environment structured power datasets. In: 6th International Conference on Intelligent Environments, pp. 29–33. IEEE Press, New York (2010)
14. Janetzko, H., Stoffel, F., Mittelstdt, S., Keim, D.A.: Anomaly detection for visual analytics of power consumption data. Comput. Graph. **38**, 27–37 (2014)
15. Kroß, J., Brunnert, A., Prehofer, C., Runkler, T.A., Krcmar, H.: Stream processing on demand for lambda architectures. In: Beltrain, M., et al. (eds.) EPEW 2015. LNCS, vol. 9272, pp. 243–257. Springer, Heidelberg (2015)
16. Lee, A.H., Fung, W.K.: Confirmation of multiple outliers in generalized linear and nonlinear regressions. J. Comput. Stat. Data Anal. **25**(1), 55–65 (1997)

17. Lee, W., Stolfo, S.J., Chan, P.K., Eskin, E., Fan, W., Miller, M., Zhang, J.: Real time data mining-based intrusion detection. In: DARPA Information Survivability Conference and Exposition II, DISCEX 2001, vol. 1, pp. 89–100. IEEE Press, New York (2001)

18. Liu, F., Jiang, H., Lee, Y.M., Snowdon, J., Bobker, M.: Statistical modeling for anomaly detection, forecasting and root cause analysis of energy consumption for a portfolio of buildings. In: 12th International Conference of the International Building Performance Simulation Association (2011)

19. Liu, G., Zhu, W., Saunders, C., Gao, F., Yu, Y.: Real-time complex event processing and analytics for smart grid. Procedia Comput. Sci. **61**, 113–119 (2015)

20. Liu, X., Iftikhar, N., Xie, X.: Survey of real-time processing systems for big data. In: 18th International Database Engineering & Applications Symposium, pp. 356–361. ACM, New York (2014)

21. Liu, X., Nielsen, P.S.: Streamlining smart meter data analytics. In: Proceedings of the 10th Conference on Sustainable Development of Energy, Water and Environment Systems, SDEWES 2015.0558, pp. 1–14 (2015)

22. Liu, X., Nielsen, P.S.: A hybrid ICT-solution for smart meter data analytics. J. Energy (2016). doi:10.1016/j.energy.2016.05.068

23. Liu, X., Golab, L., Ilyas, I.F.: SMAS: a smart meter data analytics system. In: Proceedings of the ICDE, pp. 1476–1479 (2015)

24. Liu, X., Golab, L., Golab, W., Ilyas, I.F.: Benchmarking smart meter data analytics. In: Proceedings of the EDBT, pp. 385–396 (2015)

25. Magld, K.W.: Features extraction based on linear regression technique. J. Comput. Sci. **8**(5), 701–704 (2012)

26. Martnez-Prieto, M.A., Cuesta, C.E., Arias, M., Fernnde, J.D.: The solid architecture for real-time management of big semantic data. Future Gener. Comput. Syst. **47**, 62–79 (2015)

27. Marz, N., Warren, J.: Big Data: Principles and Best Practices of Scalable Realtime Data Systems, 1st edn. Manning Publications Co., Greenwich (2013)

28. Meng, X., Bradley, J., Yavuz, B., Sparks, E., Venkataraman, S., Liu, D., Xin, D.: MLlib: Machine Learning in Apache Spark (2015). arXiv preprint: arXiv:1505.06807

29. Preuveneers, D., Berbers, Y., Joosen, W.: SAMURAI: a batch and streaming context architecture for large-scale intelligent applications and environments. J. Ambient Intell. Smart Environ. **8**(1), 63–78 (2016)

30. Schneider, M., Ertel, W., Ramos, F.: Expected Similarity Estimation for Large-Scale Batch and Streaming Anomaly Detection (2016). arXiv preprint: arXiv:1601.06602

31. Sequeira, H., Carreira, P., Goldschmidt, T., Vorst, P.: Energy cloud: real-time cloud-native energy management system to monitor and analyze energy consumption in multiple industrial sites. In: 7th IEEE/ACM International Conference on Utility and Cloud Computing, pp. 529–534. IEEE Press, New York (2014)

32. Villari, M., Celesti, A., Fazio, M., Puliafito, A.: Alljoyn lambda: an architecture for the management of smart environments in IOT. In: IEEE International Conference on Smart Computing Workshops, pp. 9–14. IEEE Press, New York (2014)

33. Zaharia, M., Chowdhury, M., Das, T., Dave, A., Ma, J., McCauley, M., Stoica, I.: Resilient distributed datasets: a fault-tolerant abstraction for in-memory cluster computing. In: 9th USENIX Conference on Networked Systems Design and Implementation, p. 2. USENIX Association (2012)

34. Zaharia, M., Das, T., Li, H., Shenker, S., Stoica, I.: Discretized streams: an efficient and fault-tolerant model for stream processing on large clusters. In: 4th USENIX Conference on Hot Topics in Cloud Computing, p. 10. USENIX Association (2012)
35. Zhang, Y., Chen, W., Black, J.: Anomaly detection in premise energy consumption data. In: Power and Energy Society General Meeting, pp. 1–8. IEEE Press, New York (2011)

Big Data Indexing and Searching

Large Scale Indexing and Searching Deep Convolutional Neural Network Features

Giuseppe Amato, Franca Debole, Fabrizio Falchi, Claudio Gennaro[(⊠)], and Fausto Rabitti

ISTI-CNR, Via G. Moruzzi 1, 56124 Pisa, Italy
{giuseppe.amato,franca.debole,fabrizio.falchi,claudio.gennaro, fausto.rabitti}@isti.cnr.it

Abstract. Content-based image retrieval using Deep Learning has become very popular during the last few years. In this work, we propose an approach to index Deep Convolutional Neural Network Features to support efficient retrieval on very large image databases. The idea is to provide a text encoding for these features enabling the use of a text retrieval engine to perform image similarity search. In this way, we built LuQ a robust retrieval system that combines full-text search with content-based image retrieval capabilities. In order to optimize the index occupation and the query response time, we evaluated various tuning parameters to generate the text encoding. To this end, we have developed a web-based prototype to efficiently search through a dataset of 100 million of images.

Keywords: Convolutional neural network · Deep learning · Inverted index · Image retrieval

1 Introduction

Deep Convolutional Neural Networks (DCNNs) have recently shown impressive performance in the Computer Vision area, such as image classification and object recognition [9,15,20]. The activation of the DCNN hidden layers has been also used in the context of transfer learning and content-based image retrieval [6,19]. In fact, Deep Learning methods are "representation-learning methods with multiple levels of representation, obtained by composing simple but nonlinear modules that each transform the representation at one level (starting with the raw input) into a representation at a higher, slightly more abstract level" [16]. These representations can be successfully used as *features* in generic recognition or visual similarity search tasks.

The first layers of DCNN are typically useful in recognizing low-level characteristics of images such as edges and blobs, while higher levels have demonstrated to be more suitable for semantic similarity search. One major obstacle to the use of DCNN features for indexing large datasets of images is its internal representation, which is high dimensional leading to the curse of dimensionality [7]. For instance,

© Springer International Publishing Switzerland 2016
S. Madria and T. Hara (Eds.): DaWaK 2016, LNCS 9829, pp. 213–224, 2016.
DOI: 10.1007/978-3-319-43946-4_14

in the well-known AlexNet architecture [15] the output of the sixth layer (fc6) has 4,096 dimensions, while the fifth layer (pool5) has 9,216 dimensions. An effective approach to tackle the dimensionality curse problem is the application of approximate access methods such as permutation-based indexes [1,5].

A drawback of these approaches is on one hand the dependence of the index on a set of reference objects (or pivots); on the other hand the need to reorder the result set of the search according to the original feature distance. The former issue requires the selection of a set of reference objects, which represents well the variety of datasets that we want to index. The latter issue forces to use a support database to store the features, which requires efficient random I/O on a fast secondary memory (such as flash-based storages).

In this paper, we propose an approach to specifically index DCNN Features to support efficient content-based on large datasets, which we refer to as *LuQ*. The proposed approach exploits the ability of inverted files to deal with the sparsity of the Convolutional features. To this end, we make use of the efficient and robust full-text search library Lucene[1]. The idea is to associate each component of the feature vector with a unique alphanumeric keyword and to generate a textual representation in which we boost the relative term proportionally to its intensity.

A browser-based application that provides an interface for combined textual and visual searching into a dataset of about 100 million of images is available at http://melisandre.deepfeatures.org.

The paper is organized as follows. Section 2 makes a survey of the related works. Section 3 presents the proposed approach. Section 4 discusses the validation tests. Section 5 concludes.

2   Related Work

Recently, a new class of image descriptors, built upon Deep Convolutional Neural Networks (DCNNs), have been used as effective alternative to descriptors built using local features such as SIFT, SURF, ORB, BRIEF, etc. Starting from 2012 [15], DCNNs have attracted enormous interest within the Computer Vision community because of the state-of-the-art results achieved in the image classification challenge ImageNet Large Scale Visual Recognition Challenge (ILSVRC). In Computer Vision, DCNN have been used to perform several tasks, including not only image classification, but also image retrieval [2,6] and object detection [8], to cite some. In particular, it has been proved that the multiple levels of representation which are learned by DCNN on specific task (typically supervised) can be used to transfer learning across tasks. This means that the activation of neurons of a specific layers, preferably the last ones, can be used as features for describing the visual content [19].

Liu et al. [17] proposed a framework that adapts Bag-of-Word model and inverted table to DCNN feature indexing, which is similar to LuQ. However,

---

[1] http://lucene.apache.org.

for large-scale datasets, Liu et al. have to build a large-scale visual dictionary that employs product quantization method to learn a large-scale visual dictionary from a training set of global DCNN features. In any case, using this approach the authors reported a search time that is one order higher than in our case for the same dataset.

3 Text Encoding for Deep Convolutional Neural Network Features

3.1 Feature Indexing

The main idea of LuQ is to index DCNN features using a text encoding that allows us to use a text retrieval engine to perform image similarity search. As discussed later, we implemented this idea on top of the Lucene text retrieval engine; however any full-text engine supporting vector space model can be used for this purpose.

In principle, to perform similarity search on DCNN feature vectors we should compare the vector extracted from the query with all the vectors extracted from the images of the dataset and take the nearest images according to L2 distance. This is sometimes called brute-force approach. However, if the database is large, brute-force approach can become time-consuming.

Starting from this approach, we observed that given the sparsity of the DCNN features, which contain mostly zeros (about 75 %), we are able to use a well-known IR technique, i.e., an inverted index.

In fact, we note that since typically DCNN feature vectors exhibits better results if they are L2-normalized to the unit length [4,19], if two vectors $\mathbf{x}$ and $\mathbf{y}$ have length equal to 1 (such as in our case), the following relationship between the L2 distance $d_2(\mathbf{x}, \mathbf{y})$ and the inner product $\mathbf{x} * \mathbf{y}$ exists:

$$d_2(\mathbf{x}, \mathbf{y})^2 = 2(1 - \mathbf{x} * \mathbf{y})$$

The advantage of computing the similarity between vectors in terms of inner product is that we can efficiently exploit the sparsity of the vectors by accumulating the product of non-zeroes entries in the vector $\mathbf{x}$ and their corresponding non-zeros entries in the vector $\mathbf{y}$. Moreover, Lucene, as other search engines, computes the similarity between documents using the cosine similarity, which is the inner product of the two vectors divided by their lengths product. Therefore, in our case, cosine similarity and inner product are the same. Ascertained this behave, our idea is to fill the inverted index of Lucene with the DCNN features vectors. For space-saving reasons, however, text search engines do not store float numbers in the posting entries of the inverted index representing documents, rather they store the term frequencies, which are represented as integers. Therefore, we must guarantee that posting entries will contain numeric values proportional to the float values of the deep feature entries.

To employ this idea, in LuQ, we provide a text encoding for the DCNN feature vectors that guarantees the direct proportionality between the feature components and the term frequencies. Let $\mathbf{w} = (w_1, \ldots, w_m)$ denote the L2-normalized DCNN vector of m dimensions. Firstly, we associated each of its component w_i with a unique alphanumeric term τ_i (for instance, the prefix 'f' followed by the numeric values corresponding to the index i). The text encoding $doc(\mathbf{w})$ corresponding to the vector $\mathbf{w}$ is given by:

$$doc(\mathbf{w}) = \bigcup_{i=1}^{m} \bigcup_{j=1}^{\lfloor Qw_i \rfloor} \tau_i$$

Where $\lfloor \rfloor$ denotes the floor function and Q is a multiplication factor > 1 that works as a *quantization factor*[2].

Therefore, we form the text encoding of w_i by repeating the term τ_i for the non-zero components a number of times directly proportional to w_i. This process introduces a quantization error due to the representation of float components in integers. However, as we will see, this error does not affect the retrieval effectiveness. The accuracy of this approximation depends on the factor Q, used to transform the vector $\mathbf{w}$. For instance, if we fix $Q = 2$, for $w_i < 0.5$, $\lfloor Qw_i \rfloor = 0$, while for $w_i \geq 0.5$, $\lfloor Qw_i \rfloor \geq 1$. In contrast, the smaller we set Q the smaller the inverted index will be. This is because the floor function will set to zero more entries of the posting lists. Hence, we have to find a good trade-off between the effectiveness of the retrieval system and its space occupation.

For example, if we set $Q = 30$ and we have for instance a feature vector with just three components $\mathbf{w} = (0.01, 0.15, 0.09)$ the corresponding integer-representation of the vector will be $(0, 4, 2)$ and its text encoding will be: $doc(\mathbf{w}) = $ "f2 f2 f2 f2 f3 f3".

Since on average the 25 % of the DCNN features are non-zero (in our specific case the fc6 layer), the size of their corresponding text encoding will have a small fraction of the unique terms present in the whole dictionary (composed of 4,096 terms). In our case, on average a document contains about 275 unique terms, which is about 6.7 % of the dictionary because of quantization that set to zero the feature components smaller than $1/Q$.

3.2 Query Reduction

When we have to process similarity search, therefore the search engine have to treat query of that size. These unusual long queries, however, can affect the response time if the inverted index contains million of items.

A quite intuitive way to overcome this issue is to reduce the size of the query by exploiting the knowledge of the *tf*idf* (i.e., term frequency * inverse document frequency) statistic of the text encoding, which comes for free in standard full-text retrieval engines. We can retain the elements of the query that exhibit greater values of *tf*idf* and eliminate the others. For instance, for a query of about 275 unique term on average, we can take the first ten terms that exhibits the highest *tf*idf*, we obtain a query time reduction of about 96 %.

---

[2] By abuse of notation, we denote the space-separated concatenation of keywords with the union operator ∪.

This query reduction comes, however, with a price: it decreases the precision of results. To attenuate this problem, for a top-k query, we reorder the results using the cosine similarity between the original query (i.e., the one without reduction) and the first $C_r \times k$ candidates documents retrieved. Where C_r is an amplification factor that we refer to as *reordering factor*. For instance, if we have to return $k = 100$ images and we set $C_r = 10$, we take and reorder the first $10 \times 100 = 1000$ candidate documents retrieved by the reduced query.

In order to calculate the cosine similarity of the original query and the $C_r \times k$ candidates, we have to reconstruct the quantized features by accessing to the

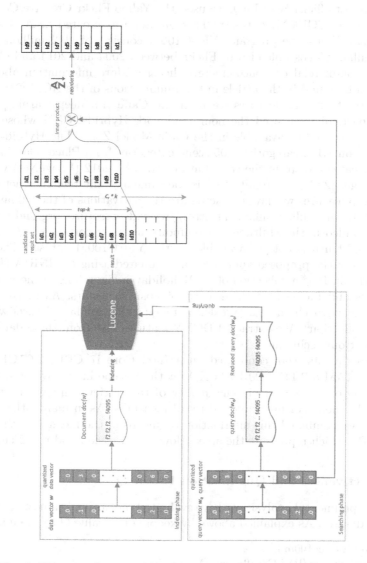

Fig. 1. Diagram showing the indexing and searching phases of LuQ.

posting list of the document returned by the search engine. As we will see, this approach does not affect significantly the efficiency of the query but can offer great improvements in terms of effectiveness.

Figure 1 summarizes the indexing and searching phases of LuQ.

4   Experiments

4.1   Setup

In order to test efficiency of LuQ, we used the Yahoo Flickr Creative Commons 100 Million (YFCC100M) dataset[3]. This dataset was created in 2014 as part of the Yahoo Webscope program. YFCC100M consists of 99.2 million photos and 0.8 million videos uploaded to Flickr between 2004 and 2014 and Creative Commons commercial or noncommercial license. More information about the dataset can be found in the article in Communications of the ACM [21].

For extracting deep features we used the Caffe [14] deep learning framework. In particular we used the neural network Hybrid-DCNN whose model and weights are public available in the Caffe Model Zoo[4]. The Hybrid-DCNN was trained on 1,183 categories (205 scene categories from Places Database and 978 object categories from the train data of ILSVRC2012 (ImageNet) with 3.6 million images [22]). The architecture is the same as Caffe reference network.

The deep features we have extracted are the activations of the fc6 layer. We have made them public available at http://www.deepfeatures.org and they will be soon included in the Multimedia Commons initiative corpus.

A ground-truth is not yet available for the YFCC100M dataset. Therefore, effectiveness of the proposed approach was evaluated using the INRIA Holidays dataset [11,13]. It is a collection of 1,491 holiday images. The authors selected 500 queries and for each of them a list of positive results. As in [10–12], to evaluate the approach on a large scale, we merged the Holidays dataset with the Flickr1M collection[5]. We extracted DCNN features also from these datasets in order to test our technique.

All experiments were conducted on a Intel Core i7 CPU, 2.67 GHz with 12.0 GB of RAM a 2 TB 7200 RPM HD for the Lucene index. We used Lucene v5.5 running on Java 8 64 bit.The quality of the retrieved images is typically evaluated by means of precision and recall measures. As in many other papers [10,11,18], we combined this information by means of the mean Average Precision (mAP), which represents the area below the precision and recall curve.

4.2   Effectiveness and Quantization Factor Q

In a first experimental analysis, we have evaluated the optimal value of Q over the Flickr1M dataset. As explained above, by keeping the value Q to the minimum,

---

[3] http://bit.ly/yfcc100md.

[4] http://github.com/BVLC/caffe/wiki/Model-Zoo.

[5] http://press.liacs.nl/mirflickr/.

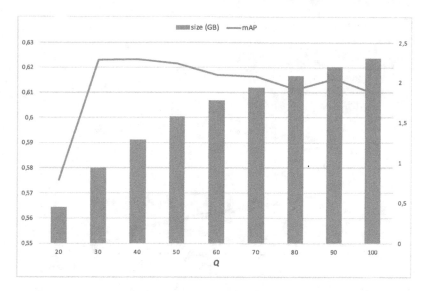

Fig. 2. Effectiveness (mAP) vs space occupation for increasing values of the quantization factor Q.

we can reduce the space occupation of the inverted index. Figure 2 shows the mAP as function of Q and the corresponding space occupation of the inverted index. From this analysis, we conclude that an optimal choice of the quantization factor Q is 30, which leads to a mAP of 0.62 and a space occupation of 2.31 GB. We stress that the mAP using the brute-force approach (on the exact DCNN feature vectors) is about 0.60. This means that our quantization error leads to a slight improvement of the precision for $Q \geq 30$. Another important aspect is that this effectiveness was obtained forcing Lucene to use the standard inner product on *tf* weight without *idf* and any other document normalization. A further improvement can be obtained using the similarity function of Lucene called LMDirichlet, which provided a mAP of about 0.64.

Figure 3 shows the document frequency distribution, in the Flickr1M dataset, of the terms τ_i (i.e., component w_i), sorted in decreasing order. As can be seen, the distribution is quite skewed and some terms are much more frequent than others ranging from 313 to 378,876 in a collection of about 1 million features. This aspect has some impact on the performance of the inverted index, since it means that the posting list of the term τ_x has 378,876 items since it appears in many image features.

The observation about document frequency leads to the idea of using *tf*idf* to reduce the query length by cutting off terms with lower *tf*idf* weight. Since in inverted files the query time is usually proportional with the length of the query [3], this approach gives a great improvement in terms of query response time.

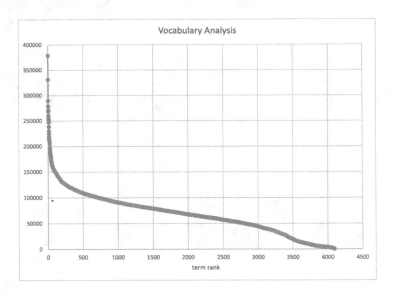

Fig. 3. Distribution of the vector components of the DCNN features ($Q = 30$).

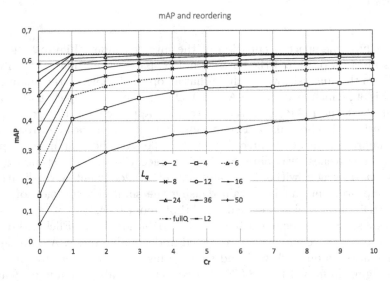

Fig. 4. Effectiveness (mAP) for various level of query lengths L_q and reordering factor C_r (with $k = 100$ and $C_r = 0$ means no reordering) vs. the query without reduction (fullQ), and the mAP obtained with sequential scan using L2.

Figure 4 shows the mAP values at different levels of reordering factors C_r and query lengths L_q. Note that, $C_r = 0$ means no reordering, $C_r = 1$, reordering of the first k candidates, $C_r = 2$, reordering of the first $2k$ candidates, and so on. Concerning, L_q, we have considered a range of values between 2 and 50. Since the average document length is about 275, this corresponds to an average length reduction from 0.3 % to 80 %. In the graph of Fig. 4, we have also plotted the mAP level obtained with query without reduction (namely fullQ) and the mAP obtained with sequential scan using L2 (brute-force). In all experiments we have set $k = 100$. As these experiments show, the configuration $C_r = 10$ and $L_q = 8$ exhibits a mAP comparable to the case of brute-force using L2.

4.3   Evaluation of the Efficiency

In order to evaluate the efficiency of LuQ, in Fig. 5, we have plotted the average search time for the same queries of the previous experiments. Please, note that the y-axis is in logarithmic scale. When there is no reordering (i.e., $C_r = 0$), the length of the query has a great impact to the search time (more than one order of magnitude). For increasing values of the the reordering factor C_r, L_q has a lower and lower influence on the search time. Clearly, increasing C_r, we increase the cost of the search. However, as we can see, there is still a big improvement in efficiency even for the case in which the reorder factor is maximum, i.e., $C_r = 10$.

In order to further validate LuQ, we test our index on the much larger dataset YFCC100M. We used the DCNN features of the fc6 layer and indexed them with Lucene using a quantization factor $Q = 30$. Since for this dataset a ground-truth is not available, we only reported the performance of the query in terms of average search time (see Fig. 6). From this experiment, we see that for instance for the configuration $L_q = 10$ and $C_r = 10$, we have an average query time of less than 4 s (without any parallelization), which is quite encouraging considering that for the same dataset the query time was of the order of 10 min using the

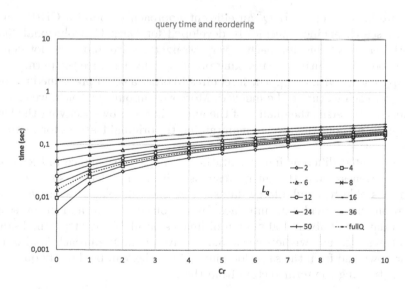

Fig. 5. Average search time (sec) for various level of query lengths L_q and reordering factor C_r (with $k = 100$ and $C_r = 0$ means no reordering) vs. the query without reduction (fullQ), and the mAP obtained with sequential scan using L2.

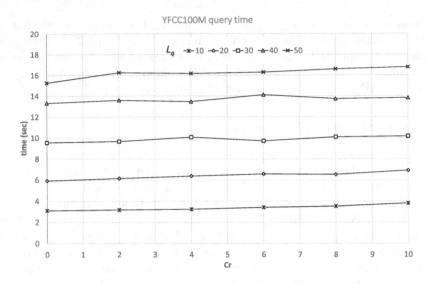

Fig. 6. Average search time (sec) for the YFCC100M dataset using lucene.

brute-force approach. In this case, we observe that C_r does not practically affect the efficiency of LuQ.

5   Conclusions and Future Work

In this work, we propose LuQ, an efficient approach to build a CBIR, on top of a text search engine, specifically developed for Deep Convolutional Neural Network Features. This approach is very straightforward and does not demand costly elaborations during the indexing process, as for instance the permutation-based approaches such as [1], which requires to order a set of predefined reference features for each feature to be indexed. Moreover, in our approach, we can tune the query costs versus the quality of the approximation by specifying the length of the query, without the need of maintaining the original features for reordering the result set.

We evaluated different implementation strategies to balance index occupation, effectiveness, and the query response time. In order to show that our approach is able to handle large datasets, we have developed a browser-based application that provides an interface for combined textual and visual searching on a dataset of about 100 million of images, available at http://melisandre. deepfeatures.org. The whole Lucene 5.5 archive of LuQ approach is also available for download from the same location. This index can be directly queried by simply extracting the term vectors from the archive.

Acknowledgments. This work was partially founded by: EAGLE, Europeana network of Ancient Greek and Latin Epigraphy, co-founded by the European Commission, CIP-ICT-PSP.2012.2.1 - Europeana and creativity, Grant Agreement n. 325122; and Smart News, Social sensing for breaking news, co-founded by the Tuscany region under the FAR-FAS 2014 program, CUP CIPE D58C15000270008.

References

1. Amato, G., Gennaro, C., Savino, P.: MI-File: using inverted files for scalable approximate similarity search. Multimedia Tools Appl. **71**(3), 1333–1362 (2014)
2. Babenko, A., Slesarev, A., Chigorin, A., Lempitsky, V.: Neural codes for image retrieval. In: Fleet, D., Pajdla, T., Schiele, B., Tuytelaars, T. (eds.) ECCV 2014, Part I. LNCS, vol. 8689, pp. 584–599. Springer, Heidelberg (2014)
3. Büttcher, S., Clarke, C.L.A.: Information Retrieval: Implementing and Evaluating Search Engines. MIT Press, USA (2010)
4. Chatfield, K., Simonyan, K., Vedaldi, A., Zisserman, A.: Return of the devil in the details: delving deep into convolutionalnets (2014). arXiv preprint arXiv:1405.3531
5. Chavez, G.E., Figueroa, K., Navarro, G.: Effective proximity retrieval by ordering permutations. IEEE Trans. Pattern Anal. Mach. Intell. **30**(9), 1647–1658 (2008)
6. Donahue, J., Jia, Y., Vinyals, O., Hoffman, J., Zhang, N., Tzeng, E., Darrell, T.: DeCAF: a deep convolutional activation feature for generic visual recognition (2013). arXiv preprint arXiv:1310.1531
7. Ge, Z., McCool, C., Sanderson, C., Corke, P.: Modelling local deep convolutional neural network features to improve fine-grained image classification. In: 2015 IEEE International Conference on Image Processing (ICIP), pp. 4112–4116. IEEE (2015)
8. Girshick, R., Donahue, J., Darrell, T., Malik, J.: Rich feature hierarchies for accurate object detection and semantic segmentation. In: Proceedings of the IEEE Conference on Computer Vision and Pattern Recognition, pp. 580–587 (2014)
9. He, K., Zhang, X., Ren, S., Sun, J.: Deep residual learning for image recognition (2015). arXiv preprint arXiv:1512.03385
10. Jégou, H., Douze, M., Schmid, C.: Packing bag-of-features. In: 2009 IEEE 12th International Conference on Computer Vision, 29 September–2 October 2009, pp. 2357–2364 (2009)
11. Jégou, H., Perronnin, F., Douze, M., Sànchez, J., Pérez, P., Schmid, C.: Aggregating local image descriptors into compact codes. IEEE Trans. Pattern Anal. Mach. Intell. **34**(9), 1704–1716 (2012)
12. Jégou, H., Douze, M., Schmid, C.: Improving bag-of-features for large scale image search. Int. J. Comput. Vis. **87**, 316–336 (2010)
13. Jégou, H., Douze, M., Schmid, C., Pérez, P.: Aggregating local descriptors into a compact image representation. In: IEEE Conference on Computer Vision and Pattern Recognition, June 2010, pp. 3304–3311 (2010)
14. Jia, Y., Shelhamer, E., Donahue, J., Karayev, S., Long, J., Girshick, R., Guadarrama, S., Darrell, T.: Caffe: convolutional architecture for fast feature embedding (2014). arXiv preprint arXiv:1408.5093
15. Krizhevsky, A., Sutskever, I., Hinton, G.E.: Imagenet classification with deep convolutional neural networks. In: Advances in Neural Information Processing Systems, pp. 1097–1105 (2012)
16. LeCun, Y., Bengio, Y., Hinton, G.: Deep learning. Nature **521**(7553), 436–444 (2015)

17. Liu, R., Zhao, Y., Wei, S., Zhu, Z., Liao, L., Qiu, S.: Indexing of cnn features for large scale image search (2015). arXiv preprint arXiv:1508.00217
18. Perronnin, F., Liu, Y., Sanchez, J., Poirier, H.: Large-scale image retrieval with compressed fisher vectors. In: 2010 IEEE Conference on Computer Vision and Pattern Recognition (CVPR), June 2010, pp. 3384–3391 (2010)
19. Razavian, A.S., Azizpour, H., Sullivan, J., Carlsson, S.: CNN features off-the-shelf: an astounding baseline for recognition. In: Proceedings of the IEEE Conference on Computer Vision and Pattern Recognition (CVPR) Workshops, June 2014, pp. 806–813 (2014)
20. Simonyan, K., Zisserman, A.: Very deep convolutional networks for large-scale image recognition (2014). arXiv preprint arXiv:1409.1556
21. Thomee, B., Elizalde, B., Shamma, D.A., Ni, K., Friedland, G., Poland, D., Borth, D., Li, L.-J.: YFCC100M: the new data in multimedia research. Commun. ACM 59(2), 64–73 (2016)
22. Zhou, B., Lapedriza, A., Xiao, J., Torralba, A., Oliva, A.: Learning deep features for scene recognition using places database. In: Advances in Neural Information Processing Systems, pp. 487–495 (2014)

A Web Search Enhanced Feature Extraction Method for Aspect-Based Sentiment Analysis for Turkish Informal Texts

Batuhan Kama[1], Murat Ozturk[1], Pinar Karagoz[1(✉)], Ismail Hakki Toroslu[1], and Ozcan Ozay[2]

[1] Computer Engineering Department, Middle East Technical University,
Ankara, Turkey
{batuhan.kama,mozturk,karagoz,toroslu}@ceng.metu.edu.tr
[2] Huawei Turkey R&D Center, Istanbul, Turkey
ozcan.ozay@huawei.com

Abstract. In this article, a new unsupervised feature extraction method for aspect-based sentiment analysis is proposed. This method improves the performance of frequency based feature extraction by using an online search engine. Although frequency based feature extraction methods produce good precision and recall values on formal texts, they are not very successful on informal texts. Our proposed algorithm takes the features of items suggested by frequency based feature extraction methods, then, eliminates the features which do not co-occur with the item, whose features are sought, on the Web. Since the proposed method constructs the candidate feature set of the item from the Web, it is domain-independent. The results of experiments reveal that for informal Turkish texts, much higher performance than frequency based method is achieved.

1 Introduction

With the increasing number of online shopping sites, people shop online rather than going to the stores [2]. Since almost all online shopping sites allow shoppers to write and read comments about the products sold, people searching for items to purchase read these comments and then decide accordingly. Furthermore, studies show that in addition to online shoppers, offline shoppers also receive support from the online shopping sites by reading comments before going to shopping. Websites providing comment support on items are not only online shopping sites. In addition, forums and blogs are huge data sources on this area. Moreover, social networking sites like Facebook and Twitter are sometimes also used for rating a service or product, too.

Although there are basically three kinds of websites to provide reviews about products, the number of such websites is considerably high. Examining such a huge number of websites with reviews requires a system that automatically finds and interprets the comments and reviews from these websites for an intended topic. One of the most popular approaches to interpret documents, sentences

© Springer International Publishing Switzerland 2016
S. Madria and T. Hara (Eds.): DaWaK 2016, LNCS 9829, pp. 225–238, 2016.
DOI: 10.1007/978-3-319-43946-4_15

and any other texts is sentiment analysis. Since it is more important to extract the views of the writers on each specific features of items, rather than the overall view on the whole item, a specific version of sentiment analysis, namely, feature based sentiment analysis is preferred.

Basically, sentiment analysis concentrates on analyzing and summarizing people's opinions, feelings and notions towards all kind of entities such as products, services and topics and their aspects. Commonly three different levels of sentiment analysis are studied: document-level, sentence-level and feature-level [13]. Document level sentiment analysis (DLSA) aims to determine whether the overall sentiment in the document is positive or negative [15]. Sentence level sentiment analysis (SLSA) focuses on determining the orientation of feeling in each sentence. Both DLSA and SLSA processes are too general to classify sufficiency of features of items/products; hence feature level sentiment analysis (FLSA) is developed to find polarity of sentiments for each different aspect of entities mentioned in a sentence or document.

All FLSA methods contain two common steps: feature extraction and feature-sentiment classification. Feature extraction step extracts features from a given text while feature-sentiment classification step matches extracted features with the sentiments in the given text. In this study, a new method to extract features of a topic from a given set of Turkish informal texts is proposed. Frequency of nouns in the item reviews and the Web search are the main components of the proposed method.

The paper is organized as follows. Section 2 summarizes the related work in the literature. Section 3 gives technical background related to the proposed method. Section 4 introduces the proposed method, and, Sect. 5 presents the results of experiments using proposed method over a dataset containing Turkish forum entries. Finally, Sect. 6 summarizes and concludes the study.

2   Related Work

In this section, novel studies based on DLSA and FLSA are explained, specifically focusing on FLSA studies.

The studies on DLSA can be divided into two categories as supervised and unsupervised approaches. In one of initial studies on supervised approaches, authors classified the movie reviews as positive and negative by running Naive Bayes and SVM algorithms with unigrams as features [15]. In [13], the proposed method includes part of speech (POS) tags of words, sentiment words, sentiment shifters and terms with their frequencies as the common features for supervised learning methods. The first known study on unsupervised DLSA uses syntactic patterns to extract phrases expressing opinions [20]. In this study, firstly, possible opinion phrases are extracted. Then, the orientation of the each extracted phrase is found out by deciding whether this phrase is closer to "excellent" or "poor" by using Pointwise Mutual Information (PMI) algorithm. Finally, aggregate score of all phrases is calculated to find sentiment orientation of the whole document.

The studies on FLSA have two main focus areas: aspect (feature) extraction and aspect-sentiment classification. Aspect-sentiment classification is the process of determining the orientation of sentiment expressed on each aspect in a sentence [13]. Aspect extraction involves finding explicit and implicit aspects from a given text. Explicit aspects are the noun or noun groups which directly exist in sentences. On the other hand, implicit aspects are the noun or noun groups which are implied by adjective, adverbs or verbs [13].

For explicit aspect extraction, in the baseline study given in [8], firstly, all words in the dataset are labeled with their POS tags. Then, each occurrence of all nouns and noun groups are counted. Finally, all nouns and noun groups with frequency over a given threshold are extracted as features. In [16], precision of the baseline algorithm is enhanced by deleting noun phrases which are predicted to be non-aspects from the aspect set. To decide for a non-aspect, the frequency of co-occurrence of this aspect with its product on the Web is used. In [4], efficiency of baseline method is further improved by applying several filters. In [8], the proposed approach can extract infrequent features which cannot be discovered by the first approach. In this study, aspects are extracted with frequency based approach firstly. Then, the sentiment words which are closest to these aspects are marked. Finally, whole dataset is rescanned and the nouns and noun groups which are not labeled as aspect yet but affected by any of these sentiment words are added to aspect list as infrequent aspects. In [18], several improvements are applied by using dependency relations as an enhancement over the method given in [8].

Supervised approaches are classified as an alternative approach to extract explicit aspects. In [10], authors propose a method where both aspects and sentiment phrases are extracted by lexicalized Hidden Markov Model (lexicalized-HMM), which uses linguistic features to learn patterns. A Conditional Random Field (CRF) based approach is proposed in [9]. A hybrid model is proposed in [11], which integrates Skip-CRF and Tree-CRF in order to train system with structural features. Yu et al. in [22] proposed a method which uses one-class Support Vector Machine (SVM), which is a method that requires only positive data to be labelled, to extract aspects. In this study, only aspects in training dataset are labelled since one-class SVM needs only positive data.

Topic models are another approach to extract explicit features. A combined model using HMM and probabilistic Latent Semantic Analysis (pLSA) is proposed for sentiment analysis in [14]. In this study, an aspect-sentiment joint model is used to train system with a small training set. In another study based on Latent Dirichlet Allocation (LDA) [12], aspects and sentiment words are extracted together without requiring any training data since a joint model is used. The authors of [5] achieved extracting features without sentiments by using local LDA.

The other direction on FLSA is extracting implicit features. In [19], in order to map implicit aspect (i.e. sentiment words) to their corresponding explicit aspects, a mutual reinforcement approach is used. The approach proposed in [17] takes a list of unique implicit features and a training data. A sentence-based co-occurrence matrix of implicit features with other words is created in the training phase. Then, this co-occurrence matrix is used to assign implicit features

Table 1. Turkish genitive and third person suffixes

Last vowel	Genitive suffix	Third person suffix
a, ı	-ın	-ı
e, i	-in	-i
o, ö	-un	-u
u, ü	-ün	-ü

to explicit features in sentences of test data. Two-step co-occurrence association rule mining approach is used in [7]. Firstly, association rules are generated between sentiment words and explicit features. Then, explicit aspects are clustered.

In this work, in order to overcome observed limitations of the previous studies on Turkish informal texts, we propose to use the web search results as an additional checking layer for feature extraction.

3   Preliminaries

3.1   Morphological Analysis

The previous researches have shown that almost all explicit features in a text are nouns or nouns groups. Therefore, analyzing sentences morphologically and extracting POS tags is needed during feature extraction process. Morphological analysis refers to the mental system involved in word formation or to the branch of linguistics that deals with words, their internal structures, and how they are formed [3]. Morphological analysis of texts is a kind of text processing process and it is highly dependent on the language used in text. In this context, since each language has its own rules, these rules have to be known to analyze sentences morphologically. Natural Language Processing (NLP) tools can be used for this purpose with high performance.

In this study, *Turkish possessive construction* is used as a part of the proposed method. In Turkish, a *possessive construction* is a noun group where the second noun is possessed by the first noun. For instance, the nouns, *telefon* (phone) and *ekran* (screen), generates the possessive construction *telefonun ekranı* (telefon-un ekran-ı), which refers to *phone's screen*. While producing possessive construction, a suffix of ownership (genitive suffix) is added to the possessor, which is the first noun and third person suffix is added to the possessed object, which is the second noun. The genitive suffix or third person suffix which will be added to a word is chosen with respect to last vowel of this word as given in Table 1. In addition, if possessor or possessed object ends with a vowel, the character n is added to beginning of the genitive or third person suffix.

3.2   Pointwise Mutual Information

Pointwise mutual information (PMI) is a statistical measure to calculate the probability of two events being together [6]. In text processing domain, the PMI score of two words is calculated by Eq. 1.

$$pmi(word_1; word_2) = \log_2 \frac{P(word_1, word_2)}{P(word_1)P(word_2)} \qquad (1)$$

The problem with calculating PMI score for two words is to decide how to find scores of probability $P(word)$ and joint probability $P(word1, word2)$. In [20], these two scores are calculated as in Eqs. 2 and 3.

$$P(word) \equiv hits(word) \qquad (2)$$

$$P(word_1, word_2) \equiv hits(near(word_1, word_2)) \qquad (3)$$

In these equations, $hits(X)$ function refers to number of occurrence of a given word X on the web and $near(Y, Z)$ returns the word groups that start with word Y and end with word Z, where at most 10 words exist between Y and Z. The study in this article uses PMI formula for texts with modifications to find co-occurrence rate of two words on the Web for feature extraction process.

3.3   Tools

In this section, the tools used for Turkish morphological analysis and the Web search are introduced. While Zemberek is chosen for Turkish morphological analysis, API support of Yandex is used for the Web search. Nevertheless, the proposed approach does not directly depend on these tools.

Zemberek is the mostly used NLP tool for Turkish. It has NLP abilities such as morphological analysis, spellchecking, word suggestion for incorrect words, sentence extraction and spelling [1]. In our methodology, Zemberek is used for two purposes: to extract sentences from given texts, and to extract part of speech tags of words by analyzing the extracted sentences morphologically.

Yandex search engine API[1] is a service that allows users to send a query to Yandex search engine and get results in XML format. In this article, search engine API is used while calculating $hits(X)$ in the PMI equation. While several different parameters are returned in the result set from the search engine, only the parameter which states frequency of search string is used.

4   Methodology

In this study, a three-step approach for extracting features from given informal text documents is proposed. In the first two steps, state-of-the-art approaches, which are frequency based feature extraction and frequency based feature extraction with sentiment word support, are applied on the informal texts. Since the accuracy results are not at acceptable level, a new method, that is web search based feature extraction method, is proposed to increase the accuracy of the feature extraction process. The overall methodology is shown in Fig. 1.

---

[1] https://tech.yandex.com/xml/.

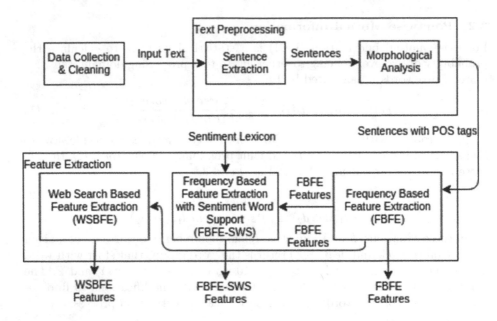

Fig. 1. The pipeline of modules in our feature extraction algorithm

4.1 Data Collection and Text Preprocessing

In this phase, the dataset used in the experiments is collected first. For this purpose, Donanimhaber[2] site, which contains user reviews and comments intensely, is selected to collect data after a thorough search on the Web. In order to crawl the comments and users' views from this site, jsoup[3] library is used. All pages under selected item's main topic are crawled. The crawled data are in XML format where each node includes user id, user's comment and date information. The crawled data are then loaded into MySQL[4]. In this regard, detailed queries can be performed on the dataset where results are used in the experiment phase.

The text data collected from the Web are analyzed and converted into format used in feature extraction methods. Firstly, all sentences in the text collection are extracted. Afterwards, these sentences are analyzed morphologically and each word is stored together with its POS tag.

4.2 Frequency Based Feature Extraction (FBFE)

In this method, all words labeled as noun or noun group (which are consecutive nouns not separated by any punctuation) are counted separately and their counts (number of occurences) in the whole data set are obtained. Among them, the nouns and noun groups with counts above a experimentally determined threshold are selected as features.

---

[2] http://www.donanimhaber.com.
[3] http://jsoup.org/.
[4] https://www.mysql.com.

Table 2. FBFE and FBFE-SWS example

Noun	Noun (English)	Feature	Dataset Freq	FBFE (T = 10)	Sent Based Freq	FBFE-SWS (T = 20)
arkadaş	friend	No	13074	F	2321	F
ekran	screen	Yes	6743	F	1261	F
teşekkür	thank	No	4830	F	258	F
ihtimal	possibility	No	884	F	106	F
ekran ışık	screen light	Yes	70	F	15	NF
ekran boyut	screen size	Yes	59	F	16	NF
işlemci	processor	Yes	14	F	2	NF
ağaç	tree	No	9	NF	N/A	N/A

4.3 Frequency Based Feature Extraction with Sentiment Word Support (FBFE-SWS)

In this method, for each sentence, firstly sentiment words are marked by using a sentiment word lexicon. Secondly, for each marked sentiment word, the noun or noun group in its neighborhood, which is the closest to the sentiment word by word distance, is determined and this noun or noun group is included into the frequency list. For example, in sentence "Gps gayet güzel çalışıyor, ancak harita uygulaması kötü (GPS works well, but map application is awful)", sentiment words are *güzel* and *kötü* and noun and noun groups are *gps* and *harita uygulaması*. Since *güzel* is closer to *gps* and *kötü* is closer to *harita uygulaması*, this noun and noun group are added to frequency list. The rest of the method is the same as in FBFE.

Table 2 gives an example on how FBFE and FBFE-SWS methods work. Assume that there is a dataset with limited number of nouns on mobile phone's. In this table, *Is Feature* column states whether corresponding noun or noun phrase is a feature or not. *Dataset Freq* gives frequencies on nouns and noun phrases in the example dataset while FBFE column labels whether they are feature (F) or not feature (NF) with respect to FBFE method.The same applies for *Sentiment Based Freq* and *FBFE-SWS* columns.

To apply FBFE method onto the example dataset, firstly, all nouns and noun groups are extracted with their frequencies and listed as sorted on their frequencies. In the example, six nouns and two noun groups given under *Noun* column are extracted and their frequencies are shown in *Dataset Freq* column. Between these nouns with frequencies higher than a threshold are selected as features. In the example, the threshold is assumed as 10, therefore, as stated in *FBFE* column, first seven nouns are labeled as features while the last noun is eliminated. Then, these seven nouns are given to FBFE-SWS method as input. In this method, each occurrence of these nouns as a neighbor of sentiment word is counted. A new threshold value is defined and nouns with occurrence counts above the threshold are returned as features. Results of FBFE-SWS on example dataset for threshold value 20 can be seen in *FBFE-SWS* column.

4.4 Web Search Based Feature Extraction (WSBFE)

WSBFE is the main novel approach proposed in this paper and given in Algorithm 1. In this method, search engine API and results of FBFE (given in Sect. 4.1) are used for feature extraction. In this method, instead of using nouns' and noun groups' frequencies in our dataset, their frequencies on the Web are used.

To apply this method, firstly, results of FBFE are determined. Then, for each noun or noun group labeled as feature by FBFE method, a search query is generated to be sent to search engine API. After these queries are sent to Web search engine, occurrence counts of this search results are stored in the list.

One of the key points of this algorithm is to generate search queries. They can be generated as stated in the Sect. 3.1. If a noun or noun group is a feature of an item whose features are sought, then it is highly possible that this noun or noun group is used together with its item name on the Web many times. For example, for the noun *ekran* search query *telefonun ekranı* is created and it is searched on the web as described in Sect. 3.3.

In the second phase, after all noun and noun groups are searched on the Web, the list that keeps these nouns and noun groups with their occurrence counts is sorted in descending order. Finally, an experimental threshold is defined and nouns and noun groups with frequencies above this threshold are returned as features. For instance, *telefonun ekranı* and *telefonun görüntüsü* have high occurrence rates on the Web and they are selected as features while *telefonun ürünü* is an infrequent phrase on the Web and not selected as feature.

In the third phase, new features are extracted from noun groups that are below the threshold. For each such noun group, it is checked whether at least one word of this noun group belong to previously returned features list or not. If at least one such word exists, then this noun group is also added to the list. For example, *telefonun ekran görüntüsü* is not a common noun phrase on the Web; however, since both *telefonun ekranı* and *telefonun görüntüsü* are common, *telefonun ekran görüntüsü* is also selected as feature by using this phase.

In Table 3, an example application of WSBFE method on the example data set given in Table 2 is presented. First two columns are the same as in Table 2. On the other hand, *Search Query* column gives noun and noun phrases which are searched on the Web with preceding *Telefonun* word. While *Search Results* column shows the search counts for each search query, last two column gives whether these nouns are features or not with respect to Phase 2 and Phase 3 of WSBFE algorithm.

All seven nouns proposed as features by FBFE are passed to WSBFE method as in FBFE-SWS. Then, in the first phase, search queries are created for these nouns. Secondly, all of these search queries are searched on the Web search engine and the occurrence count of each search query is stored as in FBFE and FBFE-SWS approaches. In Phase 2, nouns with a higher Web occurrence counts than

---

Algorithm 1. Web Search Based Feature Extraction algorithm

1: **procedure** WSBFE
Require: *item*:item in the dataset, *fbfe_features*:the list of features proposed by FBFE
2:     *feature_occurrence* ← {} //list to keep features' occurrence on the Web
3:     //phase 1 - finding occurrence count of nouns with the item on the Web
4:     **for** each feature $f \in fbfe\_features$ **do**
5:         **if** f does not contains *item* **then**
6:             *search_query* ← *item* + *Genitive Suffix* +"" + f + *Third Person Suffix*
7:         **else**
8:             *search_query* ← f
9:         *occurrence_count* ← get_occurrence_on_yandex(*search_query*)
10:        add {f, *occurrence_count*} to *feature_occurrence*
11:    //phase 2 - extracting nouns and nouns groups frequent on the Web
12:    sort(*feature_occurrence*)
13:    *threshold* ← get_experimental_threshold(*feature_occurrence*)
14:    *proposed_features* ← {}
15:    *non_features* ← {}
16:    **for** each {f, *occurrence_count*} $\in$ *feature_occurrence* **do**
17:        **if** *occurrence_count* $\geq$ *threshold* **then**
18:            add f to *proposed_features*
19:        **else**
20:            add f to *non_features*
21:    //phase 3 - extracting features that are infrequent on the Web
22:    **for** each $f \in$ *non_features* **do**
23:        **if** f is a noun group **then**
24:            *isFeature* ← False
25:            **for** each word $w \in f$ **do**
26:                **if** $w \in$ *proposed_features* **then**
27:                    *isFeature* ← True
28:                    break
29:            **if** *isFeature* is True **then**
30:                add f to *proposed_features*
            return *proposed_features*

---

the threshold value are labeled as features. In the example dataset, threshold is set as 1000 and *ekran, ekran boyutu* and *işlemci* are labeled as features as stated in *WSBFE Phase 2* column. Finally in Phase 3, each word of noun phrases labeled as Non-Feature in Phase 2 is checked whether it is a feature or not. If at least one word of a noun phrase is a feature, then this noun phrase is added to features list in Phase 3. In the example, *ekran boyut* is the only noun phrase that is non-feature with respect to Phase 2. Since, *ekran* is a feature of Phase 2, *ekran boyut* is returned as feature in Phase 3.

Table 3. WSBFE example

Noun Noun	Noun (English)	Feature	Search query (Telefonun ...)	Search results	WSBFE Phase 2 (T = 1000)	WSBFE Phase 3
arkadaş	friend	No	... arkadaşı	3	NF	NF
ekran	screen	Yes	... ekranı	28890	F	F
teşekkür	thank	No	... teşekkürü	0	NF	NF
ihtimal	possibility	No	... ihtimali	0	NF	NF
ekran ışık	screen light	Yes	... ekran ışığı	354	NF	F
ekran boyut	screen size	Yes	... ekran boyutu	1581	F	F
işlemci	processor	Yes	... işlemcisi	3952	F	F
ağaç	tree	No	... ağacı	N/A	N/A	N/A

Table 4. Dataset statistics

Total number of topics	3458
Total number of entries	179608
Total number of author	9305
Average entry per topic	51.94
Average entry per author	19.30
Total size of crawled data	93.80 MB

5   Experimental Analysis

5.1   Dataset

In order to evaluate the performance of the proposed method, we have created a data set of user reviews and comments for a selected mobile phone model from Donanimhaber[5], which is well-known and one of the biggest technology forums in Turkish.

In this forum, firstly, all pages under selected *mobile phone* main topic is crawled. Secondly, all the entries under these pages are extracted and added to our dataset. Table 4 shows the detailed statistics of our dataset.

We have compared the results of the proposed method with two methods, one baseline and another method presented in [8]. We have used precision, recall and f-score metrics for evaluation.

5.2   Experimental Analysis Results

Frequency Based Feature Extraction (FBFE). As seen in Table 5, for this baseline method, which is based on considering frequent nouns in the text as features, although very high recall is achieved, the precision is very low which

---

[5] http://www.donanimhaber.com.

Table 5. Experiment results

Method	Precision	Recall	F-score
FBFE	4.41	99.00	8.44
FBFE-SWS	4.55	99.00	8.70
WSBFE	59.24	84.90	69.79

causes very low f-score. This is an expected result since nouns are expected to cover almost all of the features, however they include a high number of false positives.

On the other hand, in [8], the same method is used as a baseline and it is reported that 56 % precision and 68 % recall rates are obtained. There are several reasons for this performance difference. The first one is due to the content of the data set. Our data set includes high number of sentences that do not refer to any feature. As the second reason, the language used in the data set is informal and hence morphological analysis fails to label some of the slang use of the words and word groups correctly. Moreover, since the approach is originally developed for English texts, linguistic features of Turkish may also affect the performance.

In addition, the points where FBFE is not sufficient can be seen in the example given in Table 2. Since the dataset used contains a lot of frequent nouns which are not features, there are a lot mislabelled samples in the example. For example, *ihtimal* is not a feature of the item but it is a highly used word in Turkish, hence, it is frequent and selected as feature by FBFE. On the other hand, since a low threshold value is selected, all features are extracted. However, even this high recall value does not lead to high f-score.

Frequency Based Feature Extraction with Sentiment Word Support (FBFE-SWS). In order to apply this method, which is presented in [8], we needed a sentiment corpus in Turkish. To this aim, we used the Turkish sentiment words collection of [21]. In their study, a subset of English sentiment lexicon of SentiStrength[6] is translated to Turkish.

As seen in Table 5, in this method, very low precision and very high recall results are obtained as in FBFE method, while much better results are obtained in [8] with 59 % precision and 80 % recall. The reasons for these results is the informal language used in our dataset and linguistic features of Turkish language, as in the results of FBFE.

As can be seen in Table 2, similar results are obtained in FBFE-SWS. Since all nouns can have neighbor sentiment words even they are not a feature of item, low precision value is achieved. Again, much higher recall value is obtained but low f-score caused this method fail.

---

[6] http://sentistrength.wlv.ac.uk.

Table 6. Effect of threshold on accuracy

Threshold	Precision	Recall	F-score
10000	90.05	22.78	36.36
5000	84.13	30.20	44.44
2000	71.80	43.84	54.44
1000	67.13	57.88	62.16
500	62.99	74.83	68.40
300	59.24	84.90	69.79
250	57.27	86.62	68.95
100	49.51	92.98	64.61

Web Search Based Feature Extraction (WSBFE). Experimental results show that (Table 5) the precision ratio is considerably improved through the proposed results. Hence, we can deduce that incorporating Web search results for the candidate terms can eliminate, especially slang use of nouns and noun groups that are not related with the item and hence are not features, effectively. On the other hand, this technique fails to recognize some of features that are specific to the item and that are not used frequently as search terms. Therefore, we observe some decrease in recall rate. However, on the overall, there is a huge increase in f-score.

As given in the experiment results, Table 3 shows that WSBFE is much more effective than other two methods. In the results of example given Table 3, Phase 2 of WSBFE gives much better results than FBFE and FBFE-SWS. However, it misses a feature which is a non-frequent noun phrase. After Phase 3 is applied, it is seen that this missing infrequent feature is extracted without any loss on Phase 2' s results. Hence, both high precision and recall values are obtained.

Table 6 shows the results of WSBFE algorithm under varying threshold values. In this table, precision, recall and F-score values are calculated for only intra-group of noun and noun phrases whose occurrence count on the Web is equal to or more than 1. Increasing threshold value results with higher precision and lower recall values. The best f-score value is achieved when threshold value is set to 300 and this value is used for the result reported in Table 5.

6   Conclusion

Sentiment analysis is one of the most basic research topics and it is an active research area due to its impact on both commercial and academical efforts. Aspect extraction is the first step of aspect-based sentiment analysis and it is important to match correct sentiment words with correct features.

In this article, we build a framework to extract features of a target item from informal text by using an unsupervised learning based solution. We adapt basic ideas from the literature and propose a new approach that is based on constructing a search term by using the candidate features of the item and using the search count for determining the features.

For the experiments, we set the target item as a mobile phone model and constructed a data set by collecting postings on a Turkish forum containing entries written about the item. The experimental analysis results show that the proposed approach improves the feature extraction performance in comparison to two basic approaches from the literature.

Although we developed and tested our methods for Turkish, since we used the linguistic property that Turkish is an agglutinative language while creating Web search queries, our approach may fit other agglutinative languages well. Moreover, no words are used between product word and feature word in Turkish, therefore, our approach can also easily be adopted to other languages having the same property.

We plan to extend the proposed approach in several directions. One research direction is to build a hybrid model that incorporates both the term frequencies in the data set and the search query result counts. Another point to work on is transferring the feature extraction knowledge among semantically similar target items in order to facilitate the process.

Acknowledgements. This work is supported by Ministry of Science, Technology and Industry with funding Project No. 0740.STZ.2014.

References

1. Akın, A.A., Akın, M.D.: Zemberek, an open source NLP framework for Turkic languages. Structure **10**, 1–5 (2007)
2. Allen, K.G., Reynolds, T.: Thanksgiving weekend shopping brings big in-store and online crowds, according to NRF survey (2015). https://nrf.com/media/press-releases/thanksgiving-weekend-shopping-brings-big-store-and-online-crowds-acco rding-nrf/. Accessed 29 November 2015
3. Aronoff, M., Fudeman, K.: Thinking about morphology and morphological analysis. In: What is Morphology?, pp. 1–31. Blackwell Publishing (2004)
4. Blair-Goldensohn, S., Hannan, K., McDonald, R., Neylon, T., Reis, G.A., Reynar, J.: Building a sentiment summarizer for local service reviews. In: WWW Workshop on NLP in the Information Explosion Era, vol. 14, pp. 339–348 (2008)
5. Brody, S., Elhadad, N.: An unsupervised aspect-sentiment model for online reviews. In: Human Language Technologies: The 2010 Annual Conference of the North American Chapter of the Association for Computational Linguistics, pp. 804–812. Association for Computational Linguistics (2010)
6. Church, K.W., Hanks, P.: Word association norms, mutual information, and lexicography. Comput. Linguist. **16**(1), 22–29 (1990)
7. Hai, Z., Chang, K., Kim, J.: Implicit feature identification via co-occurrence association rule mining. In: Gelbukh, A.F. (ed.) CICLing 2011, Part I. LNCS, vol. 6608, pp. 393–404. Springer, Heidelberg (2011)
8. Hu, M., Liu, B.: Mining and summarizing customer reviews. In: Proceedings of the Tenth ACM SIGKDD International Conference on Knowledge Discovery and Data Mining, pp. 168–177. ACM (2004)

9. Jakob, N., Gurevych, I.: Extracting opinion targets in a single-and cross-domain setting with conditional random fields. In: Proceedings of the 2010 Conference on Empirical Methods in Natural Language Processing, pp. 1035–1045. Association for Computational Linguistics (2010)

10. Jin, W., Ho, H.H., Srihari, R.K.: A novel lexicalized hmm-based learning framework for web opinion mining. In: Proceedings of the 26th Annual International Conference on Machine Learning, pp. 465–472. Citeseer (2009)

11. Li, F., Han, C., Huang, M., Zhu, X., Xia, Y.-J., Zhang, S., Yu, H.: Structure-aware review mining and summarization. In: Proceedings of the 23rd International Conference on Computational Linguistics, pp. 653–661. Association for Computational Linguistics (2010)

12. Lin, C., He, Y.: Joint sentiment/topic model for sentiment analysis. In: Proceedings of the 18th ACM Conference on Information and Knowledge Management, pp. 375–384. ACM (2009)

13. Liu, B.: Sentiment analysis and opinion mining. Synth. Lect. Hum. Lang. Technol. 5(1), 1–167 (2012)

14. Mei, Q., Ling, X., Wondra, M., Su, H., Zhai, C.: Topic sentiment mixture: modeling facets and opinions in weblogs. In: Proceedings of the 16th International Conference on World Wide Web, pp. 171–180. ACM (2007)

15. Pang, B., Lee, L., Vaithyanathan, S.: Thumbs up?: sentiment classification using machine learning techniques. In: Proceedings of the ACL 2002 Conference on Empirical Methods in Natural Language Processing, vol. 10, pp. 79–86. Association for Computational Linguistics (2002)

16. Popescu, A.-M., Etzioni, O.: Extracting product features and opinions from reviews. In: Kao, A., Poteet, S.R. (eds.) Natural Language Processing and Text Mining, pp. 9–28. Springer, London (2007)

17. Schouten, K., Frasincar, F.: Finding implicit features in consumer reviews for sentiment analysis. In: Casteleyn, S., Rossi, G., Winckler, M. (eds.) ICWE 2014. LNCS, vol. 8541, pp. 130–144. Springer, Heidelberg (2014)

18. Somasundaran, S., Wiebe, J.: Recognizing stances in online debates. In: Proceedings of the Joint Conference of the 47th Annual Meeting of the ACL and the 4th International Joint Conference on Natural Language Processing of the AFNLP, vol. 1, pp. 226–234. Association for Computational Linguistics (2009)

19. Su, Q., Xu, X., Guo, H., Guo, Z., Wu, X., Zhang, X., Swen, B., Su, Z.: Hidden sentiment association in chinese web opinion mining. In: Proceedings of the 17th International Conference on World Wide Web, pp. 959–968. ACM (2008)

20. Turney, P.D.: Thumbs up or thumbs down?: semantic orientation applied to unsupervised classification of reviews. In: Proceedings of the 40th Annual Meeting on Association for Computational Linguistics, pp. 417–424. Association for Computational Linguistics (2002)

21. Vural, A.G., Cambazoglu, B.B., Senkul, P., Tokgoz, Z.O.: A framework for sentiment analysis in Turkish: application to polarity detection of movie reviews in Turkish. In: Gelenbe, E., Lent, R. (eds.) Computer and Information Sciences III, pp. 437–445. Springer, London (2013)

22. Yu, J., Zha, Z.-J., Wang, M., Chua, T.-S.: Aspect ranking: identifying important product aspects from online consumer reviews. In: Proceedings of the 49th Annual Meeting of the Association for Computational Linguistics: Human Language Technologies, vol. 1, pp. 1496–1505. Association for Computational Linguistics (2011)

Keyboard Usage Authentication Using Time Series Analysis

Abdullah Alshehri$^{(\boxtimes)}$, Frans Coenen, and Danushka Bollegala

Department of Computer Science, The University of Liverpool,
Liverpool L69 3BX, UK
{a.a.alshehri,coenen,danushka.bollegala}@liverpool.ac.uk

Abstract. In this paper, we introduce a new approach to recognising typing behaviour (biometrics) from an arbitrary text in heterogeneous environments using the context of time series analytics. Our proposed method differs from previous work directed at understanding typing behaviour, which was founded on the idea of usage a feature vector representation to construct user profiles. We represent keystroke features as sequencing discrete points of events that allow dynamically detection of suspicious behaviour over the temporal domain. The significance of the approach is in the context of typing authentication within open session environments, for example, identifying users in online assessments and examinations used in eLearning environments and MOOCs, which are becoming increasingly popular. The proposed representation outperforms the established feature vector approaches with a recorded accuracy of 98 %, compared to 83 %; a significant result that clearly indicates the advantage offered by the proposed time series representation.

Keywords: Keystroke recognition · Keystroke time series · Typing patterns

1 Introduction

Biometrics are acknowledged to provide a robust method for authenticating users based on their personal traits, as opposed to *token-based* mechanisms (such as passwords). Personal traits can be classified as being either behavioural or physiological [14]. The usage of behavioural biometrics has received prominent attention in the context of user authentication because they offer the advantage that they do not require specialised equipment [20]. Unlike physiological biometrics (for example fingerprints or iris data) that do require such equipment. One form of behavioural biometric is keystroke patterns; the typing patterns produced when an individual uses a keyboard. Keystroke patterns are a promised behavioural biometric that can provide unobtrusive authentication to confirm the legitimate users. A frequently cited example of the use of keystroke patterns for user authentication purposes is to confirm user credentials (such as password and username). In this case, authentication process is conducted by

© Springer International Publishing Switzerland 2016
S. Madria and T. Hara (Eds.): DaWaK 2016, LNCS 9829, pp. 239–252, 2016.
DOI: 10.1007/978-3-319-43946-4_16

comparing timing features of successive keystrokes with a stored typing profile so as to authenticate the person inputting the credentials [3,9,11,13,18]. To date keystroke patterns are typically represented as feature vectors comprised of quantitative statistical values, for instance, the calculated average flight time (interval) of frequent consecutive pairs of *graphs* (keypress sequences), usually *bi*-graphs [9]. A comparison, between a learnt user profile and previously unseen profiles, is then performed using a variety of paradigms, such as classification, AI based or Neural Network, to see whether two corresponding profiles are matched. However, the process is becoming harder in the case of dealing with arbitrary (free) text where constructing the feature vector is becoming stochastic. The reason is that the typed text is expected to be different each time; and therefore, the sequencing of key presses is largely lost. There is also a great deal of variability in the statistical features used to construct the feature vectors. Consequently, the reported results to date have tended not to be as good as anticipated to apply in heterogeneous environments [1,4,8,10,12,16]. The conjecture of studying typing patterns based on free text is that keystroke can be applied to continuous surveillance in heterogeneous environments where typing patterns are extracted from the arbitrary text. For example, it is sensible to be employed for continuous authentication in online assessments and examinations frequently used in eLearning environments and MOOCs[1], which is becoming increasingly popular.

The idea of this work is directed to deal with keystroke feature representation in the context of time series paradigm rather than using feature vectors based classification approach. The intuition is that time series representation can be more readily used to identify dynamically suspicious behaviours from free text. Furthermore, time series avails to capture keystroke sequences, unlike in the case of statistical techniques. We have considered that a typing session is represented as a series of discrete points P_M expressed in the temporal domain, where M is the number of points in a keystroke time series. Each point P is defined as pairs $P = (t, k)$, where t is a time stamp or time identifier; and k is depended multi-dimensional keystroke features. The diversity of keystroke timing features allow us to implement the proposed representation in two ways: (i) one in the 2D space as Flight time F^t (interval time between consecutive keystrokes) is recorded along the y-axis, where indexing keystrokes KN is along the x-axis; and (ii) representing features in 3D space (x,y,z) where we use F^t along the y-axis, Hold time HD^t (the length time of pressing a key) over the z dimension, and indexing keystrokes KN along the x-axis ticks. The purpose of implementing the two methods of representation (2D and 3D) is that to evaluate the effectiveness of using multivariate features for keystroke time series, and to compare which one can result in a more understanding of typing patterns in time series paradigm.

The main contribution of this work is that to introduce a different representation of keystroke timing features in the context of time series analysis to extract

---

[1] Massive Open Online Course (MOOC): is a web-based teaching distance that allows users to participating a variety of learning resources including filmed lectures, board discussion, etc. It is widely becoming used in the academic teaching process. See https://www.mooc-list.com/.

meaningful patterns in heterogeneous environments. Thus, keystroke biometrics can be eligible to use in different disciplines not only for authentication purposes, such as psychological detection [2], intrusion detection [17] or deceptive writing recognition [5].

The remainder of this paper is organised as follows. In Sect. 2, related work of keystroke feature representation methods is reviewed. Section 3 introduces a description of keystroke time series representation. Similarity method of keystroke time series is then discussed in Sect. 4. The evaluation and comparison of the proposed approach are reported on in Sect. 5. Finally, the work is summarised and concluded in Sect. 6.

2   Previous Work

There is a little work in the literature that has investigated the use of typing patterns generated from free text for user authentication purposes [1]. Most studies, as noted in the introduction to this paper, have adopted a feature vector representation where the features are computed statistical measurements. For example Dowland and Furnell [7] used digraph latency for the feature vector representation from which a binary classifier was generated. The classifier operated using the mean and standard deviation of digraph occurrences in a training profile. Principal disadvantages of the approach were that to achieve a reasonable classification performance a substantial amount of data was required with which to train the classifier. Furthermore, a dedicated classifier was necessary for each individual. The approach would thus be difficult to apply in heterogeneous environments such as eLearning platforms and MOOCs. An alternative approach was presented in Gunetti and Picardi [10] where the average time for pressing frequent sequences (n-graphs) was recorded and stored in arrays, one per n-graph. Common n-graphs were extracted for corresponding samples (reference and test). The elements of the arrays were then ordered and the distance between sample pairs computed by comparing the ordering in the reference array with the order in the test array. This measure was referred to as "the degree of disorder". However, learning a reference feature sample depends on all other samples in the reference profile. This can cause an efficiency issue when dealing with large numbers of samples as would be expected with respect to heterogeneous environments. Ahmed *et al.* [1] used key-down time information and the average digraph flight time to represent feature vectors to be employed in the context of a classifier. Although they obtained good results in heterogeneous environments, the issue is that the scalability of results is largely influenced by changing the environments conditions, such as using different keyboard layout. Indeed, the demand for developing such generic mechanism that able to recognise typing patterns in heterogeneous environments is desirable. Thus, the monitoring of keystroke sequencing over the temporal domain is argued for a better understanding of the arbitrary text, unlike constructing vectors to interpret the extracted features. The concept of using time series analysis, to the best knowledge of the authors, has not been considered in the previous work on keystroke free text detection.

3   Keystroke Time Series

Time series is a sequential ordering of data points that occur within an interval time [19], as each point corresponds multiple values. We first start with providing basic definitions in regards to keystroke time series:

Definition 1. *A Keystroke Time Series K_{ts}: is an ordered discrete sequence of points P; $K_{ts} = [P_1, P_2, \ldots, P_i, \ldots, P_M]$ where $M \in \mathbb{N}$ is the length of series and P_i is a tuple corresponding pairs of dimensional features.*

Thus, different keystroke time series may have different lengths M that describing an independent typing task in the session.

Definition 2. *A point tuple P_i in K_{ts}: is dependent dimensional features consists of two instances $\langle t, k \rangle$ where: (i) t is the indexing sequence of time stamp (KN) in which keys are pressed; and (ii) k is a set of timing attributes and descriptive features including: flight time (F^t), key-hold (KH^t) and key code (K_{cod}). So each p_i can be formally written as $p_i = \langle t_i, k_i \rangle$ where:*

- $\forall p_i \in K_{ts} : p_i \leftarrow \langle t_i, k_i \rangle$
- $\forall t_i \wedge k_i \in p_i : t_i = KN; k_i = \{F_i^t, KH_i^t, K_{cod_i}\}$

Definition 3. *Keystroke time series subset S: is a set of keystroke time series with length L, generated from K_{ts}, $S = [p_1, p_2, \ldots, p_i, \ldots, p_L]$ where $L < M$.*

Based on the above definitions, we can exploit the dimensionality of keystroke features to visualise time series in different spaces (2D and 3D). This is expected to give a better explanation of the typing rhythm of free text when adopting multi-features in the temporal domain to discriminate unique patterns.

3.1   2D Keystroke Time Series Representation

Keystroke temporal events have been represented as 2D series using two features: F^t and HD^t, respectively. The indexing sequence KN has been used along the x-axis, where F^t value or HD^t are used along the y-axis. Thus, a tuple p_i is underlying the sequential ID number KN per keystroke for t_i; and F^t or HD^t as for k_i, so a keystroke sequence can be simultaneously represented as $K_{ts} = \{\langle KN_1, F_1^t \rangle, \langle KN_2, F_2^t \rangle, \ldots\}$, $K_{ts} = \{\langle KN_1, HD_1^t \rangle, \langle KN_2, HD_2^t \rangle, \ldots\}$.

Recall that the value of F^t has to meet a pre-defined threshold value θ, to ensure the fluency of sequence; otherwise, some long stops over the typing session may affect similarity measurement (as we describe later on Sect. 4). We have considered the value of 3000(ms) as a normal variation, $\theta = 3000$. In Algorithm 1, if the value of F^t is greater than the threshold θ (line 7), then reduce F^t to Zero. So, every point with Zero value is considered as a reasonable stopping of typing. Figure 1 shows F^t values in one independent task that has been taken from our dataset. It can be observed in Fig. 1(a) that outlier values of F^t can describe a spurious behaviour where Fig. 1(b) depicts keystroke series after minimising F^t values.

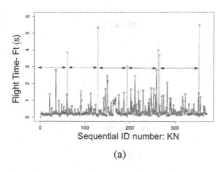
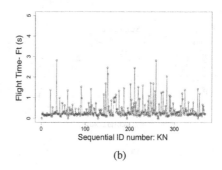

(a)   (b)

Fig. 1. Keystroke time series before removing outlier values of F^t (**a**); and keytroke time series after removing outlier values of F^t (**b**).

Thus, each typing session is represented as 2D time series that can discriminate a distinct pattern of typing. Figure 2(a) and (b) give two keystroke time series, taken from our evaluation dataset (see Sect. 5), for Subject 2 writing two different texts. It can be observed that this subject has a steady rhythm fluctuating between 0.1 and 1 ms. In contrast, Fig. 2(c) and (d) show two time series, for Subject 9, have a range between 0.1 and 1.5 with some peaks that favourably can introduce a similarity typing pattern for the same subject. From the figure, it can be seen that there are apparent dissimilarities in the keystroke pattern between the different subjects (despite writing different texts).

Algorithm 1. Removing Outlier Values of Flight Time (F^t)

Require: $K_{ts} \leftarrow$ keystroke time series, $\theta \leftarrow$ threshold outlier value.
Ensure: $\widehat{K_{ts}} \leftarrow$ Reduce outlier values in K_{ts}.
 1: $K_{ts} = (p_1, p_2, \ldots, p_i, \ldots, p_L)$
 2: $p_i \leftarrow \langle t_i, k_i \rangle$
 3: $L \leftarrow$ length of K_{ts}
 4: **for** $i = 1$ to L **do**
 5:     **for** each p_i in K_{ts} **do**
 6:        $p_i \leftarrow \langle F_i^t \rangle$ ▷ Search only for the value of F^t
 7:        **if** $p_i > \theta$: **then**
 8:           $p_i == 0$
 9:           $\widehat{K_{ts}} =$ Update (K_{ts})
10:        **end if**
11:     **end for**
12: **end for**
13: **Return** $\widehat{K_{ts}}$

3.2   3D Keystroke Time Series Representation

Further dependent features have been employed to conceptualise keystroke time series in 3D representation. Representing keystroke time series in the 2D space,

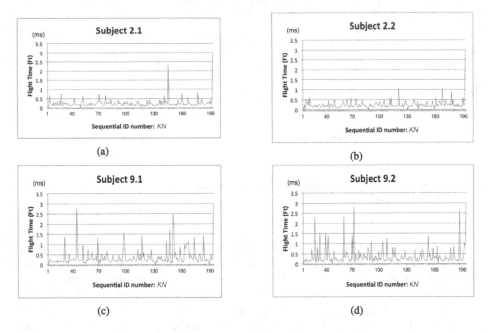

Fig. 2. Examples of keystroke time series representation: **(a)** and **(b)**, time series for Subject 2 writing two different texts; **(c)** and **(d)**, time series for Subject 9 writing two different texts.

in some cases, may affect the discrimination of patterns. For example, the time sequence may appear somewhat similar over some ticks in the series where it can influence the accuracy when calculating the similarity between two-time series. The conjuncture is then that 3D practically can show a preference to calculate the weight of series in three dimensions x, y, z rather than noticing data sequence in 2D space. To this end, we incorporated hold-time KH^t feature along the z-axis in the coordinate space. So, each point in the sequence p_i has: (i) the sequencing numbering KN over the x-axis, (ii) F^t over the y-axis, and (iii) HD^t over the z-axis. Thus, the tuple consists 3 dimensional features as $\forall p_i \in K_{ts} : p_i = \langle [t_i : KN], [k_i : F^t, KH^t] \rangle$.

4   Measuring the Similarity of Keystroke Time Series

Having represented keystroke time series, the similarity can be computed between the current series with one or more series. Given two keystroke time series $K_{ts1} = \{p_1, p_2, \ldots p_i, \ldots, p_M\}$ and $K_{kt2} = \{q_1, q_2, \ldots q_j, \ldots, q_N\}$, where M and N is the length of the two series, the simplest way to define similarity s is by directly computing the Euclidean distance between each points. However, this requires both time series to be of the same length $M = N$, where this is not necessarily the case at all time. The similarity should be performed between sequences that have varied lengths (when $M \neq N$). To this end, Dynamic Time

Warping (DTW) is the best choice that allows for non-linearity matching of two-time series with different lengths [15].

4.1 DTW Similarity in 2D Space

Lets assume that we have the above keystroke time series K_{ts1} and K_{ts2}, and also assumed that $M \neq N$. The two corresponding time series are constructed in a matrix X with $N \times M$, $X = (c_1, c_2, \ldots, c_n, \ldots, c_L)$, $N \leqslant L \leqslant M$. The elements of X are then computed by the squared distance D of F^t between the two corresponding points p_i and q_j.

$$c_{ij} \leftarrow D(c_n) = \sqrt{(F_j^t - F_i^t)^2} \tag{1}$$

The lowest cumulative distance ξ in each cell is the founded as in the following equation:

$$\xi(c_{ij}) = D + P_{min(\varepsilon(i,j-1), \varepsilon(i-1,j-1), \varepsilon(i-1,j))} \tag{2}$$

where D is the current distance of the i-th and j-th points in the cell c_{ij} and P is the lowest value obtained form: (i) the vertical cell $(i, j-1)$, (ii) diagonal cell $(i-1, j-1)$, and (iii) the horizontal cell $(i-1, j)$. The idea is then to find the lowest cost of the path (*Warping Distance* WD), where it is describing continuous cells in the matrix that mapping the alignment between K_{ts1} and K_{ts2}. The lower WD concerning the two time series being compared the similar the two time series are; if $WD = 0$, the two time series are identical.

$$WD = \min[\sum_{n=1}^{n} \xi(c_{ij})_n] \tag{3}$$

Recall that at the same manner, the value of WD is calculated when applying independently HD^t as the feature of interest in time series representation. Therefore, we compare the obtained values of WD for the both applied features, respectively, as described in Sect. 5.

4.2 DTW Similarity in 3D Space

With respect to 3D representation, we slightly modified the concept presented for 2D to perform DTW similarity. As described in Sect. 3.2, two features of interested, F^t and HD^t have represented two depended dimensions. This can affect measuring the distance for each cell in the matrix X. To avoid some computational conflicts, we simply find a weighted value w^* for each point p_i in the 3D space rather than separately computing the distance for each instances. The value of w^* is founded by computing the percentage between flight time F^t and hold time HD^t as in the following equation:

$$w_{p_i}^* = \sqrt{\log\left(\frac{F_i^t}{HD_i^t}\right)^2} \tag{4}$$

Therefore, the elements of matrix X is filled by calculating the distance between each corresponding $w_{q_j}^*$ and $w_{p_i}^*$ in the series.

$$c_{ij} \leftarrow D(c_n) = \sqrt{(w_{q_j}^* - w_{p_i}^*)^2} \tag{5}$$

By constructing the matrix X, the WD is then computed at the same method described for 2D representation.

5   Evaluation

For the evaluation propose, we have examined the proposed method to detect typing patterns by simulating the operation of on-line assessments where students were asked to respond to discussion questions. A number of experiments have been conducted by applying the proposed representation in 2D and 3D; and then to compare the operation with statistical feature vector approach similar to that used in earlier work on free text typing recognition.

5.1   Data Collection

Keystroke timing data was collected (in milliseconds) using a Web-Based Keystroke Timestamp Recorder (WBKTR) developed by the authors. WBKTR was developed in JavaScript whereas it can work on cross-platforms web browsers. There is no need to install a third party or plug-ins, so it works smoothly without annoying users with further obligations. Another advantage of using JavaScript is that to avoid any implications for network delay when passing data to the server, which can affect the accuracy of recorded time. Thus, the script function works at the end user station to record time stamp within the current accuracy of the computer clock. Ideally, this can give a reliable accuracy of the recorded time. A front-end page in HTML was showing three discussion questions, similarly to the board discussion applied in eLearning environments. The interface can be found at (http://cgi.csc.liv.ac.uk/~hsaalshe/WBKTR3.html). A total of 17 subjects at the graduate level, ages between 20–35 were asked to response questions (for simplicity we used the term subject to refer each participant). The identity of the respondents was anonymised for privacy concerns. They were asked to type at least 100 words in response to each question with no maximum limitation so that adequate numbers of keystrokes (not less than 100 keystrokes per question) could be collected. For convenience, a scripting function was used to count the number of words per question. Samples with a number of keystrokes less than 100 (per question) were discarded. The reason is that 100 keystrokes can sufficiently provide a meaningful typing pattern [10]. Figure 3 illustrates that a total number of keystrokes more than 100 gives a steady accuracy of pattern detection. During the session, flight time F^t and hold time HD^t are recorded per each keystroke. A PHP script was used to store the identified attributes in the form of a plain text file on a server side for each subject.

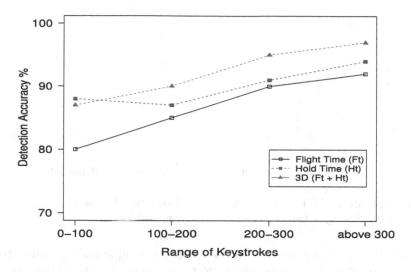

Fig. 3. The recorded accuracy with a range of keystroke number between 100 and 300 for different features in time series (Ft, Ht, 3D). The indication is that the detection accuracy is steadily performed while the number of keystrokes is above 100.

5.2   Analysis

The total number of samples (keystroke time series) N that we obtained is 17. Each sample S_i was splitted into three equal lengths of subset (keystroke subset time series) $\{s_{1i}, s_{2i}, s_{3i}\}$, as one for training and the other two for testing purpose. This results in a total number of $m = 3 \times N$ samples. This division allows to expand the comparison status by grouping samples into three main groups: (i) Group a, as $a = \{s_{11}, s_{12}, \ldots, s_{1i}\}$, (ii) Group b, as $b = \{s_{21}, s_{22}, \ldots, s_{2i}\}$; and (iii) Group c, as $c = \{s_{31}, s_{32}, \ldots, s_{3i}\}$. So, we implemented multiple experiments by swapping groups each time. This gave us also a wider comparisons each time as we simulated different (training and testing) samples for the same subject. The different groups of dataset being compared is then as follows: (i) $a. \vee \{b, c\}$, (ii) $b. \vee \{a, c\}$ and (iii) $c. \vee \{a, b\}$, the symbol $\vee$ is used to denote the versus status.

Figure 4 simplifies the idea of matching different groups. The warping distance WD is then performed as explained in Sect. 4. Figure 5(a) illustrates the WD of two samples s_{1i} and s_{2i} from the same user, whilst Fig. 5(b) shows the WD of two samples from two different users. A distinction can clearly be perceived. Lets perform the comparison in the combination group $a. \vee \{b, c\}$, for each sample s_{1i} the WD was compared with that of all the remaining samples $s_{\{2,3\}_i}$. A similarity threshold σ, for each subject, has been calculated by the average

Fig. 4. Dataset has been divided into three groups a, b, c; and multiple comparisons have been conducted between different combination of groups: $a. \vee \{b, c\}$, $b. \vee \{a, c\}$, $c. \vee \{a, b\}$.

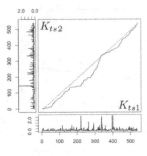
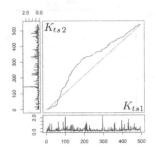

(a) WD of the same subject   (b) WD for two different subjects.

Fig. 5. Application of DTW. It can be observed that the WD is more alignment in (a) as the sequencing of two time series for the same subject.

value of WD for s_{1i}, s_{2i} and s_{1i}, s_{3i}. Whenever a sample s_{1i} was found to be the most similar to σ this was considered to be a *correct* match; otherwise, the match was deemed to be *incorrect*. Thus, each smaple has a rank value r by ordering the WDs for each corresponding sample in the ascending order. The detection accuracy was computed as the ratio between the number of incorrect matches ℓ prior to a correct match being arrived at $(\ell = \sum (r - 1))$ and the total number of test cases τ $(\tau = n \times (m - 1))$. For simplicity, we calculate the accuracy of 3D representation as: $\frac{850-17}{850} \times 100 = 98.10\%$, as 17 is the value of ℓ and 850 is the value of τ (Table 3). As the same manner, we have implemented 2D representation using F^t and HD^t features, respectively. Tables 1, 2 and 3 introduce the accuracy results obtained from each feature with respect to time series representation.

False Rejection Rate (FRR) and False Acceptance Rate (FAR): As same as other biometric applications, we have evaluated our proposed approach by calculating the percentage of False Rejection Rate (FRR); and False Acceptance Rate (FAR). According to the European Standard for access control, the acceptable rate of FRR is 1%, where the rate of FAR is 0.001% [14]. Thus, we used these metrics to measure how far our proposed, as biometric authentication, from this standard.

In each combination in our experiments, we calculated FRR by computing the number of subjects n that their samples' rank r is not equal to 1, $\sum_1^n r \neq 1$. If the equivalent sample's rank are not equal to 1, this means that sample is falsely rejected. In contrast, FAR is calculated by the number of samples that recorded a higher rank than the current equivalent samples where all samples, smaller than the current equivalent sample, supposed to be accepted as real users.

Tables 1, 2 and 3 present the recorded results of FRR and FAR obtained from our proposed representation, comparing with feature vector representation as describe later in Sect. 5.3. We can clearly observe that time series representation recorded the best values for FRR and FAR in all different combinations of experiments.

5.3 Comparison with Feature Vector Approach

To obtain a reasonable evaluation of our proposed approach, we have examined the concept of the statistical feature vector style of operation found in earlier work; and compare the performance of the two methods of representations. This was performed by computing the average flight time $\mu(f^t)$ (Eq. 6) and hold time $\mu(h^t)$ (Eq. 7), respectively, for the most frequently occurring di-graphs found in the dataset.

$$\mu(f^t) = \frac{1}{n} \sum_{i=1}^{i=n} Ft_n \tag{6}$$

$$\mu(h^t) = \frac{1}{n} \sum_{i=1}^{i=n} HD_n^t \tag{7}$$

where n is the number of identified frequent di-graphs. In this manner feature vectors could be generated for each sample. The resulting representation was thus similar to that found in more traditional approaches to free text recognition [1,7,10,16]. Each sample S_i is divided into three vectors v_{1i}, v_{2i}, v_{3i}. As the same scenario in time series representation (Subsect. 5.2), we have measured the similarity between two vectors using Cosine Similarity (CS). Thus, for two vectors v_{1i} and v_{2i}, CS calculated as:

$$CS(v_{1i}, v_{2i}) = \frac{v_{1i} \cdot v_{2i}}{||v_{1i}|| \times ||v_{2i}||} \tag{8}$$

where $v_{1i} \cdot v_{2i}$ is the dot product between two feature vectors v_{1i} and v_{2i}, and $||v_{1i}||$ ($||v_{2i}||$) is the magnitude of the vector v_{1i} (y). In the same manner, as described above, we measured the similarity of each feature vector with every other feature vector using CS. Note that using CS, unlike in the case of DTW, the feature vectors need to be of the same length. In this case, the results for each subject are listed in descending order of CS ($CS = 1$ indicates a perfect match). We also computed FRR and FAR for the feature vector as well. The results obtained are presented in Tables 1, 2, and 3. It can be observed that the accuracy, in all combinations datasets, has recorded fewer values than our proposed method. It can be also noticed from the tables that a worse performance has been recorded than when using time series approach with respect to FRR and FAR.

An alternative evaluation measure that can be used to indicate the effectiveness of the proposed approach to keystroke time series is Mean Reciprocal Rank (MRR) [6]; a measure that indicates how close the position of a desired subject of interest is to the top of a ranked list. MRR is a standard evaluation measure used in Information Retrieval (IR). MRR is calculated as follows:

$$MRR = \frac{1}{|Q|} \cdot \sum_{i=1}^{|Q|} \frac{1}{r_i} \tag{9}$$

where: (i) Q is a set of queries (in our case queries as to whether we have the correct subject or not), and (ii) r_i is the generated rank of the desired response to Q_i.

Table 1. Results obtained by applying F^t as the feature applied for the two methods of representation, *Time Series*, and *Feature Vector*.

Representation Dataset \ Metrics	2D Time Series with F^t				Feature Vector with F^t			
	FRR(%)	FAR(%)	MRR	Acc(%)	FRR(%)	FAR(%)	MRR	Acc(%)
a. ∨ {b, c}	6.11	1.52	0.438	93.88	20.58	1.64	0.283	79.41
b. ∨ {a, c}	**5.17**	1.41	**0.520**	**94.82**	21.64	1.76	0.155	78.35
c. ∨ {a, b}	6.70	**1.17**	0.454	93.29	19.17	1.64	0.225	80.82

Table 2. Results obtained by applying HD^t as the feature applied for the two methods of representation, *Time Series*, and *Feature Vector*.

Representation Dataset \ Metrics	2D Time Series with HD^t				Feature Vector with HD^t			
	FRR(%)	FAR(%)	MRR	Acc(%)	FRR(%)	FAR(%)	MRR	Acc(%)
a. ∨ {b, c}	**2.70**	1.05	0.658	**97.29**	20.35	1.64	0.199	79.64
b. ∨ {a, c}	3.64	0.94	0.666	96.35	16	1.64	0.251	84
c. ∨ {a, b}	3.52	**0.70**	**0.723**	96.47	17.17	1.64	0.242	82.82

Thus, with reference to Table 3, time series representation has recorded the best values of MRR with comparing with feature vector representation. Among different methods of time series representation, 3D representation has outperformed other features with a value of 0.801 while the best MRR value on feature vector is = 0.311 (Table 3).

Table 3. Results obtained by applying F^t and HD^t as features applied for the two methods of representation, (i) *Time Series*, and (ii) *Feature Vector*.

Representation Dataset \ Metrics	3D Time Series				Feature Vector with F^t and HD^t			
	FRR(%)	FAR(%)	MRR	Acc(%)	FRR(%)	FAR(%)	MRR	Acc(%)
a. ∨ {b, c}	2	0.70	0.745	98	15.29	1.52	0.305	84.70
b. ∨ {a, c}	**1.76**	**0.58**	**0.801**	**98.20**	17.80	1.50	0.311	82.11
c. ∨ {a, b}	1.80	0.70	0.772	98.10	17.50	1.60	0.275	82.40

For completeness, the average value has been computed for each feature applied in all combinations for the both methods of representation. Table 4 summarises the average values obtained for time series representation and feature vectors in all metrics. It can be clearly observed that 3D method outperforms other representations, including 2D keystroke time series method. A clear indicator that applying multi-variate time series has promising potential to detect typing patterns from arbitrary text.

Table 4. Summary of the average values obtained for all representations.

Representation / Applied Feature ╲ Metrics	Time Series				Feature Vector			
	FRR(%)	FAR(%)	MRR	Acc(%)	FRR(%)	FAR(%)	MRR	Acc(%)
Average results for F^t	5.99	1.37	0.470	94.00	20.46	1.68	0.225	79.53
Average results for HD^t	3.29	0.90	0.480	96.70	17.84	1.64	0.230	82.09
Average results for $3D$	**1.85**	**0.66**	**0.770**	**98.10**	16.86	1.54	0.210	83.07

6   Conclusion

An approach to recognise typing patterns in heterogeneous environments, that deal with arbitrary text of typing, has been proposed. The process operates by representing keystroke timing features as discrete points in a time series where each point has a timestamp of some kind and attribute value. The proposed representation used a sequential key-press numbering system in 2D, by applying flight time and hold time, respectively; and using both features in the 3D time series representation. DTW has been adopted to measure the similarity of keystroke time series so that practically works with non-linearity time series. By implementing the proposed approach to detect typing patterns in a simulated onLine environment, recorded results show that proposed feature representation obtained an overall accuracy of 98.10 % (coped with FRR = 1.85 %, and FAR = 0.66 %). This compared very favourably with the alternative approach using feature vector with an accuracy of 83.07 % (FRR = 16.86 % and FAR = 1.54 %) when applying classical features vector representation; a clear indication that the proposed time series based approach outperforms the vector based approach. The result demonstrated that the proposed time series based approach to keystroke authentication has a significant potential benefit in the context of user authentication in heterogeneous environments such as those used in online learning and MOOCS. The authors believe that further improvement can be realised by considering different methods of representation, such as Fast Fourier Transform (FFT). Future work will also be directed at confirming the findings using larger datasets.

Acknowledgment. We would like to express our thanks to those who participated in collecting the data and to Laureate Online Education b.v. for their support.

References

1. Ahmed, A.A., Traore, I.: Biometric recognition based on free-text keystroke dynamics. IEEE Trans. Cybern. **44**(4), 458–472 (2014)
2. Bixler, R., D'Mello, S.: Detecting boredom and engagement during writing with keystroke analysis, task appraisals, and stable traits. In: Proceedings of the International Conference on Intelligent User Interfaces, pp. 225–234. ACM (2013)
3. Bleha, S., Slivinsky, C., Hussien, B.: Computer-access security systems using keystroke dynamics. IEEE Trans. Pattern Anal. Mach. Intell. **12**(12), 1217–1222 (1990)

4. Bours, P.: Continuous keystroke dynamics: a different perspective towards biometric evaluation. Inf. Secur. Tech. Rep. **17**(1), 36–43 (2012)
5. Choi, Y.: Keystroke patterns as prosody in digital writings: a case study with deceptive reviews and essays. In: Empirical Methods on Natural Language Processing (EMNLP) (2014)
6. Craswell, N.: Mean reciprocal rank. In: Liu, L., Özsu, M.T. (eds.) Encyclopedia of Database Systems, p. 1703. Springer, US (2009)
7. Dowland, P.S., Furnell, S.M.: A long-term trial of keystroke profiling using digraph, trigraph and keyword latencies. In: Deswarte, Y., Cuppens, F., Jajodia, S., Wang, L. (eds.) Security and Protection in Information Processing Systems. IFIP, vol. 147, pp. 275–289. Springer, US (2004)
8. de Lima e Silva Filho, S.R., Roisenberg, M.: Continuous authentication by keystroke dynamics using committee machines. In: Mehrotra, S., Zeng, D.D., Chen, H., Thuraisingham, B., Wang, F.-Y. (eds.) ISI 2006. LNCS, vol. 3975, pp. 686–687. Springer, Heidelberg (2006)
9. Stockton Gaines, R., William Lisowski, S., Press, J., Shapiro, N.: Authentication by keystroke timing: Some preliminary results. Technical report, DTIC Document (1980)
10. Gunetti, D., Picardi, C.: Keystroke analysis of free text. ACM Trans. Inf. Syst. Secur. (TISSEC) **8**(3), 312–347 (2005)
11. Joyce, R., Gupta, G.: Identity authentication based on keystroke latencies. Commun. ACM **33**(2), 168–176 (1990)
12. Messerman, A., Mustafic, T., Camtepe, S.A., Albayrak, S.: Continuous and non-intrusive identity verification in real-time environments based on free-text keystroke dynamics. In: International Joint Conference on Biometrics (IJCB), pp. 1–8. IEEE (2011)
13. Ogihara, A., Matsumuar, H., Shiozaki, A.: Biometric verification using keystroke motion and key press timing for atm user authentication. In: International Symposium on Intelligent Signal Processing and Communications, ISPACS 2006, pp. 223–226. IEEE (2006)
14. Polemi, D.: Biometric techniques: review and evaluation of biometric techniques for identification and authentication, including an appraisal of the areas where they are most applicable. Reported prepared for the European Commision DG XIIIC 4 (1997)
15. Rabiner, L., Juang, B.-H.: Fundamentals of speech recognition. Prentice Hall (1993)
16. Shepherd, S.J.: Continuous authentication by analysis of keyboard typing characteristics. In: European Convention on Security and Detection, pp. 111–114. IET (1995)
17. Sridhar, M., Abraham, T., Rebello, J., D'souza, W., D'Souza, A.: Intrusion detection using keystroke dynamics. In: Das, V.V. (ed.) Proceedings of the Third International Conference on Trends in Information, Telecommunication and Computing. LNEE, vol. 150, pp. 137–144. Springer, Heidelberg (2013)
18. Syed, Z., Banerjee, S., Cukic, B.: Normalizing variations in feature vector structure in keystroke dynamics authentication systems. Softw. Qual. J. **24**(1), 137–157 (2014)
19. Wang, X., Mueen, A., Ding, H., Trajcevski, G., Scheuermann, P., Keogh, E.: Experimental comparison of representation methods and distance measures for time series data. Data Min. Knowl. Discov. **26**(2), 275–309 (2013)
20. Yampolskiy, R.V., Govindaraju, V.: Behavioural biometrics: a survey and classification. Int. J. Biom. **1**(1), 81–113 (2008)

Big Data Learning and Security

A G-Means Update Ensemble Learning Approach for the Imbalanced Data Stream with Concept Drifts

Sin-Kai Wang[✉] and Bi-Ru Dai

Department of Computer Science and Information Engineering, National Taiwan University of Science and Technology, Taipei, Taiwan, ROC
M10315076@mail.ntust.edu.tw, brdai@csie.ntust.edu.tw

Abstract. Concept drift has become an important issue while analyzing data streams. Further, data streams can also have skewed class distributions, known as class imbalance. Actually, in the real world, it is likely that a data stream simultaneously has multiple concept drifts and an imbalanced class distribution. However, since most research approaches do not consider class imbalance and the concept drift problem at the same time, they probably have a good performance on the overall average accuracy, while the accuracy of the minority class is very poor. To deal with these challenges, this paper proposes a new weighting method which can further improve the accuracy of the minority class on the imbalanced data streams with concept drifts. The experimental results confirm that our method not only achieves an impressive performance on the average accuracy but also improves the accuracy of the minority class on the imbalanced data streams.

Keywords: Data stream mining · Concept drift · Ensemble classifier · Imbalance class problem

1 Introduction

A data stream is an ordered sequence of instances that arrive at a certain rate. It is a fast, continuous, real time and unlimited stream of data. There are many important data collected or generated in the form of streams in real life, such as weather information and financial transactions. The traditional classification methods are not suitable for this type of data because it is impossible to obtain all of the data from the data streams directly since the instances are still arriving continuously when the classification model is training or testing. Moreover, the generation of the data stream is often in a non-stationary environment, so the distributions of data probably change over time. This property is called concept drift [1–3]. For example, in the recommendation systems of social networks, the users usually change their topic of interest over time. The data information from the past tends to become irrelevant—or even harmful—for the current classification model, thereby causing a decrease in prediction accuracy if the classification model cannot adapt to the new concept quickly. Although many approaches to concept drift have been proposed [6–11] in recent years, these methods

© Springer International Publishing Switzerland 2016
S. Madria and T. Hara (Eds.): DaWaK 2016, LNCS 9829, pp. 255–266, 2016.
DOI: 10.1007/978-3-319-43946-4_17

do not take into account the influence of the imbalanced class distribution, and the information of the minority class is possibly ignored. In this way, even though the overall accuracy can be fairly high, the performance of the minority class is very poor.

In recent years, the imbalanced learning problem has also drawn a significant amount of interest from both academia and industry. Most traditional classification algorithms assume or expect a balanced class distribution environment or equal misclassification costs. Therefore, when presented with imbalanced datasets, it is possible that these algorithms will fail to correctly represent the distributive characteristics of the data and provide unfavorable accuracies on the minority class of the data. Several methods dealing with the class imbalance problem have been proposed [4, 5, 12–15]. Some of these methods [4, 5] can effectively solve the problem, but only for small datasets, that is, they are not suitable for the data collected in the form of streams; the other methods [12–15] can further deal with the problem on the data streams, but they do not account for concept drifts occurring in the imbalanced data streams, which diminishes the performance of the whole system.

In real life, a data stream often has the properties of multiple concept drifts and imbalance data distribution at the same time. To deal with the combined challenges of concept drifts and the imbalanced class problem, we propose a new weighting method which can maintain the performance of the majority class and improve the accuracy of the minority class better than the other online classification algorithms. This weight mechanism allows the classification model to adapt to different types of concept drifts quickly and keep the information of the minority class to update the whole system incrementally. The experimental results show that the proposed method obtains great performance on the average accuracy and improves the accuracy of the minority class.

The structure of the paper is as follows: in the next section, the related works are discussed. In Sect. 3, the proposed algorithm is introduced. The experimental results and evaluations of the method are presented in Sect. 4. Section 5 concludes the paper and discusses some the future works.

2 Related Work

In this section, some classification approaches for data streams with concept drifts, including single or multiple concept drifts, will be introduced first. After that, we will discuss some methods dealing with the class imbalance problem.

Generally, there are many types of concept drifts, such as gradual and sudden drifts, and numerous methods have been proposed to learn from data streams containing concept drifts. Nowadays, several methods have been proposed using the technique of ensemble classifiers to solve the concept drift problem with the data stream, but these approaches [6–9] only focus on—or are only good at—dealing with one type of concept drift, such as the gradual concept drift. The Dynamic Weighted Majority (DWM) [6] extends the Weighted Majority Algorithm [7] and designs a weighted ensemble classifier to determine whether concept drifts occur. According to the global performance, it adds and removes classifiers to the ensemble. If the classifier continues with a low accuracy, as indicated by a low weight, it will be removed from the ensemble. However, when using the global performance as the weight mechanism, the information of the minority class is ignored, which leads to a poor performance on the minority class. Adaptable Diversity-based Online Boosting (ADOB) [8] is a variation of the Online Boosting [19] method which distributes instances more efficiently among classifiers by controlling the diversity through the accuracy of each classifier, aiming to more quickly adapt to the situation where concept drifts occur frequently, especially if they are abrupt. This method achieves an impressive performance on datasets with abrupt concept drifts, but has poorer results than other methods on dealing with other types of concept drifts.

In recent research, many methods that can solve multiple and different concept drifts have been proposed [9–11]. The Diversity for Dealing with Drifts (DDD) [10] uses four ensemble classifiers with high and low diversity, before and after a concept drift is detected. This approach using a drift detection method tries to select the best ensemble (or weighted majority of ensembles) suffering from abrupt and gradual concept with several speeds of change, but it does not consider the relationship between diversity and concept drifts in an imbalanced environment. The Accuracy Updated Ensemble (AUE2) [11] is an ensemble method that improves [9] and uses the Hoeffding tree as the base learner. Its strategy is to remove the classifier with the worst accuracy and replace it with a new one at every chunk. Although AUE2 is able to react to different types of concept drifts well, it is doubtful whether the candidate model is always a better choice than an existing ensemble member on the imbalanced data stream.

Traditional methods for imbalance learning are mostly sampling methods and cost-sensitive learning methods [4, 5]. The sampling methods attempt to balance distributions by considering the proportions of class examples in the distribution. The cost-sensitive learning methods consider the costs associated with misclassifying examples instead of creating balanced data distributions. However, they are generally only suitable for smaller datasets and static datasets. The Oversampling-based Online Bagging (OOB) and Undersampling-based Online Bagging (UOB) [12] are resampling-based ensemble methods for online class imbalance learning. They can effectively and adaptively adjust the learning bias from the majority to the minority class through resampling. The data distribution is a major factor affecting their performance. Although some methods [12–15] evidence good performance on imbalanced class learning, they ignore the influences of concept drifts on the data streams.

In summary, different from existing approaches, our method will take into account not only imbalanced data distribution but also multiple concept drifts in the data streams. The proposed weight mechanism can react to various types of concept drifts quickly and achieve a good performance on both majority and minority classification accuracy.

3 Proposed Method

In this section, we will introduce the proposed classification method on the imbalanced data stream with concept drift. There are two major challenges for designing this classification model. The first is that when the concept drift occurs, the model should adjust itself to adapt to new concepts on the data stream as soon as possible. The second challenge is that when the class distribution of the data streams is imbalanced, the classification model should be able to identify this situation and prevent the situation of ignoring the information of the minority class. In order to deal with these challenges, in this paper, we design a new weighting mechanism for classification models in the ensemble to react to various types of concept drifts quickly and achieve a good performance for both majority and minority classification accuracy. The proposed method is called *G-means Update Ensemble* (GUE) and will be described in detail below. Note that the symbols used in this paper are summarized in Table 1.

Table 1. The summary of symbols used in this paper

Symbols	Description
B_i	The ith data chunk
k	The maximum number of classifiers in the ensemble
ε	The ensemble
C	The candidate model
C_j	The jth base model in the ensemble
w_{ij}	The weight of the jth base model corresponding to B_i
w_c	The weight of the candidate model
Y_{maj}	The label set of the majority class
Y_{min}	The label set of the minority class
acc_maj	The accuracy of the majority class
acc_min	The accuracy of the minority class
θ	The parameter to identify whether class imbalance occurs
S	The data stream

The system architecture of our method is shown in Fig. 1. Since the instances in the data stream arrive continuously, it is impossible to directly obtain the all of the data from the data streams. Therefore, the stream is usually separated into chunks with a fixed size in order to overcome the environment of data streams and each chunk will be tested first then trained alternately. This will trigger the imbalance detection before training the current chunk into a candidate model in order to decide whether the oversampling process is required. For each chunk, a new candidate classifier will be trained and be put into the ensemble when the ensemble is not full. If the ensemble is full, the candidate model will substitute the classifier of the poorest weight in the ensemble. All the classifiers in the ensemble will be used to predict the next data chunk by a majority vote. In the following paragraphs, we will first describe our weighting mechanism in detail, and then illustrate the imbalance detection.

Weighting Mechanism of the Ensemble. We used the ensemble method that trains the current chunk into a candidate model as the main architecture and used incremental classifiers as weighted components in the ensemble to handle the concept drifts. Compared with single classifier methods, ensemble methods can usually adapt to more types of concept drifts because the ensemble can contain more concepts in the data stream to make a combined prediction. When the stream arrives and fills the chunk,

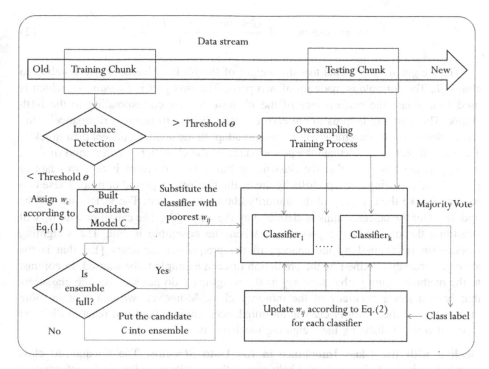

Fig. 1. System architecture of GUE

GUE will train the current chunk B_i into a candidate model C and assign a weight w_c according to Eq. (1).

$$w_c = Gmean_i, \qquad (1)$$

Where Gmean$_i$ = $\sqrt{acc\_maj_i \times acc\_min_i}$.

The $Gmean_i$ is a measure calculated from the acc_maj_i and acc_min_i in the current chunk. The acc_maj_i and acc_min_i respectively represent the accuracy of the majority class and that of the minority class in the current chunk. After the candidate model classifies the testing chunk, we can obtain the acc_maj_i and acc_min_i. The higher weight means that the candidate model can simultaneously achieve better accuracy for the majority class and the minority class. On the other hand, if the candidate model performs poorly for either the majority class or the minority class, it will get a lower weight. Most existing methods usually obtain a good performance on the overall average accuracy, whereas the accuracy of the minority class is very poor. However, we do not want to sacrifice the accuracy of the minority class. Differing from other methods, the equation takes the accuracy of the minority class into consideration. It allows the classifiers to keep more information of the minority class.

In testing phase, GUE will integrate the prediction of the classifiers in the ensemble by majority vote to classify the instances in the testing chunk. Then, GUE updates the w_{ij} for each member of the ensemble at next step according to Eq. (2).

$$w_{ij} = Gmean_{i-1,j} + \frac{Gmean_{i,j} - Gmean_{i-1,j}}{i - f + 1}. \tag{2}$$

The weight w_{ij} represents that the weight of the jth base learner corresponding to chunk B_i. The formula is made up of two parts. The first part is $Gmean_{i-1,j}$, which is used to evaluate the performance of the jth base learner corresponding to the i-1th chunk. This part can help us to preserve the classifiers with better performance at the recent chunk, so the ensemble will be able to adapt to the sudden concept drift quickly. The second part is to calculate the performance of the classifier over a period of time, where f is the time step that the classifier is built. The ensemble is able to adapt to gradual and recurring concept drifts by the evaluation over a period of time. These two parts both take the accuracy of the minority class into account. The weight w_{ij} will be updated after the current chunk is classified by the ensemble. The classifier with w_c will substitute the one with the poorest w_{ij} when the ensemble is full. The weighting mechanism is inspired by the concept of the prequential accuracy [13] that is the average accuracy obtained by the prediction of each example to be learned. In contrast to the methods using global accuracy as the weight, we do not sacrifice the classifiers that have a lower accuracy of the minority class. Moreover, we can contain more information of the minority class in an imbalanced environment and react to different types of concept drifts by the weighting mechanism.

Dealing with the Class Imbalance in the Data Streams. The imbalanced class distribution in the training chunk likely causes the classifiers to ignore the information of the minority class, thereby arriving at a worse accuracy of the prediction for the minority class. In order to solve class imbalance, one of the simplest and most effective techniques is resampling. Two major types of resampling are oversampling which increases the number of minority-class examples, and undersampling which reduces the number of majority-class examples. Some approaches [12, 14, 15] can deal with the class imbalance problem well, but only OOB [12] is designed for the environment of data streams. It is an oversampling-based method that increases the chance of training minority class examples. Therefore, OOB is integrated into the training method of our approach. However, OOB does not consider the concept drifts on the data streams, and it always performs the oversampling procedure before training classifiers every time. In real world applications, the class distribution in a chunk can vary with time and is not always imbalanced. Therefore, the process of oversampling is not always required for every chunk. Frequent use of the oversampling procedure often causes system overload and costs additional training time for each classifier in the ensemble. Furthermore, in a slightly imbalanced class distribution, the oversampling method can also cause the problem of data distortion because of increasing the chance of training minority class examples. Therefore, we design a threshold θ to identify whether the data distribution is imbalanced in the current chunk before the classifiers train with the chunk. That is, if the number of majority class examples divided by the number of minority class examples is greater than θ, the chunk will be regarded as being imbalanced and the oversampling training procedure will be executed for the purpose of increasing the chance of training minority class examples.

The pseudo-code of our method is shown in Table 2. By integrating the weighting mechanism and the adjustment of the class imbalance, our ensemble method can react to various types of concept drifts quickly and achieve a good performance on both majority and minority classification accuracy for the imbalanced data stream. In the next section, some experimental results will be illustrated to demonstrate the advantages of our method.

Table 2. Algorithm of GUE

GUE (Stream S partitioned into chunks, k: number of ensemble members)
Initialize ε is ψ
  for each chunks $B_i \in S$
C = Candidate model trained from current B_i
    If $Y_{maj} / Y_{min} > \theta$ then
    do Oversampling Training Procedures
    else
        Assign w_c according to Eq. (1)
        for each model $C_j \in \varepsilon$
        update w_o
        if $|\varepsilon| < k$ then
          $\varepsilon = \varepsilon \cup C$
    else
      if $w_{ij} < w_c$ then
        Substitute the poorest w_{ij} model with C
      end if
    end if
        for all classifier C_j except C
        train with B_i incrementally
        end for
    end for
output ε: ensemble of k weighted incremental classifiers

---

4 Experiments

In this section, we will introduce the datasets used in the experiments first. Section 4.2 will describe the setup of experiment. In Sect. 4.3, the results of experiments will be discussed to evaluate GUE against other methods.

4.1 Datasets

Generally, concept drifts can be categorized into the following four types: (1) gradual concept drift, in which the probability of the old data distribution will decrease and the probability of a new distribution will increase during a period of time until the new distribution substitutes the old one. (2) Sudden concept drift, in which the original data distribution will be changed directly to a new data distribution at a specified time. (3) Recurring class, when a class appears in the stream, then disappears for a long time, and again appears; due to the adapting mechanisms, the model will forget the class that disappears for a long time. (4) Blip, which represents a "rare event" that could be regarded as an outlier in a static distribution.

In these experiments, we used five artificial datasets generated from MOA [17], an open source framework for data stream mining, and two real world datasets that were used in [9, 11, 16–18]; these datasets contain different types of concept drifts that are briefly described below and summarized in Table 3. Note that imbalance rate means that the average of the occurrence probability of the minority class in each chunk.

Dataset Without Drift. The RBF_{ND} is a dataset without concept drift. This dataset is used to evaluate performance of our method on data streams without concept drift.

Hyp. The Hyp is a dataset with incremental gradual concept drift by slightly rotating the decision boundary of the hyper plane.

Tree. The dataset Tree contains four sudden recurring concept drifts, where recurring means that the concept will repeatedly appear in different points of time.

SEA. We use the MOA framework to create the SEA [18] dataset, which contains sudden drifts. Each concept is defined by a sum of two functions, which are both dependent on a single attribute.

RBF_{GR}. The RBF_{GR} dataset is generated by the radius basis function with a number of drifting centroids. The RBF_{GR} contains four gradual recurring concepts.

Real World Datasets. The last two datasets in Table 3 are real world datasets. The first dataset is called Electricity (Elec) and consists of the electricity market's energy prices, which are affected by market demand, supply, season, weather, and time of day. The class label identifies the price change to up or down. The second dataset, Airlines, is the prediction of whether a given flight will be delayed or not, given the information of the schedule of flights.

Table 3. Descriptions of datasets

Dataset	No. inst	No. attrs	No. class	Noise	No. drifts	Drift type	Imbalance rate
Tree	100 K	10	4	0 %	4	Sudden.recurring	0.17
SEA	100 K	3	2	10 %	3	Sudden	0.36
RBF_{ND}	100 K	20	2	0 %	0	None	0.46
RBF_{GR}	100 K	20	2	0 %	4	Gradual	0.42
Hyp	100 K	10	2	5 %	1	Gradual	0.49
Elec	45 K	7	2	–	–	–	0.41
Airlines	539 K	7	2	–	–	–	0.35

4.2   Experimental Setup

The methods in this paper were implemented in Java as part of the MOA framework [17]. The experimental environment was Intel i5-2400, 3.20 GHz processors and 8 G of RAM. The number of ensemble members was set to 10 and the chunk size was 500.

4.3   Experimental Results

The experimental results of the proposed method, GUE, and other competitors will be illustrated in this section. The performance on the classification accuracy will be discussed first, and then the execution time of the training phase and the testing phase will be presented.

Discussions on the Classification Accuracy. In this section, we compare the performance of our method GUE with AUE2 [11], ADOB [8], and OOB [12]. The accuracies obtained for each method on the artificial and real-world datasets are shown in Tables 4 and 5. Bold values represent the best results on the average accuracy of the majority and the minority class. The average rank is the average of the positions that each method achieved in different datasets. The results show that the average accuracy of the minority class of GUE outperforms other methods and obtains a good average rank on the accuracy of the majority class. Note that GUE also outperforms the other methods on the overall average accuracy and the rank. In contrast, although AUE2 achieves a good overall accuracy, its performance for the minority class is worse. OOB obtains a worse performance on the datasets with concept drifts. GUE can therefore keep the performance of the majority class and improve the accuracy of the minority class for imbalanced data streams. This is why we are able to achieve better performance than other methods on the overall average accuracy.

Table 4. Comparison of each method on average accuracy

Dataset	AUE2	ADOB	OOB	GUE
Tree	87.99 (3)	86.65 (4)	**91.87 (1)**	88.23 (2)
SEA	89.13 (2)	88.90 (3)	88.86 (4)	**89.24 (1)**
RBF$_{ND}$	97.76 (3)	96.94 (4)	**98.06 (1)**	97.78 (2)
RBF$_{GR}$	91.96 (2)	89.30 (3)	81.57 (4)	**92.18 (1)**
Hyp	**88.40 (1)**	85.90 (3)	83.54 (4)	88.13 (2)
Airlines	**66.95 (1)**	63.84 (3)	63.34 (4)	66.23 (2)
Elec	75.82 (4)	84.02 (2)	**88.72 (1)**	76.80 (3)
Rank	2.29	3.14	2.71	**1.86**
Average	85.43	85.08	85.14	**85.51**

Discussions on the Execution Time. The training time and the testing time are also important for the evaluation of the classification models on data streams due to the demand of real time response for users. The results are shown in Table 6, and Figs. 2 and 3. In order to solve the problem of imbalanced class distribution, our method

Table 5. The average accuracy of the majority and the minority class

Dataset	AUE2		ADOB		OOB		GUE	
	Majority [%]	Minority [%]	Majority [%]	Minority [%]	Majority [%]	Minority [%]	Majority [%]	Minority [%]
Tree	93.96 (3)	81.00 (3)	93.54 (4)	78.38 (4)	**94.58 (1)**	**88.06 (1)**	94.01 (2)	81.81 (2)
SEA	**94.60 (1)**	79.24 (4)	93.94 (4)	79.50 (2)	93.98 (3)	79.36 (3)	94.00 (2)	**79.83 (1)**
RBF$_{ND}$	97.64 (2)	97.91 (3)	97.31 (4)	96.51 (4)	**97.91 (1)**	**98.24 (1)**	97.58 (3)	98.01 (2)
RBF$_{GR}$	92.83 (2)	91.06 (2)	91.95 (3)	87.74 (3)	88.05 (4)	72.59 (4)	**93.18 (1)**	**91.18 (1)**
Hyp	**88.02 (1)**	88.27 (2)	85.06 (3)	86.74 (3)	83.86 (4)	83.22 (4)	87.95 (2)	**88.32 (1)**
Airlines	**83.32 (1)**	32.21 (4)	72.44 (4)	34.50 (3)	76.76 (3)	34.52 (2)	79.95 (2)	**36.00 (1)**
Elec	83.07 (3)	67.76 (4)	90.09 (2)	75.31 (2)	**91.36 (1)**	**77.26 (1)**	82.32 (4)	70.25 (3)
Rank	**1.86**	2.86	3.43	3	2.43	2.29	2.29	**1.57**
Average	**90.50**	76.78	89.20	76.95	89.50	76.18	89.86	**77.91**

Table 6. The average testing time and training time per chunk.

Dataset	AUE2		ADOB		OOB		GUE	
	Testing time [ms]	Training time [ms]	Testing time [ms]	Training time [ms]	Testing time [ms]	Training time [ms]	Testing time [ms]	Training time [ms]
Tree	8.45	**25.38**	8.24	26.01	**8.17**	28.36	8.46	25.55
SEA	2.91	10.23	2.44	10.14	**2.39**	**10.12**	2.93	10.17
RBF$_{ND}$	8.36	**22.57**	**8.22**	25.69	8.57	27.68	8.40	25.64
RBF$_{GR}$	19.36	55.83	**19.08**	55.71	19.11	56.03	19.15	**55.67**
Hyp	6.33	**24.84**	6.31	25.79	6.25	33.74	**6.19**	31.26
Airlines	**8.46**	150.73	8.77	**149.44**	8.34	151.26	8.76	150.98
Elec	**3.98**	**33.67**	4.33	39.36	4.18	39.52	4.15	37.33
Average	8.26	**46.18**	8.20	47.45	**8.14**	49.53	8.29	48.09

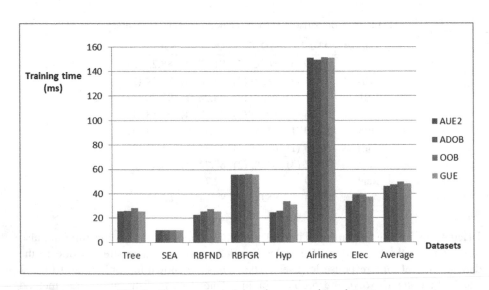

Fig. 2. The average training time on various datasets

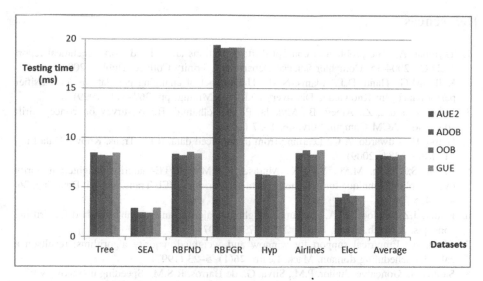

Fig. 3. The average testing time on various datasets

applies the oversampling method which leads to a slight overhead on the training time. However, this overhead is further reduced by the threshold θ we designed. In the end, the training time of our method performs similarly with other methods and is generally better than that of OOB. On the other hand, there is no significant difference on the average testing time of each method because the testing phase was not the focus of this research and was not particularly designed to minimize testing time.

5 Conclusions and Future Work

In this paper, we proposed a classification method for imbalanced data streams with concept drifts. The proposed method is called G-means Update Ensemble (GUE). GUE uses a new weighting mechanism to react to different types of concept drifts and an oversampling method to solve the class imbalance problem on data streams. The experiments showed that our method outperformed existing methods on the average accuracy and achieved better performance both for the majority and the minority class than the other methods in most of the compared datasets.

In our future work, we will try to design a method to tune the threshold θ auto-matically. This will allow our method to be suitable for different imbalanced data streams, and can enable us to observe a variety of concept drifts on the severely imbalanced data streams. It will also be a challenge to build good models which can quickly react to different types of concept drifts in such situations.

References

1. Tsymbal, A.: The problem of concept drift: definitions and related work. Technical report TCD-CS-2004-15, Computer Science Department, Trinity College, Dublin (2004)
2. Kelly, M.G., Hand, D.J., Adams, N.M.: The Impact of changing populations on classifier performance. In: Knowledge Discovery and Data Mining, pp. 367–371 (1999)
3. João, G., Indrė, Ž., Albert, B., Mykola, P., Abdelhamid, B.: A survey on concept drift adaptation. ACM Comput. Surv. **46**, 1–37 (2014)
4. Haibo, H., Edwardo, A.G.: Learning from imbalanced data. IEEE Trans. Knowl. Data Eng. **21**, 1263–1284 (2009)
5. Barua, S., Islam, M.M., Yao, X., Murase, K.: MWMOTE-majority weighted minority oversampling technique for imbalanced data set learning. IEEE Trans. Knowl. Data Eng. **26**, 405–425 (2014)
6. Kolter, J.Z., Maloof, M.A.: Dynamic weighted majority: an ensemble method for drifting concepts. J. Mach. Learn. Res. **8**, 2755–2790 (2007)
7. Blum, A.: Empirical support for winnow and weighted-majority algorithms: resultson a calendar scheduling domain. Mach. Learn. **26**(1), 5–23 (1997)
8. Santos, S., Gonçalves Júnior, P.M., Silva, G., de Barros, R.S.M.: Speeding up recovery from concept drifts. In: Calders, T., Esposito, F., Hüllermeier, E., Meo, R. (eds.) ECML PKDD 2014, Part III. LNCS, vol. 8726, pp. 179–194. Springer, Heidelberg (2014)
9. Brzeziński, D., Stefanowski, J.: Accuracy updated ensemble for data streams with concept drift. In: Corchado, E., Kurzyński, M., Woźniak, M. (eds.) HAIS 2011, Part II. LNCS, vol. 6679, pp. 155–163. Springer, Heidelberg (2011)
10. Minku, L.L., Yao, X.: DDD: a new ensemble approach for dealing with concept drift. IEEE Trans. Knowl. Data Eng. **24**(4), 619–633 (2012)
11. Brzezinski, D., Stefanowski, J.: Reacting to different types of concept drift: the accuracy updated ensemble algorithm. IEEE Trans. Neural Netw. Learn. Syst. **10**(10), 1–13 (2013)
12. Shuo, W., Leandro, L.M., Xin, Y.: A learning framework for online class imbalance learning. In: Computational Intelligence and Ensemble Learning (CIEL), pp. 36–45 (2013)
13. Ghazikhani, A., Reza, M., Hadi, S.Y.: Recursive least square perceptron model for non-stationary and imbalanced data stream classification. Evolving Syst. **4**, 119–131 (2013)
14. Mirza, B., Zhiping, L., Kar-Ann, T.: Weighted online sequential extreme learning machine for class imbalance learning. Neural Process. Lett. **38**, 465–486 (2013)
15. Shuo, W., Leandro, L.M., Xin, Y.: Resampling-based ensemble methods for online class imbalance learning. IEEE Trans. Knowl. Data Eng. **27**(5), 1356–1368 (2015)
16. Oza, N.C., Russell, S.: Experimental comparisons of online and batch versions of bagging and boosting. In: Proceedings of the Seventh ACM SIGKDD International Conference on Knowledge Discovery and Data Mining, pp. 359–364. ACM (2001)
17. Bifet, A., Holmes, G., Kirkby, R., Pfahringer, B.: MOA: massive online analysis. J. Mach. Learn. Res. **11**, 1601–1604 (2010)
18. Street, W.N., Kim, Y.: A streaming ensemble algorithm SEA for large-scale classification. In: Lee, D., Schkolnick, M., Provost, F.J., Srikant, R. (eds.) KDD, pp. 377–382. ACM (2001)
19. Oza, N.C., Russell, S.: Online bagging and boosting. In: Artificial Intelligence and Statistics 2001, pp. 105–112. Morgan Kaufmann (2001)
20. Kelly, M.G., Hand, D.J., Adams, N.M.: The impact of changing populations on classifier performance. In: Proceedings of the Fifth ACM SIGKDD International Conference on Knowledge Discovery and Data Mining, pp. 367–371. ACM (1999)
21. Harries, M., Wales, N.S.: SPLICE-2 Comparative Evaluation: Electricity Pricing (1999)

A Framework of the Semi-supervised Multi-label Classification with Non-uniformly Distributed Incomplete Labels

Chih-Heng Chung and Bi-Ru Dai[⊠]

Department of Computer Science and Information Engineering,
National Taiwan University of Science and Technology, No. 43, Sec. 4,
Keelung Road, Daan District, Taipei 106, Taiwan, ROC
D9915015@mail.ntust.edu.tw, brdai@csie.ntust.edu.tw

Abstract. In real world applications, the problem of incomplete labels is frequently encountered. These incomplete labels decrease the accuracy of the supervised classification model because of a lack of negative examples and the non-uniform distribution of the missing labels. In this paper, we propose a framework of the semi-supervised multi-label classification which can learn with the incompletely labeled training data, especially for the missing labels whose distribution is not a uniform distribution. With a modified instance weighted *k* nearest neighbor classifier, this framework recovers the labels of the training data, including both the incomplete labeled part and the unlabeled part, by iteratively updating the weight of each training instance in an acceptable execution time. The experimental results verify that the classification model trained from the recovered training data generates better prediction results in the testing phase.

Keywords: Multi-label classification · Incomplete label · Semi-supervised learning

1 Introduction

Multi-label data exist in many applications, such as the image/video annotation [1, 2], document/website categorization [3] and gene function prediction [4]. In these applications, each instance can be mapped to more than one label. For example, the landscape image in Fig. 1 could be assigned the labels {mountain, river, sky}. The multi-label classification attempts to train a model with some multi-label data, and predict the labels of other unlabeled data.

Traditional multi-label classification algorithms assume that all the labels of the training data are complete [4–10]. However, the labels of the data are usually incomplete in real applications. For example, even though trees and clouds appear in Fig. 1, the labels {tree, cloud} are not included in the given labels. Some possible reasons for these absent labels include the following: they are simply missing; the assigners considered them to be less important; or the system only collects/records the top three labels and drops the others. In this case of incomplete labels, it cannot be determined whether the absent labels in the training instances are actually negative

© Springer International Publishing Switzerland 2016
S. Madria and T. Hara (Eds.): DaWaK 2016, LNCS 9829, pp. 267–280, 2016.
DOI: 10.1007/978-3-319-43946-4_18

Assigned labels:

{mountain, river, sky}

Possible missing labels:

{cloud, rock, water, tree, blue, ...}

Fig. 1. A landscape image which is assigned the labels {mountain, river, sky}.

samples or missing labels. The traditional multi-label classification algorithms take these unknown labels as negative samples, and this assumption will lead to a worse accuracy of the prediction results [11–13].

In addition, the distribution of the missing labels is usually not uniformly distributed in real applications. For example, the images which are similar to Fig. 1 will usually be tagged with the labels {mountain, river} but without the label {rock} even though there are some rocks at the bottom left of Fig. 1. Because the importance of each label is different, some of the labels will be assigned to more data instances than the others. In many applications, such as the image/video annotation, document categorization and website categorization, the users usually assign only a few of the topmost important labels. For example, Table 1 shows an example dataset with complete labels, where the symbol 'v' denotes the corresponding image on the row being assigned the corresponding label on the column. Table 2 illustrates the label assignments of this dataset in the situation that the system only extracts the top 3 labels for each image. In this example, the number of the assigned label 'rock' is reduced from 3 to 1, but the labels 'river' and 'sky' are kept intact. The missing ratios of all labels are not uniform. In addition, the distribution of the missing label assignments in each single label is not uniform either. This prevents some methods, such as [14], from being able to precisely estimate the missing rate of the labels.

Another common problem in real applications is the lack of labeled data. Because it is impossible to manually assign labels to large amounts of data, we usually have a small number of labeled data with a large number of unlabeled data for the learning. The unlabeled data cannot be used in the traditional multi-label classification algorithms, and for this reason, semi-supervised learning has been proposed [15, 16]. Semi-supervised learning attempts to make use of unlabeled data for training and expects to train a model which is better than the model trained by only the labeled data.

However, typical semi-supervised learning algorithms still regard the labels of training data as being complete. In other words, no matter whether they are actually negative or missing because of some reasons, the absent labels are all treated as negative. Consequently, when the labels of training data are incomplete, typical

Table 1. An example dataset with complete labels.

	Mountain	River	Sky	Cloud	Water	Rock
Image 1	v	v	v	v	v	
Image 2		v	v	v		
Image 3	v	v			v	v
Image 4		v	v	v	v	v
Image 5	v		v	v		v

Table 2. An example dataset with incomplete labels. Note that the dataset is the same as the one in Table 1 but only the top 3 labels are kept.

	Mountain	River	Sky	Cloud	Water	Rock	Missing rate
Image 1	v	v	v				0.4
Image 2		v	v	v			0
Image 3		v			v	v	0.25
Image 4		v	v		v		0.4
Image 5	v		v	v			0.25
Missing rate	0.333	0	0	0.5	0.333	0.667	

semi-supervised learning algorithms generate poor results [16]. Therefore, the semi-supervised classification with incomplete labels is a difficult but significant problem in real world applications. Although some algorithms have been proposed to deal with this issue [16], the extremely long execution time for optimization is usually not acceptable in many real applications.

In summary, the challenges of semi-supervised multi-label classification with incomplete labels are as follows:

1. Fewer labeled training data with many unlabeled training data
2. Only positive samples and unknown samples, no negative samples
3. The distribution of the missing labels is not a uniform distribution, thus preventing the missing rates of the labels from being estimated precisely
4. The execution time is expected to be short.

In this paper, we propose a framework of semi-supervised multi-label classification which can learn with the incompletely labeled training data, especially for the missing labels whose the distribution is not uniform. This framework is based on training label recovery. With a modified instance weighted k nearest neighbor classifier, this framework recovers the labels of the incomplete labeled training data and the unlabeled training data by iteratively updating the weight of each training instance in an acceptable execution time. The classification model which is trained from the recovered training data by existing multi-label classification algorithms can generate better prediction results in the testing phase.

The rest of this paper is organized as follows. In the next section, we introduce the related works, and then our framework is presented in detail in Sect. 3. In Sect. 4, the experiment results are described, and finally, we share conclusions in Sect. 5.

2 Related Work

In this section, we introduce the related work of this paper. These related works are organized according to three major issues: multi-label classification, incomplete label and semi-supervised learning.

Multi-label classification learns the model of more than one label at the same time. There are several categories of multi-label classification approaches. The first category is to decompose the multi-label classification into some binary classification works where each one classifies a single label [5–7]. The ML-kNN [8] algorithm utilizes the maximum a posteriori (MAP) principle on the k-nearest neighbors of the instances to determine the labels of testing data. Another category of multi-label classification approaches learns the correlation between labels [4, 9, 10].

Although these approaches focus on multi-label learning, they cannot be directly used on incomplete label data because they consider the labels of training data as totally complete. They will assume the missing labels of incomplete label data to be all negative samples. This will result in some evaluation errors in the training phase and lead to a poorly performing classification model. There has, however, been research which focuses on the multi-label learning with incomplete labels. In [11], an image annotation algorithm with incomplete labels was devised. It proposes a ranking-based multi-label learning framework and exploits the group lasso technique to combine ranking errors. In [12], an algorithm FastTag which also focuses on image tagging was presented. This algorithm builds two classifiers, one that reconstructs the incomplete labels, and the other that maps the image features to the reconstructed label set; these two classifiers are further combined by a joint convex loss function via coregularization. In [17], an algorithm named Multiview Imperfect Tagging Learning (MITL) was introduced. MITL solves the problem of incomplete labels by extracting the information of the incomplete labeled training dataset from multiple views in order to differentiate the data points in the stage of classification. However, MITL requires the existence of different views to a dataset, and, ideally, these views are conditionally independent. In [13], the WEak Label Learning (WELL) algorithm was designed to solve the incomplete label problem. WELL formulates the objective function as a convex optimization problem which can be solved efficiently. It also exploits the correlation between labels by assuming a group of low-rank base similarities, and then derives the appropriate similarities between instances for different labels from these base similarities. Although these methods attempt to solve the incomplete label problem, they do not exploit the unlabeled data.

In many cases, the labeled data are few, thus making it very difficult to learn a high quality model only from the labeled data. Furthermore, the labeled data are usually incomplete in real world applications. In the problem of semi-supervised multi-label classification with incomplete labels, there is the attempt to learn the model not only from the incomplete labeled data but also from the unlabeled data. There are several works which focus on the problem of semi-supervised multi-label classification with incomplete labels. In [15], the Semi-Supervised Correspondence Hierarchical Dirichlet Process (SSCHDP) algorithm was proposed to develop an inductive semi-supervised learning statistical model to complete the missing labels in the incomplete training data.

In [16], low-rank label matrix recovery was integrated into the manifold regularized vector-valued prediction framework, and the global optimal solution was solved by a proximal gradient descent with continuation algorithm. The quality of the prediction results of this algorithm is higher than other above methods, but the running time of this algorithm is extremely long and would not normally be considered acceptable in practice.

3   Method

In this section, we present our proposed method in detail. Our method attempts to recover the incomplete and unlabeled training labels. According the result of the base classifier, the labels will be iteratively recovered. The recovered labels can be used for training a better model to predict the labels of the testing data.

We define the problem of semi-supervised multi-label classification with incomplete labels in Sect. 3.1. The method of the label recovery and the selected base classifier, weighted kNN classifier, is shown in Sect. 3.2.

3.1   Problem Definition

In this subsection, we define the problem of semi-supervised multi-label classification with incomplete labels. Firstly, suppose all instances have d-dimension features and there are m possible labels. Consider the training dataset $D_{Train} = \{(x_i, y_i)\}_{i=1}^{n_{Train}}$, where $x_i \in \mathbb{R}^d$ is the i-th training feature vector and $y_i \in \{+1, -1\}^m$ is the i-th training label vector. For each y_{ij} in the vector $y_i = (y_{i1}, y_{i2}, \ldots y_{im})$, the $+1$ value indicates that the i-th instance is assigned the j-th label, and the -1 value indicates that the assignment of the j-th label for the i-th instance is unknown.

In addition, the training dataset D_{Train} can be divided into the labeled part D_L and the unlabeled part D_U, where

$$\forall (x_i, y_i) \in D_L, \exists 1 \leq j \leq m, y_{ij} = +1$$

and

$$\forall (x_i, y_i) \in D_U, \forall 1 \leq j \leq m, y_{ij} = -1.$$

In other words, D_L collects the labeled training instances which are assigned at least one label, and D_U collects the unlabeled training instances whose labels are totally unknown. Furthermore, the number p is given to denote the maximum number of assigned labels for an instance. Each labeled instance in D_L is assigned no more than p labels. That is, the number of assigned labels for each instance in D_L is an integer value in the interval $[1, p]$. In most previous research, the distribution of the missing labels of the training data is close to a uniform distribution. However, in many applications, the system only collects/records several top labels and drops the others.

Therefore, we focus on the training data with the constraint of limited labels, which will lead to a non-uniform distribution of the missing labels.

Table 3 shows an example training dataset with the number of instance $n = 9$, the number of features $d = 4$, the number of labels $m = 3$ and the maximum number of assigned labels $p = 2$ for each instance. In this example, the labeled part $D_L = \{(x_i, y_i)\}_{i=1}^4$ and the unlabeled part $D_U = \{(x_i, y_i)\}_{i=5}^9$ because the instances 1 to 4 are assigned at least one label and the instances 5 to 9 are assigned no label. The p value restricts the number of assigned labels to be in the interval [1, 2], and therefore, the instances 1 to 4 should be assigned either one or two labels. Note that the values -1 of the label assignments do not mean negative samples, but instead mean the unknown assignments, which may be either positive or negative samples.

Table 3. An example of a training dataset with the number of instance $n = 9$, the number of features $d = 4$, the number of labels $m = 3$ and the maximum number of assigned labels p is 2 for all instances. Accordingly, the training dataset can be divided into the labeled part $D_L = \{(x_i, y_i)\}_{i=1}^4$ and the unlabeled part $D_U = \{(x_i, y_i)\}_{i=5}^9$.

	Feature 1	Feature 2	Feature 3	Feature 4	Label 1	Label 2	Label 3
Instance 1	x_{11}	x_{12}	x_{13}	x_{14}	$y_{11} = -1$	$y_{12} = -1$	$\mathbf{y_{13} = 1}$
Instance 2	x_{21}	x_{22}	x_{23}	x_{24}	$\mathbf{y_{21} = 1}$	$y_{22} = -1$	$\mathbf{y_{23} = 1}$
Instance 3	x_{31}	x_{32}	x_{33}	x_{34}	$\mathbf{y_{31} = 1}$	$\mathbf{y_{32} = 1}$	$y_{33} = -1$
Instance 4	x_{41}	x_{42}	x_{43}	x_{44}	$y_{41} = -1$	$\mathbf{y_{42} = 1}$	$y_{43} = -1$
Instance 5	x_{51}	x_{52}	x_{53}	x_{54}	$y_{51} = -1$	$y_{52} = -1$	$y_{53} = -1$
Instance 6	x_{61}	x_{62}	x_{63}	x_{64}	$y_{61} = -1$	$y_{62} = -1$	$y_{63} = -1$
Instance 7	x_{71}	x_{72}	x_{73}	x_{74}	$y_{71} = -1$	$y_{72} = -1$	$y_{73} = -1$
Instance 8	x_{81}	x_{82}	x_{83}	x_{84}	$y_{81} = -1$	$y_{82} = -1$	$y_{83} = -1$
Instance 9	x_{91}	x_{92}	x_{93}	x_{94}	$y_{91} = -1$	$y_{92} = -1$	$y_{93} = -1$

The objective of the problem of semi-supervised multi-label classification with incomplete labels is training a model $f : X \mapsto Y$ which maps the feature space X to the label space Y.

Overall, the framework of the problem of semi-supervised multi-label classification with incomplete labels is shown in Fig. 2. According to the input labeled and unlabeled training data, we try to build a new training dataset where the instances in this dataset are all labeled and more complete. In other words, the aim of this process is to recover some missing labels in the training dataset. The multi-label classification model is trained with this new training dataset by a multi-label classifier, and then it is applied to classify the labels of the testing data.

3.2   Label Recovery

Because the labels of the training data in the problem of the semi-supervised multi-label classification with incomplete labels are lacking, the traditional multi-label

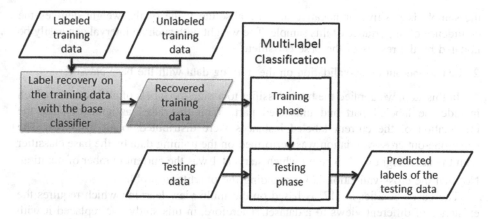

Fig. 2. The framework of the semi-supervised multi-label classification with incomplete labels.

classification algorithms cannot generate good results with such training data. Therefore, if we are able to correctly recover the labels of the incomplete labeled training data and unlabeled data, the prediction results can usually be improved. In this subsection, we will describe the process of label recovery in detail.

Our label recovery process is modified from the method called "weight the training data for missing tags" in [17]. This method was selected because it can distribute the information of labeled data to all instances, can enhance the information with high confidence and can converge quickly. It gives the weight for each training instance, and then iteratively updates the weight by a base classifier. The final result of the base classifier will be used as the recovered labels. However, [17] is based on multi-view learning, which requires the existence of different views to a dataset. Therefore, we changed the base classifier of this method for the dataset with only a single view. In this way, the label recovery process will be more general and more applicable to various types of datasets. The steps of this method are listed below.

1. Initialization

This method tries to distribute the positive and negative information from labeled instances to all instances and optimize this information during the iterative steps. However, there is no negative sample in the training data, which means it has to add some negative information into the training data and make a distinction between unknown and negative samples.

Firstly, we changed all y_{ij} with the unknown value from -1 to 0. Secondly, for each label j, p_j instances with $y_{ij} = 0$ were chosen randomly and the values of y_{ij} for the selected instances were modified to -1, where p_j is the total number of positive instances of label j. In summary, the value set of all y_{ij} was changed from $\{+1, -1\}$ to $\{+1, 0, -1\}$, where $+1$ represents positive samples, 0 represents unknown samples and -1 represents negative samples.

Finally, the initial weight was initialized as $w_{ij}^0 = y_{ij}$. These weights were updated to the interval $[-1, 1]$ in the following steps, where the sign of the weight denotes that the sample is positive or negative and the absolute value of the weight denotes the confidence of importance of this sample. The weight with such an interval can easily be updated by the results of the base classifier.

2. Leave-one-out cross-validation on the training data with the base classifier

In this step, we applied the base classifier to assign the labels of all instances which include the labeled part and unlabeled part. With this label assignment step, the information of the current labeled instances were distributed to all instances. The leave-one-out cross-validation was performed on the training data by the base classifier with the last weight w_{ij}^{t-1}, where t which starts at 1 was the current number of iteration. Not that the cross-validation is for avoiding the overfitting problem.

The base classifier in [17] is based on the multi-view learning which requires the existence of different views to a dataset. Therefore, in this study, we replaced it with another base classifier which is more suitable for various types of datasets. However, not all of the classification algorithms are suited for this method. In our opinion, a suitable base classifier for the problem studied in this paper has to satisfy several requirements.

Firstly, the base classifier should be able to learn with the training data with weighted instances. The importance of each instance is represented by the weight. The weights are optimized iteratively according to the prediction results of the base classifier. Therefore, the base classifier should change the prediction result with different weights to provide the needed information for updating the weights.

Secondly, because the interval of the updated weights is $[-1, 1]$, the output of the base classifier has to be in a fixed interval, and the threshold between positive and negative results also has to be fixed. For example, if the output value is in $[0, 1]$ and is considered as positive when the value is greater than 0.5, we can easily map it to $[-1, 1]$ and set the threshold between positive and negative results to 0.

In addition, this base classifier is required to be able to generate available classification results when the training labels are lacking. Some strong classifiers, such as LIBSVM [18], tend to predict all label assignments as negative when the training labels are lacking, which makes them inappropriate choices for our base classifier.

The execution time of the base learning is also an important consideration. The base classifier will be executed many times because the process iteratively performs the cross-validation. The total execution time will become unacceptable if it costs a lot of time for a single execution.

In order to satisfy all the requirements of the base classifier, the instance weighted k nearest neighbor (instance WkNN) algorithm was selected. The basic concept of the instance WkNN is introduced below. Given the training data and the weight of each training instance, the Euclidean distance between training instances and testing instances are calculated. For a testing instance I^{Test}, it finds the k nearest neighbors $KNN(I^{Test})$ in the training data, where the distances between the target instance and the k nearest neighbors are the smallest. Finally, the predicted label vector $q = \{q_1, q_1, \ldots, q_m\}$ of the testing instance I^{Test} is calculated as the weighted average of the label vectors of the k nearest neighbors $KNN(I^{Test})$:

$$q_j = \frac{\sum_{i \in KNN(I^{Test})} \left| w_{ij}^{t-1} \right| \cdot y_{ij}}{\sum_{i \in KNN(I^{Test})} \left| w_{ij}^{t-1} \right|}, \text{ for each } j = 1, 2, \ldots, m,$$

where the weight w_{ij}^{t-1} is the last weight of the i-th instance on the j-th label.

The instance WkNN can satisfy the requirements of the base classifier discussed above. Obviously, the predictions of the instance WkNN are based on the weighted training instances. The interval of the output predictions of the instance WkNN is $[-1, 1]$, which can directly be used in the updating step without additional mapping function. This allows the generation of available classification results when the training labels are lacking because the information will be distributed and combined between the nearest neighbors. Therefore, the instance WkNN can produce available classification results using the few labeled training instances and then distribute the information to similar neighbors for the next optimization step. In addition, the instance WkNN can easily perform the leave-one-out cross-validation by ignoring each instance itself when it finds the k nearest neighbors. It is also applicable to the iterative execution because the distances between the instances only need to be calculated once and do not change in the iterations.

However, the typical instance WkNN optimizes the weights of instances by the gradient descent. However, instead of the gradient descent, we used the weight updating step in [17] with the instance WkNN for a faster convergence that still ensures good information distribution.

3. Updating the weights

In this step, the current weight w_{ij}^t is updated for each instance i and label j. This updating step is for enhancing the information with high confidence and reducing the information with low confidence. The updating value of w_{ij}^t is

$$w_{ij}^t = \begin{cases} sgn\left(q_{ij}^t\right) \cdot 1 & if sgn\left(q_{ij}^t\right) = sgn\left(w_{ij}^{t-1}\right) \text{ and } q_{ij}^t > e \text{ and } w_{ij}^{t-1} > e \\ c \cdot \frac{q_{ij}^t - Min_j}{Max_j - Min_j} & if sgn\left(q_{ij}^t\right) = -1 \text{ and } sgn\left(y_{ij}\right) = 1 \\ q_{ij}^t & otherwise \end{cases}$$

where $sgn()$ is the sign function, e is a high confidence threshold which is set in $(0.5, 1)$, c is a low confidence update range which is set in $(0, 0.5)$ and

$$Max_j = max_i\left(q_{ij}^t\right),$$

$$Min_j = min_i\left(q_{ij}^t\right).$$

4. Iteratively execute steps 2 and 3

It continues steps 2 and 3 iteratively until a given maximum number of iterations.

With these iterative steps, the information of label assignments is distributed and converges fast. Therefore, the labels of all training data can be recovered fast by these steps.

4 Experiment

In this section, we present the experimental results. The datasets used in the experiment are introduced in Sect. 4.1, and the experimental results are shown in Sect. 4.2.

4.1 Datasets

We used six datasets in the experiment: yeast [4], CAL500 [19], birds [20], emotions [21], scene [7] and mediamill [1]. The statistical information of the datasets is summarized in Table 4. The label cardinality in Table 4 represents the average number of assigned labels for each instances.

Each dataset was partitioned into a training part and testing part. For the training part, we further partitioned it into a labeled part and an unlabeled part. All the label assignments of the unlabeled part were removed. For the labeled part, the maximum number of the sampled labels was set to the half of the number of the label cardinality of this dataset. For each instance in the labeled part, if the number of labels of this instance was more than the maximum number of the sampled labels, we randomly removed its label assignments until the number of the remaining label assignments was equal to the maximum number of the sampled labels. For example, suppose the maximum number of the sampled labels was set to 1. If an instance was assigned only one label, this label assignment would kept. If an instance was assigned more than one labels, only one label assignment would be kept and the others would be removed. By this process, the label assignments were not removed uniformly.

4.2 Experimental Results

In this section, we introduce the framework of our experiment and present the experimental results.

The framework of the experiment is illustrated in Fig. 2. We applied our proposed method and the comparison method WELL [13] to recover the labels of the training data. Next, two commonly used multi-label classification algorithms, MLkNN [8] and LIBSVM [18], were applied to the recovered training data to train the classification models. Finally, we used the models to predict the labels of the testing data, and evaluated the results of the predictions. MLkNN and LIBSVM were also applied on the training data before the label recovery as the baselines. In addition, [16] was not included in our experiment because the execution time of [16] was too lengthy to be considered acceptable. For example, we applied [16] on the yeast dataset where the number of the training instances is only 1500, and found that the execution time was more than 5 h. Compared with other methods, as will be shown in Table 5, the execution time of [16] cannot be realistically applicable in practice.

Table 4. Statistical information of the datasets.

Dataset	#instances of total/training/testing	#features	#labels	Label card.	#labeled/unlabeled training instances	#max sampled labels
Yeast	2417/1500/917	103	14	4.237	750/750	2
CAL500	502/402/100	68	174	26.044	201/201	13
Birds	645/322/323	258	19	1.014	161/161	1
Emotions	593/475/118	72	6	1.869	237/238	1
Scene	2407/1211/1196	294	6	1.074	605/606	1
Mediamill	43907/4390/39517	120	101	4.376	2195/2195	2

Table 5. The experimental results.

Dataset		Proposed-MLkNN	Proposed-LIBSVM	WELL-MLkNN	WELL-LIBSVM	MLkNN	LIBSVM
Yeast	HammingLoss	0.2393	**0.2264**	0.2919	0.2985	0.3037	0.3037
	MicroF1	0.5638	**0.5697**	0.0912	0.0386	0	0
	Execution time (s)	**7.005179**		108.5215			
CAL500	HammingLoss	0.1611	0.1533	0.2869	0.2894	**0.153**	0.1531
	MicroF1	0.3134	0.3117	**0.4153**	0.4115	0.0008	0.0008
	Execution time (s)	**0.61997**		1171.003			
Birds	HammingLoss	0.0574	**0.0508**	0.4215	0.4494	0.0515	0.051
	MicroF1	0.12	0.0064	**0.156**	0.1519	0.0063	0.0063
	Execution time (s)	**0.513747**		125.12			
Emotions	HammingLoss	0.2768	**0.2556**	0.6822	0.6201	0.3178	0.3178
	MicroF1	0.5421	**0.5639**	0.4823	0.4613	0	0
	Execution time (s)	**0.404335**		28.6281			
Scene	HammingLoss	0.1834	0.164	**0.1632**	0.1743	0.1742	0.1724
	MicroF1	0.6186	**0.6393**	0.5044	0.4006	0.0836	0.1081
	Execution time (s)	**14.54812**		63.8112			
Mediamill	HammingLoss	0.0385	**0.0333**	0.1241	0.124	0.0433	0.0433
	MicroF1	0.4417	**0.4847**	0.3529	0.3529	0.0002	0.0002
	Execution time (s)	**528.2417**		29305.14			

In our work, we set the high confidence threshold $e = 0.8$ and the low confidence update range $c = 0.2$. In addition, the number of iterations was set to 5 because [17] shows that the result of this method converges quickly, in 5 iterations. The k value of the instance WkNN was set to 10, where it is the same as the suggested value of MLkNN [8]. All the parameters of WELL were set to the suggested values in [13].

We measured the classification results using two standard multi-label evaluation criteria: Hamming Loss and Micro-F1. Hamming Loss directly evaluates the ratio of the misclassified instance-label pairs, where a smaller Hamming Loss is better. Micro-F1 calculates the F1-measure on the predicted results of different labels as a whole, where both the precision and the recall are taken into consideration, and a greater Micro-F1 is better. The details of these two criteria can be found in [22]. In addition, we also compared the execution time of the label recovery.

The experimental results are listed in Table 5. The bold fonts represent the best results. Our proposed method achieves better results on most datasets. In the cases that the information is lacking, the predictions of the baselines are almost all negative. As observed in Table 5, their Micro-F1 is close to or equal to 0. In these cases, after the label recovery of our proposed method, the classification results are significantly improved.

There are two exceptions in these results: CAL500 and birds. In the results of these two datasets, the Micro-F1 of WELL is better, but the Hamming Loss of WELL is the worst. The reason is that WELL predicts too much "positive" in the results. This increases both the number of true positives and the number of false positives, and the ratio of false positives is larger than the ratio of true positives. Therefore, the Hamming Loss of WELL is the worst, but the Micro-F1 of WELL gets better because of a much higher recall value. The results of true positive, false positive, true negative, false negative, Micro-Precision and Micro-Recall of CAL500 and birds dataset are shown in Table 6.

Table 6. The experimental results of true positive, false positive, true negative, false negative, Micro-Precision and Micro-Recall of the CAL500 and birds datasets.

Dataset		Proposed-MLkNN	Proposed-LIBSVM	WELL-MLkNN	WELL-LIBSVM	MLkNN	LIBSVM
CAL500	#True positive	640	604	1773	1761	1	1
	#False positive	781	609	4102	4134	0	2
	#True negative	13956	14128	10635	10603	14737	14735
	#False negative	2023	2059	890	902	2662	2662
	MicroPrecision	0.4504	0.4979	0.3018	0.2987	1	0.3333
	MicroRecall	0.2403	0.2268	0.6658	0.6613	0.0004	0.0004
birds	#True positive	24	1	239	247	1	1
	#False positive	63	0	2513	2692	4	1
	#True negative	5761	5824	3311	3132	5820	5823
	#False negative	289	312	74	66	312	312
	MicroPrecision	0.2759	1	0.0868	0.084	0.2	0.5
	MicroRecall	0.0767	0.0032	0.7636	0.7891	0.0032	0.0032

Another observation is that the Micro-F1 of the Proposed-LIBSVM on the birds dataset is poor. The reason is that the label assignment information of the birds dataset is very lacking. Even after the label recovery, the information was still not enough for LIBSVM to produce good classification results, but it is enough for MLkNN to produce available results.

In summary, our proposed method can quickly recover the training labels under the problem of semi-supervised multi-label classification with incomplete labels. The recovered training data can be used with traditional multi-label classification algorithms to generate good classification results.

5 Conclusion

Is this paper, we proposed a framework of the semi-supervised multi-label classification which can learn with the incompletely labeled training data, especially for the missing labels whose the distribution is not uniform. The proposed framework recovers the labels of the incomplete labeled training data and the unlabeled training data using a modified instance weighted k nearest neighbor classifier, and is able to iteratively update the weight of each training instances for further optimizing the recovered labels. The experimental results verified that the classification model trained from the recovered training data generates better prediction results in the testing phase, and the execution time is more acceptable.

References

1. Snoek, C.G., Worring, M., Van Gemert, J.C., Geusebroek, J.M., Smeulders, A.W.: The challenge problem for automated detection of 101 semantic concepts in multimedia. In: Proceedings of the 14th Annual ACM International Conference on Multimedia, pp. 421–430. ACM (2006)
2. Huiskes, M.J., Lew, M.S.: The MIR Flickr retrieval evaluation. In: Proceedings of the 1st ACM International Conference on Multimedia Information Retrieval, pp. 39–43. ACM (2008)
3. Srivastava, A., Zane-Ulman, B.: Discovering recurring anomalies in text reports regarding complex space systems. In: Proceedings of the IEEE Aerospace Conference, pp. 55–63 (2005)
4. Elisseeff, A., Weston, J.: A kernel method for multi-labelled classification. In: Advances in Neural Information Processing Systems, pp. 681–687 (2001)
5. Joachims, T.: Text categorization with support vector machines: learning with many relevant features. In: Nédellec, C., Rouveirol, C. (eds.) Machine Learning: ECML-98. LNCS, vol. 1398, pp. 137–142. Springer, Heidelberg (1998)
6. Yang, Y.: An evaluation of statistical approaches to text categorization. Inf. Retrieval 1(1–2), 69–90 (1999)
7. Boutell, M.R., Luo, J., Shen, X., Brown, C.M.: Learning multi-label scene classification. Pattern Recogn. 37(9), 1757–1771 (2004)
8. Zhang, M.-L., Zhou, Z.-H.: ML-kNN: a lazy learning approach to multi-label learning. Pattern Recogn. 40(7), 2038–2048 (2007)
9. Guo, Y., Schuurmans, D.: Adaptive large margin training for multilabel classification. In: Proceeding of AAAI (2011)
10. Minh, H., Sindhwani, V.: Vector-valued manifold regularization. In: Proceeding of ICML (2011)
11. Bucak, S.S., Jin, R., Jain, A.K.: Multi-label learning with incomplete class assignments. In: Computer Vision and Pattern Recognition (CVPR), pp. 2801–2808. IEEE (2011)
12. Chen, M., Zheng, A., Weinberger, K.: Fast image tagging. In: Proceedings of the 30th International Conference on Machine Learning, pp. 1274–1282 (2013)
13. Sun, Y.Y., Zhang, Y., Zhou, Z.H.: Multi-label learning with weak label. In: Proceedings of 24th AAAI Conference on Artificial Intelligence (2010)
14. Elkan, C., Noto, K.: Learning classifiers from only positive and unlabeled data. In: Proceedings of the 14th ACM SIGKDD International Conference on Knowledge Discovery and Data Mining, pp. 213–220. ACM (2008)
15. Qi, Z., Yang, M., Zhang, Z.M., Zhang, Z.: Mining partially annotated images. In: Proceedings of the 17th ACM SIGKDD International Conference on Knowledge Discovery and Data Mining, pp. 1199–1207. ACM (2011)

16. Zhao, F., Guo, Y.: Semi-supervised multi-label learning with incomplete labels. In: Proceedings of the 24th International Conference on Artificial Intelligence, pp. 4062–4068 (2015)
17. Qi, Z., Yang, M., Zhang, Z.M., Zhang, Z.: Multi-view learning from imperfect tagging. In: Proceedings of the 20th ACM International Conference on Multimedia, pp. 479–488. ACM (2012)
18. Chang, C.C., Lin, C.J.: LIBSVM: a library for support vector machines. ACM Trans. Intell. Syst. Technol. (TIST) 2(3), 27 (2011)
19. Turnbull, D., Barrington, L., Torres, D., Lanckriet, G.: Semantic annotation and retrieval of music and sound effects. IEEE Trans. Audio Speech Lang. Process. 16(2), 467–476 (2008)
20. Briggs, F., Huang, Y., Raich, R., Eftaxias, K., Lei, Z., Cukierski, W., Hadley, S.F., Hadley, A., Betts, M., Fern, X.Z., Irvine, J., Neal, L., Thomas, A., Fodor, G., Tsoumakas, G., Ng, H. W., Nguyen, T.N.T., Huttunen, H., Ruusuvuori, P., Manninen, T., Diment, A., Virtanen, T., Mar-zat, J., Defretin, J., Callender, D., Hurlburt, C., Larrey, K., Milakov, M.: The 9th annual mlsp competition: new methods for acoustic classification of multiple simultaneous bird species in a noisy environment. In: 2013 IEEE International Workshop on Machine Learning for Signal Processing (MLSP), pp. 1–8. IEEE (2013)
21. Trohidis, K., Tsoumakas, G., Kalliris, G., Vlahavas, I.P.: Multi-label classification of music into emotions. In: ISMIR, vol. 8, pp. 325–330 (2008)
22. Zhang, Y., Zhou, Z.H.: Multilabel dimensionality reduction via dependence maximization. ACM Trans. Knowl. Discovery Data (TKDD) 4(3), 14 (2010)

XSX: Lightweight Encryption for Data Warehousing Environments

Ricardo Jorge Santos[1], Marco Vieira[1], and Jorge Bernardino[1,2(✉)]

[1] CISUC – Centre of Informatics and Systems of the University of Coimbra
FCTUC – University of Coimbra, 3030-290 Coimbra, Portugal
`lionsoftware.ricardo@gmail.com, mvieira@dei.uc.pt`
[2] ISEC – Superior Institute of Engineering of Coimbra Polytechnic of Coimbra,
3030-190 Coimbra, Portugal
`jorge@isec.pt`

Abstract. The use of data security solutions like encryption significantly reduces database performance in data-intensive processing environments such as Data Warehouses. To overcome this issue this paper proposes a transparent lightweight encryption technique based on alternating sequences of eXclusive Or (XOR) and bit switching operations. The technique aims to achieve high performance while providing significant levels of security strength, attaining better security-performance tradeoffs than most commercial encryption solutions and those proposed by the research community. The experimental evaluation conducted using the TPC-DS benchmark shows that the proposed technique adds lower database performance overhead when comparing with the AES standard encryption algorithm and with other encryption solutions such as OPES, Salsa20, TEA and SES-DW.

Keywords: Data security · Database security · Confidentiality · Encryption · Data warehousing · Decision support systems

1 Introduction

Encryption is widely used and accepted as the best method to protect sensitive data in databases against undesirable or unauthorized disclosure. However, the use of encryption introduces considerable performance costs in data-intensive environments that require frequent decryption of large sets of data, creating high security-performance tradeoffs [11]. Data Warehouses (DWs) are a foremost example of environments in which this typically occurs. DWs typically store millions of sensitive numerical values that reveal the history and trends of a business and thus should be protected against unauthorized access [8]. In practice, DWs are core systems that hold the secrets of the business and are used by business managers to gain decision support information and business knowledge.

Given their decision support nature, DW tables store extremely sensitive historical and current data and frequently require many hundreds or thousands of gigabytes of storage space. Most DW users' queries typically need to access and process a large amount of that data, frequently taking a considerable amount of time to process (from

© Springer International Publishing Switzerland 2016
S. Madria and T. Hara (Eds.): DaWaK 2016, LNCS 9829, pp. 281–295, 2016.
DOI: 10.1007/978-3-319-43946-4_19

a few seconds up to many minutes or several hours) [8]. When using encryption to protect the DW's sensitive values, the time required to decrypt those values in order to process and respond to user queries can easily increase the queries' response time by several orders of magnitude as demonstrated in [11], leading users to consider it unacceptable and thus, making the use of data encryption infeasible.

Most encryption solutions are designed mainly with the goal of achieving high security strength rather than achieving high execution performance. However, designing encryption solutions that account for acceptable security-performance tradeoffs, i.e., that ensure strong security while keeping database performance acceptable, is critical in data-intensive environments such as Data Warehouses.

In this paper we propose a lightweight encryption technique based on a cipher using alternating sets of eXclusive Or (XOR) and bit switching operations, i.e., a XOR-Switch-XOR (XSX) sequence. The main focus is on achieving high execution performance and on enabling each implementation to define distinct ways of performing the bit switching in each intermediate round. The key characteristic is that it introduces small database performance overhead for intensive data processing systems that handle large amounts of sensitive data, while providing at the same time a significant level of security strength, thus representing a viable alternative solution for such environments.

Although the security strength is a central aspect of any encryption solution, the technique proposed in this paper focuses on leveraging security-performance tradeoffs that make it feasible for data warehousing environments. Therefore, it is not within the scope of this paper to discuss how secure the solution is when compared with standard or alternative encryption algorithms, but rather to demonstrate that it ensures considerable security strength while introducing low database performance overhead. An experimental evaluation using the TPC-DS benchmark [13] is presented to compare the technique's performance against the Advanced Encryption Standard (AES) algorithm [2] as well as with other encryption solutions, namely Salsa20 [4, 5], OPES [3], SES-DW [12] and TEA [15, 16].

The remainder of this paper is organized as follows. Section 2 presents background and related work on using encryption standards and other alternative solutions in databases. Section 3 describes the XOR-Switch-XOR (XSX) encryption technique and Sect. 4 discusses its security. Section 5 presents the experimental evaluation of the proposed technique against AES, Salsa20, OPES, SES-DW and TEA. Finally, Sect. 6 presents our key conclusions and ideas for future work.

2   Background and Related Work

Several types of encryption solutions can be used in databases. Encryption standards are provided in built-in packages available in most commercial DataBase Management Systems (DBMS) and in the recent past several approaches on using encryption in databases have been proposed by the database and security research communities. This section describes the encryption standards, how they are implemented in DBMS.

2.1 Encryption Standards

The Data Encryption Standard (DES) is a 64 bit block Feistel scheme cipher which is designed to encrypt and decrypt blocks of 64 bits of data using a 64 bit key [6]. The least significant (right-most) bit in each byte is a parity bit, and should be set so that there is always an odd number of 1 s in every byte. These parity bits are ignored, so only the seven most significant bits of each byte are used, resulting in a key length of 56 bits. The algorithm goes through 16 iterations that interlace blocks of plaintext with values obtained from the key. The algorithm transforms a 64 bit input into a 64 bit output performing a series of steps. The same steps with the same key are used for decryption.

Since DES was found insecure, the Triple DES (3DES) solution was proposed as an enhancement of DES and has replaced it as an encryption standard [1]. 3DES applies the original DES three times to increase the encryption level using three different 56 bit keys (K1, K2, K3, resulting in an effective key length of 168 bits) in an Encrypt-Decrypt-Encrypt (EDE) mode (*i.e.*, the plain text is encrypted with K1, then decrypted with K2, and then encrypted again with K3). The main performance issue that affects 3DES is that since it increases the number of cryptographic operations it is one of the slowest block ciphers.

The Advanced Encryption Standard (AES) [2] is currently the most used encryption standard [11]. AES is a substitution-permutation network cipher that uses 128 bit data blocks. The data blocks are treated as an array of bytes and organized as a 4×4 column-major order matrix of bytes. AES allows three key lengths: 128, 192 and 256 bits, executing 10 rounds for 128 bit keys, 12 rounds for 192 bit keys, and 14 rounds for 256 bit keys. For both encryption and decryption, the cipher begins with an *AddRoundKey* stage. The output of this stage then goes through several rounds and in each of those rounds four transformations are performed: (1) *Sub-bytes*, (2) *Shiftrows*, (3) *Mix-columns*, (4) *Add round Key*. In the final round, there is no Mix-column transformation. Decryption is the reverse process of encryption and uses inverse functions: (1) *Inverse Substitute Bytes*, (2) *Inverse Shift Rows* and (3) *Inverse Mix Columns*. AES is faster and able to provide stronger encryption when compared with DES [11]. Nowadays, there are no known practical attacks that could feasibly break the security of correctly encrypted data with AES.

2.2 Using Encryption in Database Systems

Major commercial database suppliers such as Oracle and Microsoft provide encryption packages in the native versions of their DBMS, such as the Oracle Advanced Security Transparent Data Encryption [10] and Microsoft Transparent Data Encryption services [9] available in the latest versions of these DBMS. Such encryption packages usually only include the 3DES and AES algorithms, since they represent the only solutions that are generally accepted as standard and officially approved by governmental entities and national/global security institutions.

There are mainly two ways of applying encryption in DBMSs [7]:

(1) Tablespace/datafile encryption, which encrypts all the content within a database tablespace or datafile; and

(2) Column-based encryption, which individually encrypts each targeted column of a given table.

Tablespace/datafile encryption has severe security issues in untrusted servers. It relies on only one or two keys to encrypt and decrypt all the data, and it also leaves decrypted clear data stored in the database memory cache, therefore making it vulnerable to RAM attacks (the attacker may retrieve the true data values from the machine's memory). Contrarily, column-based encryption allows encrypting each column with its own key (or keys) and never discloses clear text true values in the database cache [10].

In what concerns database performance, queries running against data encrypted by tablespace/datafile encryption process faster than those against column-based encrypted data. Nevertheless, column-based encryption avoids decryption overhead in queries that process data from non-encrypted columns, contrarily to tablespace/datafile encryption where all data is encrypted and thus needs to be decrypted in order to process any query.

A survey on the use of standard DBMS encryption packages against DWs was published in [11]. This survey presents and discusses the impact on performance caused by the use of encryption in the decision support benchmark TPC-H [14] and a real-world sales DW and the results demonstrate performance penalties of such magnitudes that can make users consider encryption unusable in such environments.

3   XSX: XOr-Switch-XOr ENCRYPTION

The foundations for the cipher proposed in this paper are based on the requirements of leveraging execution performance with security strength. In order to accomplish this, we chose to design a cipher based on bitwise exclusive Or operations, which are typical in most encryption algorithms, and bit switching steps. In this section we describe the cipher's design and its implementation in databases.

3.1   The XSX Cipher

The design of the Xor-Switch-Xor (XSX) cipher focuses on attaining the highest amount of "data mix" with the minimum number of steps possible. By "data mix" we mean the scrambling of data bits in a way that minimizes information leakage between the cipher's intermediate states, using two 64-bit keys for XOR operations and an additional key that defines how to perform the bit switching. Considering x the plaintext value to encrypt and y as the encrypted ciphertext, Fig. 1 shows the external view of the XSX cipher for encrypting a 64-bit input x, considering the following assumptions regarding the referred keys:

- *RKey* and *CKey* represent distinct 64-bit keys;
- *SKey* represents a key value that defines the bit swapping between each pair of bits in each 64-bit state. It is up to the cipher's implementer to define which bits are switched, according to each possible value of *SKey*;

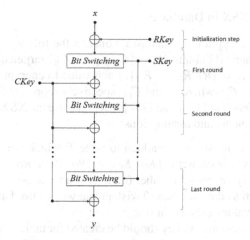

Fig. 1. XSX encryption

The encryption process starts by XORing the plaintext x with the *RKey*, producing an initial 64-bit output (corresponding to the *Initialization Step*). The cipher then processes a sequence of *NR* rounds. In each round, the intermediate 64-bit value first suffers a *Bit Switching* process that makes the bits of predetermined pairs of bits to switch places with each other, for all 64 bits. The value of *SKey* determines the bit pairing process.

Each bit only switches place once in each round. After the bit switching process finishes, the resulting 64-bit value is then XORed against the *CKey*, completing the round. The output produced by this XORing operation in the final round is the ciphered value y.

The decryption process is executed by reversing the encryption process, as shown in Fig. 2.

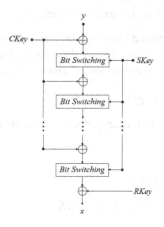

Fig. 2. XSX decryption

3.2 Implementing XSX in Databases

To implement XSX encryption in databases, consider the following: For a given table T with a set of N numerical columns $C_i = \{C_1, C_2, ..., C_N\}$ targeted for encryption and a total set of M rows $R_j = \{R_1, R_2, ..., R_M\}$. Each value to encrypt in the table will be identified as a pair (R_j, C_i), where R_j and C_i respectively represent the row and column to which the value refers ($j = \{1..M\}$ and $i = \{1..N\}$). To use the XSX cipher, we propose the following requirements and assumptions:

- An extra 64-bit column should be added to table T, which we will name $RowKey$. The value of $RowKey$ in each row j $RowKey_j = \{RowKey_1, RowKey_2, ..., RowKey_M\}$ is a 64-bit randomly generated number (e.g. using a pseudo-random generator with a long period, such as the Mersenne Twister) that will be used as the cipher's $RKey$ for all C_i column values existing in row j;
- A unique 64-bit private master key should be created for table T, which we will name as $MasterKey$. This master key is kept in a safe place outside the database and only made available to authorized users;
- To protect the $RKeys$ used in the cipher, each generated $RowKey_j$ value is stored in its respective row j after being XORed with the $MasterKey$. The original true value of each $RKey$ for each row j can then be retrieved whenever it is required by XORing the row's stored $RowKey_j$ against the $MasterKey$;
- A 64-bit random value should be generated for each of the cipher's $CKeys$. One $CKey$ should be used for each C_i column, $CKey_i = \{CKey_1, CKey_2, ..., CKey_N\}$. All defined $CKeys$ are private, kept together with the $MasterKey$ outside the database and only made available to authorized users, and should have distinct values;
- As described in the previous subsection, it is the cipher's implementer who defines the possible values for the $SKey$. Each C_i column should have one distinct $SKey$, $SKey_i = \{SKey_1, SKey_2, ..., SKey_N\}$. As the $CKeys$, all $SKeys$ are private, kept outside the database together with the $MasterKey$ and only made available to authorized users, and should have distinct values;
- We propose a minimum of four rounds to be executed by the cipher.

Based on these requirements and assumptions, and considering the use of the encryption and decryption functions presented in the general implementation shown in the previous subsection and that the cipher will perform four rounds:

$$\text{Encryption of } (R_j, C_i)$$
$$= \text{Encrypt}\left((R_j, C_i), CKey_i, RowKey_j \oplus MasterKey, SKey_i, 4\right)$$
$$= (R_j, C_i)'$$

$$\text{Decryption of } (R_j, C_i)'$$
$$= \text{Decrypt}\left((R_j, C_i)', CKey_i, RowKey_j \oplus MasterKey, SKey_i, 4\right)$$
$$= (R_j, C_i)$$

XSX can function transparently by adding a middleware mechanism that acts as a broker for user queries. This mechanism should receive all user queries, authenticate the user and replace each C_i column in each SQL query with the respective encryption or decryption function, accordingly to what is intended, if s/he is an authorized user. Finally, the broker will supply the processed results back to the user.

4 Security Analysis

The security strength provided by a cipher is one of the main aspects that leads to the acceptability of an encryption solution. In this section we analyze the security of XSX, presenting the assumed threat model and general security aspects, considerations that lead to its design and attack costs against it.

4.1 Threat Model and Overall Security Aspects of XSX

XSX operates in the same fashion as SES-DW [12], thus the threat models are similar. As previously mentioned, all user instructions are to be managed by the query broker middleware, which transparently rewrites them to query the DBMS and retrieve the results. The users never see the rewritten instructions. For security purposes, the middleware should shut off database historical logs on the DBMS before requesting execution of the rewritten instructions so they are not stored in the DBMS, since this would disclose the encryption keys. All communications between user applications, the XSX middleware and the DBMS should be done through encrypted SSL/TLS connections.

In what concerns the storage of the private keys *MasterKey*, *CKeys* and *SKeys*, these should be encrypted using the AES 256 algorithm and stored separately from the DW database(s). This would make them as secure in this aspect as any other similar solution, such as the Oracle Wallet in Oracle 11g TDE. The access to these keys is completely and uniquely managed by the middleware, once it receives a query to be processed from an authenticated and authorized user.

We assume that the DBMS is an untrusted server such as in the Database-As-A-Service (DAAS) model and that the "adversary" is someone that manages to bypass network and DBMS access/authentication controls, gaining direct access to the encrypted values stored in the DW database. We also assume that the XSX algorithm is public, so the attacker can replicate the encryption and decryption functions, meaning that the goal of the attacker is to obtain the keys in order to break security.

By making the values of $CKey_i$ distinct between columns, we also make encrypted columns independent from each other. Even if the attacker breaks security of one column in one table row, the information obtained from discovering the remaining encryption keys is limited. Thus, the attacker cannot infer information enough to break overall security; in order to succeed, s/he must recover all the *CKeys* for all columns. In the same way, using a different $RKey_j$ XORed against the table's *MasterKey* for each row also increases the key search space.

4.2 Bit Switching Versus Bit Shifting and S-Boxes

As discussed in [12], although executing complex operations such as the use of Substitution Boxes (S-boxes) provides a large amount of data mixing at reasonable speed on several CPUs - thus achieving stronger security strength faster than simple operations - the potential speedup is fairly small and is accompanied by huge slowdowns on other CPUs. In fact, it is not obvious that a series of S-box lookups (even with large S-boxes, as in AES, increasing L1 cache pressure on large CPUs and forcing different implementation techniques on small CPUs) is faster than a comparably complex series of integer operations. In contrast, simpler operations such as bit additions, bit shifting, bit switching and XORs are consistently fast, independently from the CPU. Therefore, for the design of the XSX cipher we chose a set of simpler operations: XOR and bit switching.

The reason why we chose to use bit switching instead of typical bit shifting operations is that bit switch operations introduce a higher amount of data transformation - similar to an exclusive OR - than bit shifting operations. Bit shifting only changes each bit's position but maintains the same data bits in the same sequence, therefore allowing information leakage. Additionally, allowing the cipher's implementer to define the way the bit switching occurs and how many different ways are there to accomplish this also adds complexity to any attacker's attempts to crack the keys. Besides increasing the key search space, since the way how the bit-switching occurs is defined by the cipher's implementer, different implementations of XSX will most likely have distinct ways of performing the bit-switching although the used *SKeys* between them could be the same.

4.3 Attack Costs

To break security by key search in a given column C_i, the attacker needs to have at least one pair (plaintext, ciphertext) for a row j of C_i. In this case, taking that known plaintext, the respective known ciphertext, and the respective $RowKey_j$ value (storing $RKey_j$ XOR *MasterKey*, as explained in Subsect. 3.2), s/he may then execute an exhaustive key search.

Considering that the total number of possible bit switching combinations using 64 bits is approximately $1,123 \times 10^{44}$, and that $CKey_i$, $RKey_j$ and *MasterKey* are 64-bit values, the worst scenario for the attacker is to execute an exhaustive search on a key space of $2^{64} \times 2^{64} \times 2^{64} \times 1,123 \times 10^{44} = 2^{192} \times 1,123 \times 10^{44} \cong 7,05 \times 10^{101}$ possibilities.

Note however, that $7,05 \times 10^{101}$ is the worst-case complexity and it is possible for the attacker to reduce the key search space by chosen plaintext attacks. Considering $y_1 = RowKey_j$ and $y_2 = RowKey_{j+1}$, which are known by the attacker, and $x_1 = RKey_j$ and $x_2 = RKey_{j+1}$, and that the same *MasterKey* is used for encrypting all *RKeys* ($RowKey_j = RKey_j \oplus MasterKey$ and $RowKey_{j+1} = RKey_{j+1} \oplus MasterKey$), there is information leakage given by:

$$y_1 \oplus y_2 = (x_1 \oplus MasterKey) \oplus (x_2 \oplus MasterKey) \Leftrightarrow$$
$$y_1 \oplus y_2 = (x_1 \oplus x_2) \oplus (MasterKey \oplus MasterKey) \Leftrightarrow$$
$$y_1 \oplus y_2 = x_1 \oplus x_2$$

This implies that $RKey_j \oplus RKey_{j+1} = RowKey_j \oplus RowKey_{j+1}$, reducing the possible search space for $RKey$ to 2^{32} instead of 2^{64} in each row. This reduces the key search space complexity to $2^{160} \times 1,123 \times 10^{44} \cong 1,64 \times 10^{92}$, which remains a considerable measure of security strength.

Considering NR the number of rounds to be executed by the cipher, note that a slight change in the algorithm's complexity such as the use of distinct $SKey_i$ subkey values in each round of XSX can improve XSX's security strength, increasing the number of possible bit switching combinations to $NR \times 1,123 \times 10^{44}$.

5 Experimental Evaluation

To execute the experimental evaluation, we used a data schema proposed by the TPC-DS benchmark [13]. The TPC-DS is the latest and commonly used benchmark for measuring the performance of Decision Support Systems (DSS). This benchmark has been mapped to a typical business environment as described in [8] and claims to significantly represent DSS that:

- Examine large volumes of data;
- Give answers to real-world business questions;
- Execute queries of various operational requirements and complexities (*e.g. ad-hoc* instructions, reporting actions, data mining operations, etc.);
- Are characterized by high CPU and I/O load;
- Are periodically synchronized with transactional source databases through database maintenance functions.

To execute the experiments, we chose to use the TPC-DS store sales star schema, illustrated in Fig. 3. We chose this particular schema because it represents a common business DW scenario for many enterprises, within the set of proposed data star schemas in TPC-DS. Moreover, the *Store_Sales* fact table is the biggest sized fact table of all generated tables in the complete TPC-DS database, storing a total of almost three million rows in the 1 GB scale size of the database. As shown in Fig. 3, it is composed of the *Store_Sales* fact table and ten dimension tables. The thirteen business facts in the fact table are all stored in thirteen numerical datatype columns, which is typical [8]. This implies that the 1 GB sized database of TPC-DS stores a total of approximately $13.000.000 \times 13 = 169.000.000$ encrypted values, representing a small volume of data that allows us to extrapolate how the performance could potentially be for a larger volume of data given its results.

The *Store_Sales* schema of TPC-DS with 1 GB of scale size was implemented using the Microsoft SQL Server 2012 DBMS on an Intel Core i3–3217U 1.8 GHz CPU with a 500 GB SATA3 hard disk and 8 GB DDR3L 1600 MHz SDRAM (with 2 GB devoted to database memory cache) running Windows Server 2008.

In all experiments we compare XSX with the SQL Server's TDE column-based AES128 and AES256, as well as with OPES, Salsa20, SES-DW and TEA. XSX and all other non-native DBMS solutions were implemented using C#. Each test setup comprised the execution of given workloads using one of the chosen encryption algorithms.

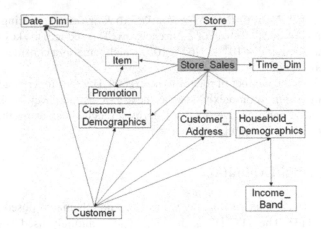

Fig. 3. TPC-DS store sales E-R diagram [13]

5.1 Analyzing Fact Table Loading Time

Figure 4 shows the total execution time (i.e., the sum of all individual query processing times) for loading all the *Store_Sales* fact table rows in each setup. Notice that standard execution time (*i.e.*, without any sort of encryption) takes 130,3 s. Figure 5 shows the total execution time overhead percentage in each encryption setup, relatively to the standard execution time.

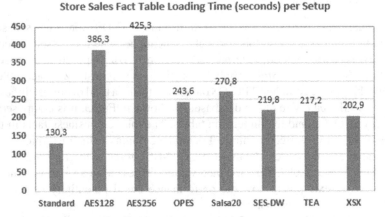

Fig. 4. Store_Sales fact table loading time (s) per setup

Observing the figures, XSX is the fastest encryption setup, taking a 202,9 s to load all *Store_Sales* rows, resulting in an absolute overhead of 72,6 s, or percentage overhead of 56 %. Relatively close, TEA and SES-DW are next in speed, both taking nearly the same time to load *Store_Sales* (217,2 and 219,8 s, respectively, adding 67 % and 69 % overhead), followed by OPES (243,6 s, adding 87 % overhead) and Salsa20 (270,8 s, adding 108 % overhead to standard execution time). The standard encryption algorithms

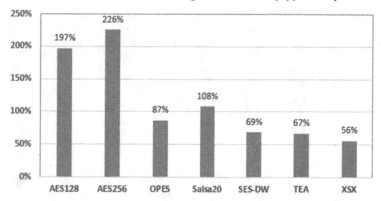

Fig. 5. Store_Sales fact table loading time overhead (%) per setup

are clearly the worst in performance, taking 386,3 and 425,3 s with AES128 and AES256, respectively, corresponding to an overhead of 197 % and 226 %.

Conclusively, XSX requires approximately 50 % more than the standard execution time, with TEA and SES-DW requiring roughly 70 % more and OPES requiring nearly 90 % more. Salsa20 takes twice the time and the standard encryption algorithms take roughly three times the standard execution time to complete the full load of the *Store_Sales* fact table.

5.2   Analyzing Query Performance

The query workload chosen to analyze database query performance in each setup included all the TPC-DS queries that process *Store_Sales* data and produce significant impact on execution time due to the use of encryption. Therefore, the query workload used in each setup was composed by the following set of thirteen TPC-DS queries: {Q8, Q9, Q28, Q36, Q43, Q44, Q47, Q48, Q59, Q65, Q67, Q70, Q89}.

Figure 6 shows the total query workload execution time in each setup. Notice that standard query workload execution time (i.e., without any encryption) takes 117,1 s. Figure 7 shows the total query workload execution time overhead percentage in each encryption setup, relatively to the standard query workload execution time. Observing the figures, XSX is once more the fastest encryption setup, taking a 168,6 s to execute all queries, resulting in an absolute overhead of 51,5 s, or percentage overhead of 44 %. TEA and SES-DW are next in speed, respectively taking 185,7 and 205,5 s, respectively adding 59 % and 76 % overhead, followed by Salsa20 (261,8 s, adding 124 % overhead). OPES and the standard encryption algorithms are clearly the worst in performance, taking 452,0, 535,5 and 560,6 s with AES128, OPES and AES256, respectively, corresponding to an overhead of 286 %, 357 % and 379 %.

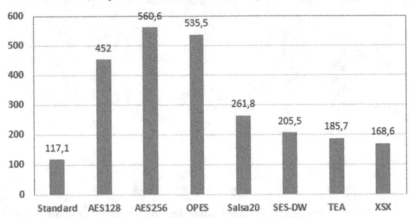

Fig. 6. Total query workload execution time (s) per setup

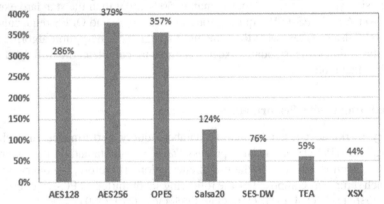

Fig. 7. Query workload execution time overhead (%) per setup

Conclusively, XSX requires approximately 50 % more than the standard query workload execution time, with TEA and SES-DW requiring roughly 60 % and 75 % more. Salsa20 takes more than twice the time, where OPES and the standard encryption algorithms taking roughly around three times the standard execution time to complete the execution of the full query workload.

Figure 8 shows the CPU execution time overhead for the full query workload per setup. As seen, the ordering of the relative speed results between the encryption solutions is similar to those already observed in Fig. 8. XSX has the best response and CPU time overhead (184 %), followed by TEA (212 %), SES-DW (261 %) and Salsa20 (310 %), while the standard encryption algorithms and OPES add between 361 % and 457 % of overhead. This means that the XSX cipher executes slightly faster than TEA (13 % less overhead) and significantly faster than SES-DW (42 % less overhead).

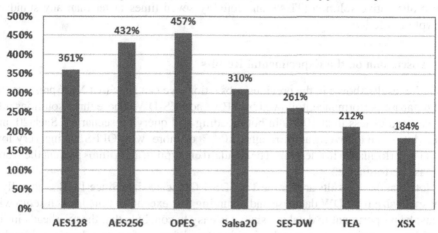

Fig. 8. Query workload CPU time overhead (%) per setup

Figure 9 shows the results for each individual query's execution time overhead, given the most significant queries in the workload (i.e., those that take the most time or that introduce the higher execution time overheads). It can be seen that they generically confirm the same tendency of the overall results of the full query workload.

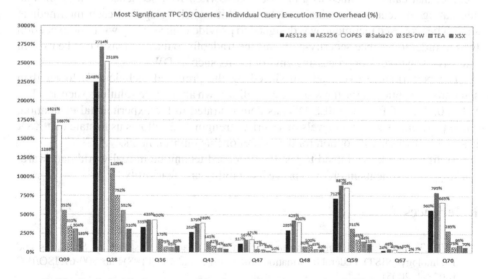

Fig. 9. Individual query workload execution time overhead (%) in each setup

For example, by observing the most significant query in terms of response time overhead (Q28) it can be seen that XSX added 310 % of overhead to the original standard execution time without encryption, while TEA, SES-DW and Salsa20 respectively added 552 %, 752 % and 1105 %, and the standard encryption algorithms added more than 2000 % of overhead. This means that XSX was almost twice as fast as the next fastest alternative solution (TEA), and roughly seven times faster than any standard encryption solution.

5.3 Discussion on the Experimental Results

From the results shown in the previous subsections we can state that XSX presents the best database performance, followed by TEA and SES-DW. These three solutions add an overhead of less than 100 % in both loading and querying scenarios. Salsa20 and OPES introduce an overhead of roughly 100 % or more, with OPES having very low database performance in querying. The standard encryption algorithms present the worst database performance.

Note that these results are for the TPC-DS 1 GB scale size. Since 1 GB is actually a very small size for a DW database and assuming that execution time in each setup will at least be proportional to database size, it is easy to conclude that the overheads introduced by encryption are extremely significant and may in fact introduce considerable performance costs.

6 Conclusions and Future Work

In this paper we have proposed a transparent lightweight encryption technique based on a cipher that can be defined as a bitwise XOr-Switch-XOr sequence, as it applies an alternating sequences of exclusive ORs with bit switching against each intermediate state. The proposed technique was designed to provide high security with low execution time in order to balance security-performance tradeoffs to make it feasible to be used in intensive sensitive data processing environments such as DWs.

The performance overhead introduced by the proposed technique is lower than standard encryption algorithms and other well-known alternative solutions such as AES, Salsa20, TEA, OPES and SES-DW, as demonstrated in the experimental evaluation, while providing significant levels of security strength. This allows us to state that XSX is a viable data security option for data warehousing environments.

As future work, we intend to test the proposed technique in real-world data warehousing environments in order to prove its viability and feasibility.

References

1. DES, Triple DES, National Bureau of Standards, National Institute of Standards and Technology (NIST), Federal Information Processing Standards (FIPS) Pub. 800-67, ISO/IEC 18033-3 (2005)
2. AES, Advanced Encryption Standard, National Institute of Standards and Technology (NIST), FIPS-197 (2001)
3. Agarwal, R., Kiernan, J., Srikant, R., Xu, Y.: Order-preserving encryption for numeric data. In: ACM SIG International Conference on Management of Data (SIGMOD) (2004)
4. Bernstein, D.J.: Snuffle 2005: The Salsa Encryption Function. http://cr.yp.to/snuffle.html

5. Bernstein, D.J.: The Salsa20 family of stream ciphers. In: Robshaw, M., Billet, O. (eds.) New Stream Cipher Designs. LNCS, vol. 4986, pp. 84–97. Springer, Heidelberg (2008)
6. DES, Data Encryption Standard, National Bureau of Standards, National Institute of Standards and Technology (NIST), Federal Information Processing Standards (FIPS) Pub. 46 (1977)
7. Huey, P.: Oracle Database Security Guide 11g. Oracle Corporation (2008)
8. Kimball, R., Ross, M.: The Data Warehouse Toolkit, 3rd edn. Wiley, Hoboken (2013)
9. Microsoft Corporation. Transparent Data Encryption (TDE), Microsoft Developer Network (MSDN) (2014). https://msdn.microsoft.com/en-us/library/bb934049.aspx
10. Oracle Corporation. Oracle Advanced Security Transparent Data Encryption Best Practices, Oracle White Paper (2012). http://www.oracle.com/technetwork/database/security/twp-transparent-data-encryption-bes-130696.pdf
11. Santos, R.J., Bernardino, J., Vieira, M.: Evaluating the feasibility issues of data confidentiality solutions from a data warehousing perspective. In: Cuzzocrea, A., Dayal, U. (eds.) DaWaK 2012. LNCS, vol. 7448, pp. 404–416. Springer, Heidelberg (2012)
12. Santos, R.J., Rasteiro, D., Bernardino, J., Vieira, M.: A specific encryption solution for data warehouses. In: Meng, W., Feng, L., Bressan, S., Winiwarter, W., Song, W. (eds.) DASFAA 2013, Part II. LNCS, vol. 7826, pp. 84–98. Springer, Heidelberg (2013)
13. Transaction Processing Council. The TPC Benchmark™ DS (TPC-DS): The New Decision Support Benchmark. http://www.tpc.org/tpcds/
14. Transaction Processing Council. The TPC Benchmark™ H (TPC-H): Decision Support Benchmark. http://www.tpc.org/tpch/
15. Wheeler, D.J., Needham, R.M.: TEA, a tiny encryption algorithm. In: Preneel, B. (ed.) Fast Software Encryption (FSE). LNCS, vol. 1008, pp. 363–366. Springer, Heidelberg (1994)
16. Wheeler, D.J., Needham, R.M.: Correction to XTEA. Technical report, Cambridge Computer Laboratory, Cambridge University, England (1998)

Graph Databases
and Data Warehousing

Rule-Based Multidimensional Data Quality Assessment Using Contexts

Adriana Marotta[1] and Alejandro Vaisman[2](✉)

[1] Universidad de la República, Montevideo, Uruguay
amarotta@fing.edu.uy
[2] Instituto Tecnológico de Buenos Aires, Buenos Aires, Argentina
avaisman@itba.edu.ar

Abstract. It is an accepted fact that a value for a data quality metric can be acceptable or not, depending on the context in which data are produced and consumed. In particular, in a data warehouse (DW), the context for the value of a measure is given by the dimensions, and external data. In this paper we propose the use of logic rules to assess the quality of measures in a DW, accounting for the context in which these measures are considered. For this, we propose the use of three sets of rules: one, for representing the DW; a second one, for defining the particular context for the measures in the warehouse; and a third one for representing data quality metrics. This provides an uniform, elegant, and flexible framework for context-aware DW quality assessment. Our representation is implementation independent, and not only allows us to assess the quality of measures at the lowest granularity level in a data cube, but also the quality of aggregate and dimension data.

1 Introduction and Background

In typical models for Online Analytical Processing (OLAP) [18], data are represented as a set of *dimensions* and *facts*. Facts are seen as points in a multidimensional (MD) space with coordinates in each dimension and an associated set of *measures* Dimensions provide appropriate contextual meaning to facts, and are usually organized as hierarchies, supporting different levels of data aggregation. *Dimension schemas* are represented as directed acyclic graphs, where the nodes are the dimension levels, with a unique bottom level, and a unique top level, denoted *All*. Each level is associated with a set of coordinate values, denoted *members*. A *dimension instance* is given by a functional relationship defined extensionally between the members in two consecutive levels.

Since a data warehouse (DW) is assumed to be a source of reliable data, which allows taking timely decisions, data quality (DQ) becomes a crucial issue. Data whose quality must be assessed includes members, measures, and aggregations. Quality dimensions typically include data completeness, consistency, accuracy, and timeliness, and for them, a wide range of metrics have been proposed [7]. Contexts, relevant to adapt system requirements to users' characteristics (e.g., location, job position), also play an important role in the assessment

© Springer International Publishing Switzerland 2016
S. Madria and T. Hara (Eds.): DaWaK 2016, LNCS 9829, pp. 299–313, 2016.
DOI: 10.1007/978-3-319-43946-4_20

of DW quality, since the latter cannot be separated from the context where data are produced and consumed [10]. We address this problem in the present work.

Motivation. DQ is crucial in the different stages of a DW system. Even though the Extract-Transform-Load (ETL) process is supposed to guarantee an appropriate DQ, it is not always the case that this quality fulfills the requirements in all possible contexts. Moreover, in many current applications particularly in a Big Data scenario, where the "load first, model later" approach is used (e.g., in near real-time DW), DQ tends to be volatile. It follows that an agile and expressive approach to DQ is needed. To address this problem, in this paper we present an approach based on logic rules, which builds from the representation of an MD data model as a collection of axioms and rules, based on the idea proposed by Minuto et al. [12], where facts are described as abstract entities (ground atoms), and dimensions give context to these facts. Previous efforts for representing contexts in MD databases using rules have been carried out. In [10], description logics was used, and DQ was addressed in a limited way. In [11], the MD model and context rules are represented as an ontology expressed in Datalog$^\pm$, although this proposal does not tackle DQ dimensions, and it is focused on query answering. Moreover, it does not consider aggregation, a key issue in DW.

Running Example. A supermarket chain maintains sales information in a DW, such that each tuple in a fact table contains a product identifier, the date of a sale, the branch identifier, and the number of units sold. The fact table is depicted in Fig. 1. Here, productId, date and branchId represent the bottom levels of the dimensions Product, Time and Branch, respectively; quantity is a measure. The attribute sId identifies a particular sale. The dimension hierarchies are the following: For the Time dimension we have date $\rightarrow$ month $\rightarrow$ year; for Branch, branchId $\rightarrow$ neighbourhood $\rightarrow$ city $\rightarrow$ country; finally, for Product, productId $\rightarrow$ family $\rightarrow$ type $\rightarrow$ category. Based on the model introduced in [12], the first fact in Fig. 1 can be represented as $AFact_{qty}(Sales, s_1, 50)$, a predicate called an *abstract fact*, whose meaning is given by the dimensions. For this, the hierarchies and their instances are represented as aggregation rules. For example, to say that "All sales of chocolate bars (a member of dimension level productId) should be assigned the family 'chocolate' ", we can write the Datalog rule [1]:

$aggr(X, \text{Product}, \text{family}, chocolate) \leftarrow aggr(X, \text{Product}, \text{productID}, choco\_bars)$

We remark that, although normally each level in a dimension is described by *attributes*, for simplicity, we will just work directly with member identifiers, e.g., we will use *choco_bars* as identifier of this member, instead as the value of an attribute name of the product identified as p_1.

Although based on [12], the model we propose in this paper differs from the former in many ways. First, the abstract facts now include a new sort, whose carrier set is a set of fact table names, allowing handling more than one fact table. Second, the above modification also allows us to express rules over aggregated abstract facts (i.e., materialized aggregate views). Finally, we simplify aggregation rules to express them in Datalog, instead of default rules.

Once the MD model is represented as described above, it is natural to define contexts as rules. For example, a context rule for measure quantity can state that

sld	productId	date	branchId	quantity
s_1	choco_bars	3-1-2014	tata_pocitos	50
s_2	choco_assortment	3-1-2014	tata_pocitos	20
s_3	choco_truffles	3-1-2014	tata_union	35
...	...	...	..	..
s_{50}	candies	5-4-2015	tata_union	35

Fig. 1. Fact table Sales

"The values for measure quantity are affected by the values of members in the dimension level family and the month of the sale." DQ metrics can be defined also as rules, e.g., a quality rule for assessing the accuracy of the measure quantity. This rule can also consider the context defined above. This provides *a uniform framework for defining the MD model, the contexts, and DQ metrics.* In the paper we explain how these rules are produced and used. It is also worth noting that our MD model allows providing context not only to measures, but also to dimension members. Further, context for measures can be given not only by dimensions (which is the natural situation), but also by external data, e.g., tables that we do not need to include in the DW process.

Contributions and Paper Outline. Summarizing, in this paper we (a) present an MD model which uses Datalog rules to represent facts, dimensions, and aggregate data at any granularity level; (b) use Datalog rules to represent contextual information in a DW; (c) apply the model to the case of DQ assessment in DW, such that DQ metrics are also defined by rules. However, our proposal can, of course, be applied in many contexts, for example, in recommender systems, or in data integration systems.

The remainder of this paper is organized as follows: in Sect. 2 we discuss related work. Section 3 presents, informally, the ideas we develop in the paper. In Sect. 4 we present the formal MD model on which we base our proposal, and in Sect. 5 we present the rule-based approach to data quality assessment in multidimensional databases. We conclude in Sect. 6.

2   Related Work

Ensuring the quality of data in databases has long been a research topic in the database community, resulting in the definition of DQ dimensions and metrics [19]. However, in classic research, DQ is considered as a concept that is independent of the context in which data are produced and used. This assumption is clearly not enough to solve complex problems. This was realized by Strong et al. [17], who thus claimed that data quality is highly dependent on the context, which became an accepted fact thereon. We next briefly review research in DW quality, context, and the combination of both concepts.

DW quality. One of the first efforts in this field, the DWQ project [9], proposed techniques to support a rigorous design and operation of DWs, mainly based on well-defined DQ factors. More recently, Daniel et al. [7] studied DQ dimensions and metrics appropriate for Business Intelligence (BI), proposing an architecture for quality-aware BI. They identified measurable quality properties that may appropriately characterize data for warehousing and mining. Gongora de Almeida et al. [2] proposed a taxonomy for DQ issues in MD data models, building from taxonomies already proposed for general DQ problems. They finally identified five DQ dimensions: Completeness, Timeliness, Uniqueness, Accuracy, and Consistency, studying each of them in detail.

Context and Context Models. There is a large corpus of work regarding data context management with a wide variety of uses. Contexts are mainly proposed for addressing personalized access to data, increasing the precision of information retrieval, automatic recommendation, and many other application fields. It has been widely accepted that most modern applications, particularly over the web, are required to be context-aware. Bolchini et al. [5] presented a survey of context models, with a well-defined structure, that identified some important aspects of context models. In particular, they remark that models must account for *space, time, context history, subject,* and *user profile.* Finally, user preferences in databases have been extensively studied, and there are basically two approaches for this: (a) Quantitative, where a numerical score is associated with tuples [6]; (b) Qualitative, where a partial order relation is defined on tuples [16].

Context and DW Quality Assessment. Several works relate data context to DW systems and MD data. Poreppelmann et al. [15], proposed a framework that allows context-sensitive assessment of DQ, through the selection of adequate dimensions for each particular decision-maker context, and her information demand. Bolchini et al. [4] define context-aware views over large information systems, one of whose components is a DW. For this, they propose a Context Dimension Tree (CDT), to model context in terms of a set of dimensions, each capturing a different characteristic of such context. Pérez et al. [13], proposed to contextualize a traditional DW through documents gathered from web forums. Pitarch et al. [14] use context in order to deduce a new classification of the DW content. They call this a "context-aware hierarchization" of measures.

Related to the work we present in this paper, some efforts proposed the use of logic rules to represent contexts. In [10], the authors propose to use DW dimensions to provide context to fact data, and represented dimensions using description logics, an approach which is known to be computationally expensive, in general. As an evolution of this work, Milani et al. [11], represented the MD model as an ontology expressed in Datalog$^\pm$, as well as context rules. However, the work does not address DQ dimensions or metrics, and it is focused on improving the quality of query answering. Moreover, it does not consider aggregation, a key issue in DW, which we also consider in our work. We built on these ideas, and propose a rule-based model for assessing DW quality accounting for the context.

3   Context-Dependent Data Warehouse Quality

We now present, informally, the notion of contexts, and how they can be applied to assess the quality of data at each stage in a typical DW architecture.

3.1   Data Quality in a Data Warehouse Architecture

The analysis of DQ in data warehousing depends on the stage of the DW lifecycle we are considering for such analysis. In this work we consider three specific stages, each of them corresponding to a particular component of the DW system architecture: (i) Source data extraction; (ii) DW loading; and (iii) DW exploitation. We claim that for each one of them, different DQ models may be defined, and different context information should be taken into account.

In the *data extraction stage*, the context for DQ assessment highly depends on the type of the data source, as discussed in [8], where three types of data sources are defined: human-sourced, process-mediated, and machine-generated data sources. Human-sourced data sources are all kinds of data generated by humans in different places (e.g., social networks, blogs, etc.). Process-mediated sources are traditional business data, usually stored in relational databases. Machine-generated sources are data generated by sensors, machines, and the kind. The authors defined specific DQ dimensions for each type of source. Along these lines, we consider that for the *data extraction stage*, the *context* consists of information about the source type, e.g., for machine-generated sources, the context would be given by the environment where the measures are taken, and the underlying measurement process of the sensors.

In the *DW loading stage*, data are structured following a logical model, where relational tables represent dimensions or facts, groups of attributes represent hierarchies and other attributes represent DW measures (or indicators). To assess the quality of these data, the *context* to be considered is related to the role of each data item in the MD model. For example, the quality of DW measures may be assessed considering the dimension data that influence the measure.

In the *DW exploitation stage*, the user becomes a first-class citizen in the DQ assessment scenario, following the well-known *fitness for use* approach. Therefore, the context in this case is given by information about the user, such as user environment, user profile, and user quality requirements.

3.2   Data Quality Dimensions

A *quality dimension* [3,17] is a concept that captures a particular facet of DQ; A *quality metric* is a quantifiable instrument that defines the way a dimension is measured, and indicates the presence or absence of a DQ problem. Many different metrics can be associated to the same DQ dimension, to measure several different aspects of the dimension. While a large set of DQ dimensions were proposed in the literature, there is a basic set of them, which are widely accepted [3]. This set includes: accuracy, completeness, consistency, and freshness. In what follows, we will focus on *accuracy* and *consistency*, which we consider enough to convey the idea of our proposal. Further, both dimensions are relevant to assess the quality of DW measures (the business indicators for decision makers) **Accuracy** specifies how accurate and valid the data are. We talk about *semantic accuracy* to indicate how close is a real world value to its representation in the database; *syntactic accuracy*, on the other hand, tells if a value belongs to a valid domain. **Consistency** indicates the satisfaction of semantic rules defined over two or more data items that belong to the database. Thinking in relational terms, these items may belong to the same tuple, to different tuples of the same table or to different tables of the same database.

3.3   Contexts and Data Quality

As mentioned above, context affects DQ in many ways, and through many dimensions. Contextual data may be internal, external, or both. We denote *internal context* the one that consists only of data from the DW, and *external context* the one which consists only of data from outside the DW. *Internal-external contexts* include both kinds of data. We formalize these notions in Sect. 5.

We illustrate the above using our running example. We propose some quality metrics (that we will formalize in Sect. 5), that account for the context, to be applied at different stages of a DW architecture. We start with the *accuracy* dimension, and define the metrics *sales_bound_I* and *sales_bound_IE* as follows:

(a) *sales_bound_I*. This metric rates as inaccurate a sale whose quantity is greater than a certain upper bound, given a certain *internal* context. The context consists of a product or group of products, a time period and a place defined by selecting a level for each of the dimensions **Product**, **Time** and **Branch**. For example, suppose we want to detect if the sales quantity of a product in the *chocolate* family in *January* in *Uruguay* is greater than 50, which probably is an inaccurate value, given our domain knowledge. In this case, the context is composed of the levels (family, month, country), that will be instantiated by the member values (*chocolate*, *January*, *Uruguay*), when the formula for the metric is evaluated.

(b) *sales_bound_IE*. This metric rates as inaccurate a sale whose value is greater than an upper bound, given a certain *internal-external* context. The *internal* context is given by dimensions **Product** and **Time**. The *external* context is given by a table containing external data. For example, suppose we want to verify if during unusually high temperatures in the winter, the sales of chocolate bars appear to be increasing, which may reveal some data error. We instantiate the metric setting the upper bound to 200. The context is formed by the dimension levels product_id, year, the table **Temperatures** shown in Fig. 2, and the condition $maxTemp > 1.4 * avgTemp$. In this table, maxTemp indicates the maximum temperature of the day, and avgTemp the historical average of maximum temperatures for that day. We can instantiate the formula for the metric, to verify if the sales quantity of the product *choco_bars*, during days in 2014 when the maximum temperature was unusually high, was greater than 200. These measures will be suspected of being incorrect.

day	maxTemp	avgTemp
3-7-2014	12	7
4-7-2014	14	7
5-7-2014	15	7
...	...	...

Fig. 2. External table Temperatures

Let us now consider *consistency*. The metric sales_consistency_IC checks the consistency between the sale quantity values of two different facts corresponding to two different moments, given a certain *internal* context, which consists of a product or group of products and two time periods. This metric is instantiated defining a level and a member for the dimension Product, and a level and two members for the dimension Time. For example, suppose we want to check if the sale quantity of chocolates in certain year is smaller than in the previous year, which would be suspicious, since we intuitively know that sales have been increasing year after year. We instantiate the metric such that the context is formed by the levels (family, year), and the members are (*chocolate, 2013, 2014*).

Note that the three metrics proposed, are applied at the *loading stage* of the DW architecture explained in Sect. 3.1. In Sect. 5 we also propose metrics to be applied at the exploitation stage.

4   Rule-Based Warehouse Model

MD databases are usually presented as collections of concrete facts and dimension instances, stored in relational implementations called *fact tables* and *dimension tables* [18], respectively. We now introduce the MD model from a formal, many-sorted, axiomatic and proof-theoretic perspective. Let us consider the following sorts and their associated finite sets: a sort **D** of dimension names, a sort **L** of level names, and a sort **C** of constants.

Definition 1 (Dimension Schema). *Given a dimension name $d \in$ **D**; a set LevelSet(d) included in* **L***; a relation on $LevelSet(d)^2$ denoted $\preceq_d$, such that $\preceq_d^*$, its transitive closure, is a partial order on LevelSet(d), with a unique minimum element, denoted l_{bottom}, and a unique top element, denoted All, such that $l_{bottom} \neq$ All. An element $l \in LevelSet(d)$ is called a level of d, and it is denoted $d : l$. The pair $(LevelSet(d), \preceq_d)$ is called a dimension schema for d.*   □

Definition 2 (Instance Sets of Levels). *Given LevelMember, a relation in* **D** × **L** × **C***, that associates constants in* **C** *(which we will denote coordinates) with levels in a dimension; and a tuple (d, l, c), called a dimension level member, meaning that c is a coordinate in level l of dimension d (for clarity, in what follows we will denote this tuple as $d : l : c$). The set $Iset(d : l) = \{c \mid (d : l : c) \in LevelMember\}$ is called the Instance Set of level l in dimension d. Also, there exists a constant all $\in$ **C** such that $Iset(d : All) = \{all\}, \forall d \in$ **D***.   □

Example 1. The pair $(LevelSet(\mathsf{Product}), \preceq_{\mathsf{Product}})$, is a schema for dimension Product in Sect. 1. $LevelSet(\mathsf{Product}) = \{\mathsf{productId}, \mathsf{family}, \mathsf{type}, \mathsf{category}, \mathsf{All}\}$ and $\preceq_{\mathsf{Product}}$ is defined according to the hierarchies in Sect. 1.   □

Definition 3 (Space). *Let $d_1 : l_1, \ldots, d_k : l_k$ be dimensions in $\mathbf{D}$, and let $l_1, \ldots, l_k$ be levels in $\mathbf{L}$, such that for all $1 \le i \le k$, $l_i \in LevelSet(d_i)$. The tuple $(d_1 : l_1, \ldots, d_k : l_k)$ is called an MD space.*   □

Definition 4 (Cell). *Let $(d_1 : l_1, \ldots, d_k : l_k)$ be an MD space, and let $c_1, \ldots, c_k$ be coordinates such that for all $1 \le i \le k$, $c_i \in ISet(d_i : l_i)$. The tuple $(d_1 : l_1 : c_1, \ldots, d_k : l_k : c_k)$ is called a cell in the MD space.*   □

Facts are abstract entities (regardless their nature and the measures that apply to them), identifying instances of some class (like object IDs in object-oriented models). We define a general sort $\mathbf{A}$ for abstract facts, and different types $\mathbf{M}$ for measures. $\mathbf{F}$ is a set of fact table names.

Definition 5 (Fact Valuation). *Facts and measures are associated through a family of predicates AFact_M, with signature $\mathbf{F} \times \mathbf{A} \times \mathbf{M} \to boolean$. A fact valuation is a ground atom formed upon predicate AFact_M for some measure.* □

Example 2. In the example in Sect. 1, values s_1, s_2, s_3 in column sId of table Sales are abstract facts, i.e., members of the carrier set of sort $\mathbf{A}$. A sort $\mathbf{Qty}$ represents the measure quantity, and values 50, 20, and 35 in column quantity are members of the carrier set that sort. Fact valuations for rows in table Sales are: $AFact_{\mathbf{Qty}}(\mathsf{Sales}, s_1, 50), AFact_{\mathbf{Qty}}(\mathsf{Sales}, s_2, 20), AFact_{\mathbf{Qty}}(\mathsf{Sales}, s_3, 35)$.   □

Facts are given their real meaning when they are associated with the coordinate of a cell, by attaching abstract facts to level members. The following definition allows performing this association.

Definition 6 (Fact Aggregation). *Consider a predicate aggr with signature $\mathbf{A} \times \mathbf{D} \times \mathbf{L} \times \mathbf{C} \to Boolean$, associating an abstract fact with a level member. Ground atoms for aggr are denoted fact aggregations. The informal meaning of an abstract aggregation (a, d, l, c) is that an abstract fact a contributes to the content of some cell with coordinate c belonging to level l of dimension d.*   □

Example 3. The following atoms are examples of fact aggregations, corresponding to the first row in Table Sales in Sect. 1: $aggr(s_1, \mathsf{Product}, \mathsf{productId}, choco\_bars)$; $aggr(s_1, \mathsf{Branch}, \mathsf{branchId}, tata\_pocitos)$. Note that the notion of fact aggregation allows defining facts at any granularity level, therefore we can represent materialized views. For instance, $AFact_{\mathbf{Qty}}(\mathsf{SalesByFamily}, pf_1, 105)$, represents the aggregation by product family, and its meaning is given by the aggregated abstract fact $aggr(pf_1, \mathsf{Product}, \mathsf{family}, chocolate)$.   □

MD databases are built from collections of base facts (i.e., facts with attributes and measures), stored in relational implementations called fact tables. From a logical perspective, base facts are sets of statements denoted axiom bases, providing a basis for logical inference. We formalized this next.

Definition 7 (Axiom set). *Let d be a dimension in* **D**. *We define the* axiom set *as the set* $W(d) = \{\mathrm{aggr}(X, d, l_{bottom}, \mathbf{b})$, *such that* l_{bottom} *is the bottom level of dimension* d, *and* $\mathbf{b} \in Iset(d : l_{bottom})\}$. $W(d)$ *contains fact aggregation schemas (with abstract fact variable X) for the bottom level of dimension* d.   □

Data in an MD database are usually summarized according to extensions of the hierarchies of dimensions and those extensions are called *rollups*, basically functions that relate coordinates in a level of a dimension hierarchy with coordinates in all of its immediate successor levels. In our approach, these rollups are represented as rules. For instance, the rollup: $d : l_2 : c_1 \mapsto d : l_1 : c_2$, relating two coordinates c_1 and c_2 belonging to levels l_1 and l_2, such that $l_1 \preceq l_2$ in a dimension d respectively, is represented as an *aggregation rule* of the form:

$$aggr(X, d, l_2, c_2) \leftarrow aggr(X, d, l_1, c_1)$$

Variable X will be substituted by values of sort **A** (abstract facts) when evaluating the rule. Intuitively, the meaning of an aggregation rule is: if an abstract fact X aggregates over a coordinate c_1 in level l_1 of dimension d, we infer that X should be aggregated over coordinate c_2 in level l_2. Rule schemas of the form $aggr(X, d, \mathsf{All}, \mathsf{all}) \leftarrow aggr(X, d, l_{bottom}, a)$, where l_{bottom} is the bottom level of dimension d, are called implicit aggregation rule schemas.

Example 4. The rollup Product : productID : *choco_bars* ↦ Product : family : *chocolate*, is mapped to the aggregation rule:
$aggr(X, \mathsf{Product}, \mathsf{family}, chocolate) \leftarrow aggr(X, \mathsf{Product}, \mathsf{productID}, choco\_bars)$ □

Definition 8 (Dimension Instance). *A set of aggregation rules for a dimension d is called a dimension instance of dimension d.*

The notions of fact valuation and fact aggregation (Definitions 5 and 6, respectively), capture the concept that the context of a fact is given by the dimensions, which is implicit in the MD model. We also point out that: (a) Our rule-based model is completely independent of the actual representation of the data in a DW. For example, the first row in the fact table in Fig. 1, can be considered a synthesis of the following assertions: $AFact_{\mathbf{Qty}}(\mathsf{Sales}, s_1, 50)$, $aggr(s_1, \mathsf{Product},$ productId, *choco_bars*), $aggr(s_1, \mathsf{Branch}, \mathsf{branchId}, tata\_pocitos)$, $aggr(s_1, \mathsf{Time},$ date, *3-1-2014*); (b) The model supports incomplete information in a given fact. For example, if the product, the date of a sale, or even the quantity sold are unknown, we could still represent this fact; (c) The model supports aggregated materialized views (summary tables), e.g., an abstract fact may refer to aggregated facts, like in $AFact_{\mathbf{Qty}}(\mathsf{SalesByFamily}, pf_1, 105)$, whose context is given by $aggr(pf_1, \mathsf{Product}, \mathsf{family}, chocolate)$, where SalesByFamily is a summary table.

5   Rule-Based Data Quality

We now define a context model for MD data, that fits the model presented in Sect. 4, and apply this model to the assessment of DW quality. The main idea is to represent DQ metrics and contexts as Datalog rules. In this work, we will consider only rules expressed as conjunctive queries and unions, expressing the theories in non-recursive Datalog [1]. The following definitions formalize the concepts introduced in Sect. 3.

Definition 9 (Context and Context Rules). *Let $\mathcal{M}$ be an MD model, consisting in a collection of fact valuations (Definition 5), dimension schemas (Definition 1), and dimension instances (Definition 8); a collection of data objects, which for simplicity we assume to be a set of tables $\mathcal{T} = T_1, \ldots, T_k$. A context rule is a rule defined over the elements in $\mathcal{M}$ and $\mathcal{T}$.* □

Definition 10 (Data Quality Rules). *Given an MD model $\mathcal{M}$, a collection of tables $\mathcal{T}$, and a set of context rules $\mathcal{C} = \mathcal{C}_1, \ldots, \mathcal{C}_n$. There is also a set M_q of data quality metrics. A data quality rule $\mathcal{Q}$ is a rule defined over the elements in $\mathcal{M}$, $\mathcal{T}$, and $\mathcal{C}$, expressing a data quality metric in M_q.* □

In what follows, we present some example cases that characterize the use of our proposal, starting from the general case, and then showing other situations where our approach allows a general management of DQ using rules. In the general case, context and rules are defined over base fact data, only considering data in the DW. Then we extend this, to show that external data can be included into the context. We continue showing that quality of aggregate data can also be accounted for. We conclude the study allowing data users to be part of the context, therefore showing how we can use our approach to customize data quality values depending on the characteristics of the user. We next show a portion of our running example, in our rule-based MD formalism.

(1) Rules defining the rollups for the Product dimension:

$aggr(X, \mathsf{Product}, \mathsf{family}, chocolate) \leftarrow aggr(X, \mathsf{Product}, \mathsf{productID}, choco\_bars)$

$aggr(X, \mathsf{Product}, \mathsf{family}, caramel) \leftarrow aggr(X, \mathsf{Product}, \mathsf{productID}, candygum)$

(2) Rules defining the rollups for the Branch dimension:

$aggr(X, \mathsf{Branch}, \mathsf{country}, uruguay) \leftarrow aggr(X, \mathsf{Branch}, \mathsf{branchID}, tata\_union)$

(3) Rules defining the rollups for the Time dimension:

$aggr(X, \mathsf{Time}, \mathsf{month}, 1 - 2014) \leftarrow aggr(X, \mathsf{Time}, \mathsf{date}, 3 - 1 - 2014)$

$aggr(X, \mathsf{Time}, \mathsf{year},' 2014') \leftarrow aggr(X, \mathsf{Time}, \mathsf{month}, 1 - 2014)$

(4) Rules defining dimension instances (internal context values):

$aggr(s_1, \mathsf{Time}, \mathsf{date}, 3\text{-}1\text{-}2014)$.

$aggr(s_1, \mathsf{Product}, \mathsf{productID}, choco\_bars)$.

$aggr(s_1, \mathsf{Branch}, \mathsf{BranchID}, tata\_pocitos)$.

$aggr(s_3, \mathsf{Time}, \mathsf{date}, 3\text{-}1\text{-}2014)$.

$aggr(s_3, \mathsf{Product}, \mathsf{productID}, choco\_truffles)$.

$aggr(s_3, \mathsf{Branch}, \mathsf{branchID}, tata\_union)$.

(5) Rules defining abstract facts:

$AFact_{\mathbf{Qty}}(\mathsf{Sales}, s_1, 50)$

$AFact_{\mathbf{Qty}}(\mathsf{Sales}, s_2, 20)$

$AFact_{\mathbf{Qty}}(\mathsf{Sales}, s_3, 35)$

Basic Case: Factual Warehouse Data. We start showing how *Sales_bound_I*, the metric presented in Sect. 3 can be expressed in our rule formalism. We want to assess the *accuracy* of the measure quantity, for a given context. For this, we will define a quality rule, that will be evaluated against a theory formed by the union of the rules defining the MD model, and the rules defining the context. In this example we just consider the *internal context*. We assume that the context for the measure quantity is given by the family of the product sold, the country, and the month of the sale. We can express this context as follows:

$$context(X, m, y, z, w) \leftarrow AFact_{\text{Qty}}(\text{Sales}, X, m), aggr(X, \text{Product}, \text{family}, y),$$
$$aggr(X, \text{Time}, \text{month}, z), aggr(X, \text{Branch}, \text{country}, w).$$

Of course we can define more than one context for the same measure, and use them in several quality rules. We are interested in the accuracy dimension. We thus define the following metric: *A measure is likely to be inaccurate if the quantity sold of a product in the 'chocolate' family, in January in Uruguay is greater than 50.* This rule is written as:

$$sales\_bound\_I(X, m) \leftarrow context(X, m, chocolate, 1 - 2014, uruguay), m > 50.$$

This rule is evaluated in the usual way. For example, we replace the predicate *context* by the right hand side of the rule defining it, and find a match for the first *aggr* predicate. Note that this predicate does not match any ground atom *aggr* (the bottom levels of the dimensions), since family is not a ground predicate (a logic fact). Thus, we need to look into the rollup rules. We match the predicate with the head of the first, i.e., with $aggr(s_1, \text{Product}, productID, choco\_bars)$, by instantiating X with s_1. Then, we match the second *aggr* predicate, e.g., against the ground rule $aggr(s_1, \text{Time}, date, 3\text{-}1\text{-}2014)$. We proceed analogously for the third *aggr* predicate. Finally, we instantiate X with s_1 in the first *aggr* predicate, obtaining: $aggr(s_1, \text{Product}, family, chocolate)$, replace s_1 in the $AFact$ predicate in the context rule, match with $AFact_{\text{Qty}}(\text{Sales}, s_1, 50)$, instantiate variable m with value 50, producing the rule $context(s_1, 50, chocolate, 1 - 2014)$. In this case, we do not find a match, because $m > 50$ does not hold, e.g., *the first fact satisfies the quality rule.*

External Databases. We now add external information, in order to give context to a DQ metric. These data are not part of the DW, since, e.g., it implies an increase in the maintenance costs, and their usage may be sporadic. We have called this, the *external context*. Consider for example, the *Sales_bound_IE* metric of Sect. 3. As we explained, sales can be affected by abnormal temperatures. Thus, we extend the previous example checking if, during days with unusually high temperatures (e.g., in the winter), there are facts indicating that the sales of chocolate bars are increasing, which may appear, *prima facie*, potentially erroneous. For that, we use external information about daily and average temperatures, shown in table Temperatures in Fig. 2. We can express table Temperatures as a collection of ground predicates, of the form:

$$temperature(3 - 7 - 2014, 12, 7), temperature(4 - 7 - 2014, 14, 7)...$$

We now define that the context for the metric is given by the product and the temperature at the date of the sale. We also include the condition that the

sld	family	year	sumA
v_1	chocolate	2014	105
v_2	pasta	2014	200
...	...	...	..

Fig. 3. Aggregate view SalesbyYrFam

maximum temperature at such date is 40 % higher that the historical average temperature for the day. The rule that defines this context reads:

$$context(X, m, y, d) \leftarrow AFact_{\mathbf{Qty}}(\mathsf{Sales}, X, m), aggr(X, \mathsf{Product}, \mathsf{productId}, y),$$
$$aggr(X, \mathsf{Time}, \mathsf{date}, d), temperature(d, z, w), z > 1.4 * w.$$

Finally, we can write the context-aware quality rule stating that, given the context above, if a sale of chocolate bars exceeds 200 units, it should be considered suspicious of being wrong:

$$sales\_bound\_IE(X, m) \leftarrow context(X, m, chocolate\_bars, d), m > 200.$$

Aggregation. Our proposal allows to include aggregate materialized views. Consider, e.g., the view SalesbyYrFam in Fig. 3, containing the total sales by year and product family. Considering each element in this view as an abstract (aggregated) fact allows us to treat this as a regular fact table. For example, for the aggregate measure sumQuantity (abbreviated sumQ), we can define a context given by the product family and the year, as follows:

$$context(X, m, d, y) \leftarrow AFact_{\mathbf{sumQ}}(\mathsf{SalesbyYrFam}, X, m),$$
$$aggr(X, \mathsf{Product}, \mathsf{family}, y),$$
$$aggr(X, \mathsf{Time}, \mathsf{year}, d).$$

We can now formalize the *Sales_consistency_IC* metric of Sect. 3 (accounting for consistency), writing the rule stating that: *An aggregated yearly sale for the family of chocolates can be suspected to be incorrect, if is less than in the previous year.* The intuition behind this rule is that our business knowledge tells that sales have been increasing year after year. The quality rule reads:

$$sales\_consistency\_IC(X, m) \leftarrow context(X, m, d, chocolate),$$
$$context(X_1, m_1, d_1, chocolate),$$
$$X \neq X_1, m < m_1, d > d_1.$$

Accounting for Users' Context. We now present a metric to be used at the *exploitation stage* in our DW architecture of Sect. 3.1. More than often, different users require different quality thresholds. For example, a high-level user may require a less precise value, while a user closer to the production line may need more accurate records. Our approach is flexible enough to handle this situation.

Assume, for instance, a base predicate $user(x, pos)$, representing the final users of a DW system. For example,

$$user(u_1, CEO), \ user(u_2, CFO), \ user(u_3, Director), \ user(u_4, SalesRep)$$

Now, assume that we want to customize the quality metrics depending on the user. Thus, we define a base predicate (that will be instantiated with ground facts), $threshold(p, t)$, defining a threshold value for each position. We will have ground atoms like: $threshold(CEO, 20), threshold(SalesRep, 10)$. We are ready now to reformulate the quality rule $Sales\_bound\_I$, in a way such that we can obtain, for each user, the facts that do no satisfy the quality conditions, considering the corresponding position's threshold. The rule would read:

$$sales\_bound\_I(X, m, p) \leftarrow context(X, m, chocolate, 1 - 2014, uruguay),$$
$$m > t, threshold(p, t), user(u, p)$$

Remark 1. All queries in this section have been run, just as a poof-of-concept, using a logic-rule engine, writing the rules in Prolog, with a data sample of ten thousand abstract facts. The results were very promising, although in this paper we do not focus in performance, but in showing the viability of our approach.

Discussion. Note that in this paper we have limited ourselves to nonrecursive datalog, and we did not even use negation. This is because our goal at this point is to show the applicability of the approach. Note that we could have also used negation, to express more complex queries, and still remain within polynomial complexity, since it has been proved that for range-restricted queries, nonrecursive datalog (with negation) is equivalent to relational algebra. All in all, even the most basic Datalog version allows us to: express context for MD data, not only using DW dimensions but external data too; express and formalize data quality metrics over both, base and aggregate fact tables; and define context-dependent quality rules. Finally, our intuition is that for realistic, practical cases, this theoretical machinery is more than enough.

6   Final Remarks and Future Work

We have addressed the problem of DW quality taking into account the context where data are produced and consumed. We studied this problem from a logic-based perspective. To achieve this goal, we first represented the MD model using Datalog rules, adapting, and simplifying, the model in [12]. We then showed how contexts for MD data can be represented also as logic rules. The picture gets completed by defining DQ metrics also by means of rules. In this way, we provided a uniform framework for giving context to warehouse data, supporting internal and external contextual data. To the best of our knowledge, this is the first proposal of this kind that applies context to DQ metric in a DW. The above does not mean that the proposal is deemed to remain as a theoretical discussion. On the contrary, we are providing an implementation-independent framework, which can be easily turned into a practical solution, since we just use non-recursive Datalog, whose complexity is the same as the relational algebra, so we do not add additional complexity to the picture. Our future steps (on which we are currently working)include the implementation of this proposal (so far, we have worked with proof-of-concept implementations of our approach), and the application of the framework to real-world case studies.

References

1. Abiteboul, S., Hull, R., Vianu, V.: Foundations of Databases. Addison-Wesley, Boston (1995)
2. Gongora de Almeida, W., de Sousa, R., de Deus, F., Amvame Nze, G., Lopes de Mendonca, F.: Taxonomy of data quality problems in multidimensional Data Warehouse models. In: Proceedings of CISTI, pp. 1–7 (2013)
3. Batini, C., Scannapieco, M.: Data Quality: Concepts, Methodologies and Techniques. Data-Centric Systems and Applications. Springer, Heidelberg (2006)
4. Bolchini, C., Curino, C.A., Orsi, G., Quintarelli, E., Rossato, R., Schreiber, F.A., Tanca, L.: And what can context do for data? Commun. ACM **52**(11), 136–140 (2009)
5. Bolchini, C., Curino, C.A., Quintarelli, E., Schreiber, F.A., Tanca, L.: A data-oriented survey of context models. SIGMOD Rec. **36**(4), 19–26 (2007). http://doi.acm.org/10.1145/1361348.1361353
6. Ciaccia, P., Torlone, R.: Modeling the propagation of user preferences. In: Jeusfeld, M., Delcambre, L., Ling, T.-W. (eds.) ER 2011. LNCS, vol. 6998, pp. 304–317. Springer, Heidelberg (2011)
7. Daniel, F., Casati, F., Palpanas, T., Chayka, O.: Managing data quality in business intelligence applications. In: Proceedings of the International Workshop on Quality in Databases and Management of Uncertain Data, pp. 133–143, Auckland (2008)
8. Firmani, D., Mecella, M., Scannapieco, M., Batini, C.: On the meaningfulness of "big data quality" (invited paper). Data Science and Engineering pp. 1–15 (2015). http://dx.doi.org/10.1007/s41019-015-0004-7
9. Jarke, M., Jeusfeld, M.A., Quix, C., Vassiliadis, P.: Architecture and quality in data warehouses. In: Pernici, B., Thanos, C. (eds.) CAiSE 1998. LNCS, vol. 1413, p. 93. Springer, Heidelberg (1998)
10. Malaki, A., Bertossi, L.E., Rizzolo, F.: Mutidimensional contexts for data quality assessment. In: Proceedings of AMW, pp. 196–209, Ouro Pretol (2012)
11. Milani, M., Bertossi, L., Ariyan, S.: Extending contexts with ontologies for multidimensional data quality assessment. In: Proceedings of ICDE Workshops, pp. 242–247 (2014)
12. Minuto, M., Vaisman, A., Terribile, L.: Revising data cubes with exceptions: a ruled-based perspective. In: Proceedings of DMDW 2002, CEUR-WS, vol. 58, pp. 72–81 (2002)
13. Perez, J., Berlanga, R., Aramburu, M., Pedersen, T.: Towards a data warehouse contextualized with web opinions. In: Proceedings of IEEE-ICEBE 2008, pp. 697–702 (2008)
14. Pitarch, Y., Favre, C., Laurent, A., Poncelet, P.: Context-aware generalization for cube measures. In: Proceedings of DOLAP, pp. 99–104 (2010)
15. Poeppelmann, D., Schultewolter, C.: Towards a data quality framework for decision support in a multidimensional context. IJBIR **3**(1), 17–29 (2012)

16. Stefanidis, K., Pitoura, E., Vassiliadis, P.: Managing contextual preferences. Inf. Syst. **36**(8), 1158–1180 (2011)
17. Strong, D.M., Lee, Y.W., Wang, R.Y.: Data quality in context. Commun. ACM **40**(5), 103–110 (1997). http://doi.acm.org/10.1145/253769.253804
18. Vaisman, A., Zimányi, E.: Data Warehouse Systems: Design and Implementation. Data-Centric Systems and Applications. Springer, Heidelberg (2014)
19. Wang, R.Y., Strong, D.M.: Beyond accuracy: what data quality means to data consumers. J. Manage. Inf. Syst. **12**(4), 5–33 (1996)

Plan Before You Execute: A Cost-Based Query Optimizer for Attributed Graph Databases

Soumyava Das[✉], Ankur Goyal, and Sharma Chakravarthy

IT Laboratory, Department of Computer Science and Engineering,
The University of Texas at Arlington, Arlington, TX 76019, USA
{soumyava.das,ankur.goyal}@mavs.uta.edu, sharma@cse.uta.edu

Abstract. Proliferation of NoSQL and graph databases indicates a move towards alternate forms of data representation beyond the traditional relational data model. This raises the question of processing queries efficiently over these representations. Graphs have become one of the preferred ways to represent and store data related to social networks and other domains where relationships and their labels need to be captured explicitly. Currently, for querying graph databases, users have to either learn a new graph query language (e.g. Metaweb Query language or MQL [6]) for posing their queries or use customized searches of specific substructures [14]. Hence, there is a clear need for posing queries using the same representation as that of a graph database, generate and evaluate alternate plans, develop cost metrics for evaluating plans, and prune the search space to converge on a good plan that can be evaluated directly over the graph database.

In this paper, we propose an approach for effective evaluation of queries specified over graph databases. The proposed optimizer generates query plans systematically and evaluates them using appropriate cost metrics gleaned from the graph database. For the time being, a graph mining algorithm has been modified for evaluating a given query plan using constrained expansion. Relevant metadata pertaining to the graph database is collected and used for evaluating a query plan using a branch and bound algorithm. Experiments on different types of queries over two graph databases (Internet Movie Database or IMDB and DBLP) are performed to validate our approach. Experimental results show that the query plan generated by our system results in exploring significantly fewer portions of the graph as compared to any other query plan for the same query.

1 Introduction

Relational database management systems (RDBMSs) are good at managing transactional data. With the proliferation of applications rich in relationships (e.g., social networks) graphs are becoming the preferred choice as the data model for representing/storing data with relationships. Ability to efficiently query this representation using intuitive ways of querying is central for its ubiquitous usage. Given a graph G and a user-defined query Q, we want to retrieve

© Springer International Publishing Switzerland 2016
S. Madria and T. Hara (Eds.): DaWaK 2016, LNCS 9829, pp. 314–328, 2016.
DOI: 10.1007/978-3-319-43946-4_21

a set of subgraphs of G that are isomorphic to Q. Graph query finds use in a plethora of domains. For example, in a bibliography network such as DBLP [1], users are eager to extract coauthor and paper information across years; and in a movie database like Internet Movie Database (IMDB) [2], movie enthusiasts tend to look for movies or series belonging to particular genres or containing specific actors or directors. Note that queries can contain various relational conditions on node and edge labels and unbounded variables to be instantiated to appropriate values during query evaluation. Queries can also include logical operators. With growing graph sizes, number of answers to queries typically increase, making a clear case for the need of query optimization for graph databases.

Graph query answering problem has three major challenges: (1) graph query answering involves graph isomorphism which is NP Complete [8]; (2) querying on graphs need users to either learn a specialized query language (like Metaweb Query Language for Freebase [6] and Cypher Query Language [12] for Neo4j [3]) or use customized query patterns; and (3) large graphs suffer from random access problem and hence need to take advantage of indexing techniques. Note that, graph query answering typically starts from a set of matching vertices in the query, expand them in constrained ways until all edges in the query are expanded. This results in the generation of a set of intermediate results. Intuitively, lesser the number of intermediate results, lesser is the query response time. Given a vertex/edge, indexing helps to quickly find it in a graph. However, indexing does not guide the choice of choosing an initial set of nodes or the set of subsequent nodes and edges needed for query expansion. Index-based approaches do not handle unbound variables and relational conditions. Therefore there is a growing need and strong motivation to take advantage of well known database query optimization techniques to address the problem of reducing the number of intermediate results in graph query answering.

Query optimization on a relational database uses metadata extracted from the database [13]. This metadata can be effectively used to estimate the cost of a query plan to help minimize the desired cost (disk I/O and cpu cost) of processing a query in an RDBMS. Similarly, a query plan in a graph database can be thought of as a sequence of vertex/edge expansions to cover the query. The choice of starting point(s) and the sequence of subsequent vertices/edges expanded, affect the intermediate results generated following *that* particular query plan. Hence, given a query plan, the goal is to *estimate* the number of intermediate results generated while answering the query. Note that determining the exact set of intermediate results is possible only when the query is executed. However, we need a technique to quickly estimate the size of these intermediate results while generating a query plan. Graph metadata can play a crucial role in approximating the size of these intermediate results. Label frequency and average node degree are examples of metadata specific to a graph database. It is also not useful to collect a large amount of metadata as it takes space, time to compute, and resources to update. Therefore, one of the challenges is to identify a relatively small set of graph metadata that can help to generate "good" plan for a given query. In this paper we identify the metadata needed and used by a graph query plan generator. We also discuss the use of metadata for computing the cost of a partial query plan. The contributions of the paper are:

- Identifying a set of metadata and algorithms for extracting and storing them in a graph catalog. We show its applicability in handling graph queries containing both logical and relational operators.
- Cost formulas to estimate the average number of intermediate substructures generated by a partial plan. This is used to prune the search space of query plans.
- Modification of the SUBDUE mining algorithm [11] to accept a query plan and evaluate that query in a constrained manner without generating all substructures as is typically done in mining.
- Experimental validation of the cost estimate and evaluation of plans consisting of relational and logical operators on different real world databases.

The rest of the paper is organized as follows. Section 2 discusses the related work. Section 3 introduces the preliminaries and problem definition. Section 4 discusses the catalog, cost formulas and the query plan generator, and Sect. 5 shows the related experiments while Sect. 6 concludes the paper.

2   Related Work

Graph databases have received a lot of attention over the recent years due to proliferation of data with relationships such as bioinformatics, social networks, web and telecommunication networks. Hence it becomes necessary to manage large graphs in relational DBMSs [5]. However, existing database models and query languages lack the native support for (large) graphs. The popularity of graph data calls for newer techniques to tackle graph data and search for user defined patterns in them. Typically, there are two types of graph databases. The transaction graph database consists of a large number of small graphs while a single-graph database is a large single graph. In all of these graph databases, the common problem is to answer a user defined query, which can be formulated as a combination of logical and relational operations on graph databases.

In any graph database, the graph query problem is to find the set of subgraphs isomorphic to the input query graph. The major challenge is to reduce the number of pairwise graph comparisons and a number of graph indexing techniques have been developed for it [7,9,15,16]. A graph database can also be comprehended as a forest of disconnected components, so solving graph querying for a single graph would also fit into the graph transaction setting. A straightforward approach is to store the underlying graph as relational tables and make use of SQL queries to address graph querying. However, SQL queries are not the best for graph querying as evaluating queries leads to a large number of self-joins. Even though database specific optimizations, like query plan generator tries controlling join order, the number of intermediate results can grow excessively large, especially when the graphs are large. Hence there is a clear motivation to query a graph using graph traversal techniques and devise optimizing techniques for querying in graph databases.

Indexing on a single graph has received some attention in recent times. GraphGrep [9] proposed a path length based indexing scheme. In TALE [15], the authors made use of a neighborhood based indexing technique to match nodes. GRAY [16] used a label propagation based indexing scheme. The common aspect among all these indexing schemes is given a node they quickly find matches of that node to give a head start into query answering. However, none of them uses an educated approach to decide the choice of a start node and the subsequent edges. We demonstrate that metadata can be more useful than indexing for queries with logical and relational operators.

3   Preliminaries and Problem Definition

This section presents the key concepts, notations and terminologies used in this paper.

3.1   Attributed Graph

A graph is attributed where each vertex belongs to a category/type. Formally, an attributed graph is defined as $G = (V, E)$ with V vertices, E edges. The vertices are again categorized into two types: instance vertices (I) and type attributes (T) where $V = I \bigcup T$. There exists a mapping $\phi: I \rightarrow T$ such that, for each v in I, $\phi(v)$ is the category or type node connected with that v. Each instance vertex is connected to only one type node and to multiple instance vertices. Optionally, edges can have edge labels. All edges going out of a type node has the same label. This is typical of many real-world graph databases with multiple labels in some cases and absence of type nodes in some cases.

Figure 1 shows an example of an attributed graph. The colored circles (Person, movie, and year) are the three type nodes in the graph. Type nodes are connected to instances which are represented using white circles. Note that type nodes are connected to instance nodes containing numeric types (e.g. year) or non numeric types (e.g. Movie). The graph is typically represented as a set of vertices with vertex labels followed by a list of edges between vertices taking $O(V + E)$ space. In case of a directed graph, the edges showcase the source destination relation-

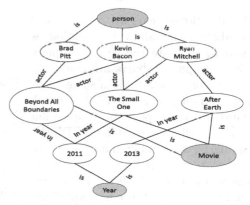

Fig. 1. Example attributed graph from IMDB (Color figure online)

ship. Any other representation that captures this information can also be used.

3.2   Graph Query

A graph query on an attributed graph is defined as $Q = (V_q, E_q)$ where V_q is the set of vertices, E_q is the set of edges and $V_q = I_q + T_q$ where T_q is the set of type nodes associated with the I_q instance vertices. The user can optionally leave vertex labels unbound (or unspecified) in the query. Following the graph representation, queries are also represented as a set of V_q vertices followed by a set of E_q edges. In graph databases some interesting queries that can be specified are: (i) Queries with one or more type nodes (ii) Queries where instance nodes can have multiple relational operators (*e.g. year* $>$ 1980, *actor* $\neq$ *Brad Pitt etc.*) (iii) Queries with unbounded vertex and edge labels (iv) Queries with logical AND/OR operators on instance node values.

4   Methodology and Approach

For a query on a relational database, the query optimizer generates a "good" plan using the database catalog and the cost formulas for estimating the execution cost of *that* query. We model our system along similar lines. However, the cost in the case of graph databases need to be properly identified. Figure 2 discusses the architecture of our query optimizer. We extract metadata from the graph and store it in a graph catalog. For any given query, the *partial plan generator* generates alternative plans and uses the *cost estimator* to estimate the cost of evaluating that plan. The *pruner* helps control the search space of query plans. Below, we explain in detail our three major contributions: the graph catalog, the partial query plan generator and the cost formulas used by the cost estimator.

4.1   Graph Catalog

As mentioned earlier, in relational databases, metadata information is used to estimate the cost of a query plan. Graph databases are analogous to relational databases in some ways. For example, in an attributed graph database, the type nodes are similar to the attribute names in an RDBMS. Similarly, the number of instances connected to type node is analogous to the attribute cardinality. The connectivity between instance nodes of different types, with different edge labels,

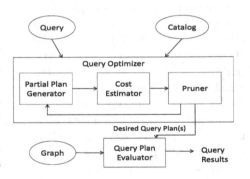

Fig. 2. Graph Database Query Optimizer (GDQP)

estimates the connection cardinality which is similar to the join cardinality in a database. Depending on how a graph is used (whether materialized in memory or not), the cost that needs to be taken into account varies. In this paper, since we are using Subdue which materializes the graph, we do not consider the I/O cost but only the number of intermediate substructures generated. The cost formulas may change under a different assumption. We here discuss the relevant graph statistics which can be gleaned from the graph database for catalog creation. Table 1 introduces the notations that we use throughout the paper.

Table 1. Symbols

Symbol	Description
$I(t)$	Number of instances connected with type t
n_i^t	i^{th} instance of type t where $1 \leq i \leq I(t)$
$edge(n_i^{t1}, n_j^{t2}, el)$	edge between n_i^{t1} and n_j^{t2} with edge label el
$Max(t)$	Maximum value across all instances of type t (if numeric)
$Min(t)$	Minimum value across all instances of type t (if numeric)

I **Type Cardinality (TC):** Type cardinality is defined as the number of instances of a particular type node. Intuitively, if query answering starts from a type node, the type cardinality helps estimate the number of substructures generated after exploring the type node. For type node t, the type cardinality is $I(t)$. Given the type node, degree of that type node is its type cardinality.

II **Average Instance Cardinality (AIC):** Once the query answering reaches an instance node, the number of substructures generated in the next round depends on the degree of the instance node. Since there are multiple instances of a particular type node, with different degrees, we introduce average instance cardinality to estimate the average number of neighbors to be searched from an instance of type t. For a given instance node of a specific type, the average instance cardinality is computed by taking the average of degrees of instances of that type. Average instance cardinality of an instance node with type t is

$$AIC(instance\ of\ type\ t) = \frac{1}{I(t)} \times \sum_{i=1}^{i=I(t)} degree(n_i^t) \qquad (1)$$

III **Average Connection Cardinality (ACC):** Query answering explores the neighborhood of an instance but only chooses a few connected instances, based on the query connectivity (edge label and connected instance node) for expansion. Since multiple instances of type $t1$ can be connected with multiple instances of type $t2$ by the same edge label el, we again derive the average connectivity information for our catalog. Intuitively, this value gives us the expected number of expansions from an instance node to another with a particular edge label. The average connection cardinality starting from an instance node of type $t1$ and reaching another instance of type $t2$ with edge label el is defined as

$$ACC(t1, t2, el) = \frac{1}{I(t1)} \times \sum_{i=1,j=1}^{i=I(t1),j=I(t2)} edge(n_i^{t1}, n_j^{t2}, el) \qquad (2)$$

IV **Min and max values of type nodes:** Note that some special type nodes are only connected to numeric instance nodes. This gives user the opportunity to ask range based queries using these attributes. Hence for selectivity, we keep the minimum and maximum numeric value associated with such special type nodes. This intuitively hints at the number of substructures to be generated following user-defined selectivity criteria. For a type node t with numeric or categorical attributes the selectivity is defined as

$$
Selectivity = \begin{cases} \frac{1}{I(t)} & node = value \\ 1 - \frac{1}{I(t)} & node \neq value \\ \frac{Max(t) - value}{Max(t) - Min(t)} & node > value \\ \frac{value - Min(t)}{Max(t) - Min(t)} & node < value \end{cases} \tag{3}
$$

The three major operations to answer a graph query are: (1) choosing a starting node (type or instance); (2) expanding the neighborhood; and (3) retaining substructures using query conditions. Type cardinality and instance cardinality help choose start nodes and estimate neighborhood expansion cost. Average connection cardinality, analogous to joins helps determine the number of expansions. Min and max values estimate number of intermediate substructures based on query conditions on numeric or categorical attributes. Table 2 shows a graph catalog created from the example graph in Fig. 1. **AIC** is depicted as an instance followed by "*" while **ACC** is between two instances with defined labels. Fields starting with a type node and ending with an instance node indicate **TC**.

Table 2. Graph catalog extracted for graph in Fig. 1

Node and its type	Node and its type	Edge label if used	Avg cardinality
person	person instance	is	3
person instance	person	is	1
person instance	movie instance	actor	5/3
person instance	*	*	8/3
movie	movie instance	is	3
movie instance	movie	is	1
movie instance	person instance	actor	5/3
movie instance	year instance	in year	1
movie instance	*	*	11/3
year	year instance	is	2
year instance	movie instance	in year	3/2
year instance	year	is	1
year instance	*	*	5/2
Node	Min	Max	#Unique values
year instance	2011	2013	2

Catalog generation for the system is outlined in Algorithm 1. We make a single pass over the input graph data file using very little memory unlike other

Algorithm 1. *Graph Catalog Generator*

1: **function** CATALOGGENERATOR(V,E)
2:    vertexTypeMap = null
3:    **for** each vertex v in V **do**
4:       if v is instance
5:          vertexTypeMap[v]=instance
6:       else
7:          vertexTypeMap[v]=type
8:    **end for**
9:    **for** each edge e in E **do**
10:       source = vertexTypeMap[e.source]
11:       dest = vertexTypeMap[e.dest]
12:       source==type and dest==instance
13:          update type cardinality
14:          update min and max value if e.dest has numeric type attribute
15:       if source==instance and dest==instance
16:          update average instance cardinality
17:          update average connection cardinality
18:    **end for**
19: **end function**

---

algorithms that require graph construction in main memory and its traversal. The catalog is created by reading one input graph line at a time and maintaining a few counters. Processing of vertices provide information on the instance and type nodes. Processing of the edges is used to populate the type cardinality, instance cardinality and connection cardinality. If the instance node has a numeric or categorical attribute, the min and max value associated with its type is computed accordingly. The number of lines in the graph file is equal to the number of vertices and edges hence catalog creation requires $O(|V| + |E|)$ time. Hence the catalog generation time scales linearly with graph size. The number of counters depends on the unique type nodes and unique edge labels present in the graph, which are orders of magnitude lesser than $|V|$ and $|E|$. Since the underlying graph does not change, the catalog incurs an one time generation cost. Once created, the catalog can be used to answer any number of queries. If new vertices and edges are added, the catalog need not be recreated but incrementally updated.

4.2   Cost Formulas

We propose cost formulas for estimating the number of intermediate results generated during query execution as the cost of that plan. Typically, answering a query requires expanding a node to connected nodes in the graph matching edge and node label conditions from the query. Our total estimated cost of a plan is a cumulative cost of partial query plans generated for evaluating a query. A partial query match grows by expanding matched node(s) incurring an expansion cost. Out of these expansions, only the ones matching the query conditions are retained, hence controlling the number of substructures. We introduce two parameters to keep track of each query plan in any iteration i: (i) $cost_i$ and (ii) $currSub_i$. The $currSub_i$ parameter keeps track of estimated number of substructures generated until iteration i. The $cost_i$ parameter estimates the cumulative neighborhood expansion cost till the ith iteration. A query must have at least one node. The node can be a type node or an instance node of a known type with or without relational operators. Given a start node, the initialization of cost and currSub in iteration 1 is computed as shown below:

$$cost_1 = I(t) \times Selectivity \ (from \ Eq.3); \quad currSub_1 = cost_1 \qquad (4)$$

For subsequent iterations (i+1 where ($i \geq 1$)), the current number of substructures is updated based on the query expansion conditions on edge labels and the relational operator on the connecting instance node. This involves both connection cardinality (for edge label) and selectivity (for connecting instance node).

$$currSub_{i+1} = currSub_i \times ACC \times Selectivity \qquad (5)$$

In each iteration, partially completed substructures incur a neighborhood expansion cost for satisfying the query conditions. Note that in any iteration, we are adding only one edge on a particular node (type or instance) as the underlying system is implemented that way. Hence $cost_{i+1}$ captures the cumulative cost of neighborhood expansion for the $currSub_i$ matched patterns and the previous cost ($cost_i$). Mathematically,

$$cost_{i+1} = cost_i + currSub_i \times node \ cardinality \qquad (6)$$

$$where \ node \ cardinality = \begin{cases} TC & if \ expanding \ on \ type \ node \\ AIC & otherwise \end{cases}$$

4.3   Query Plan Generator

The query plan generator takes a graph query and the metadata generated for that graph as input and generates the best cost query plan using a branch and bound algorithm. The cost formulas are used as a heuristic to guide the branch and bound algorithm to limit the search space. The generator uses each node in the query as a start node for the generation of alternate query plans. From each start node partial query plans are generated by adding an edge from the query and computing cost of partial plans using the formulas discussed earlier. At every iteration, top-k plans (that correspond to the lowest costs) are considered as candidates for the next iteration. When a complete plan is generated (number of iterations equal to the number of edges in the query), all plans that have a cost higher than the completed plan are pruned. Note that partial plans with a cost lesser than the cost of the current completed plan are still expanded to guarantee an optimal plan. The plan generator emits the optimal completed plan. Note that, in case of relational queries (like OR) the emitted query plan correspond to multiple plans and the query result is an union of results of those plans.

Algorithm 2. *Query Plan Generator*

1: **Input:** Query graph G_q, Catalog C, user defined k
2: **Output:** k alternative plans with their estimated cost
3: planList = NULL
4: **for** each node q in G_q **do**
5:      add q as start point of plan
6:      initialize q.cost using Eq. 4
7:      initialize q.currSub
8: **end for**
9: **while** number of plans < k **do**
10:      **for** each plan p in k lowest cost plans **do**
11:          expand plan by adding an edge
12:          update p.cost using Eq. 6
13:          update p.currSub using Eq. 5
14:      **end for**
15:      update k lowest cost plans
16:      update planList on plan completion
17: **end while**
18: emit k lowest cost plans

Algorithm 2 outlines the algorithm used by our generator. The user can generate top-k plans by explicitly specifying a k value. Obviously, $k = 1$ emits the best plan and k with a large value generates all plans along with their estimated cost information.

Figure 3 shows how above-described plan generator works on a query on the IMDB graph, along with the catalog used by the plan generator. We show all plans and their final costs to demonstrate the cost differences between the best and worst plans. For this example, it is clear that there is orders of magnitude difference between the costs. Once a good (or best) query plan is generated, any system can be used to execute the query plan on the graph. A good query plan would require the least execution time in any graph querying system assuming node-wise expansion of the plan.

We have modified SUBDUE [4,11], a popular graph mining system into a querying system, Query Processor-Subdue (QP-Subdue) [10] by making the following changes. Instead of starting from all nodes (as in the case of mining), QP-Subdue starts from node(s) specified by the plan. We replace the unconstrained expansion strategy in SUBDUE by a constrained expansion strategy which supports checks of relational query conditions on node labels and edge labels. Moreover, support for logical operations (AND/OR) on node values is also added to QP-Subdue. We believe that any other querying system will show the same trend as QP-Subdue. However, QP-Subdue carries forward the advantages and disadvantages of SUBDUE. Typically SUBDUE expands one edge at a time and hence requires multiple traversals of the adjacency list of the same vertex. This can be avoided by performing simultaneous expansions on a node. The cost formulas need to be adjusted to reflect the approach used.

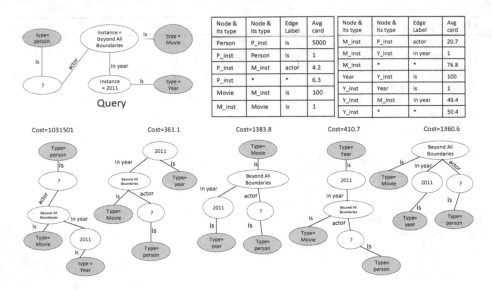

Fig. 3. Query plan generation steps for an example query

Note that, in Fig. 3 the plans with lower costs have always expanded to known nodes before unknown nodes thereby achieving better pruning during the expansion phase. This is analogous to query plans in a relational model where selections and projections are pushed down the query plan tree. Intuitively, expanding to smaller number of nodes (similar to lower intermediate result cardinality in an RDBMSs) achieves lesser query computation cost. We believe that the actual running time of these plans on the graph will have a positive correlation with the estimated query cost. We shall verify our conjecture in Sect. 5 on a diverse set of queries to show the effectiveness of our approach.

5   Experimental Analysis

In this section, we present the results of our experimental analysis performed on various queries over different databases. The experimental results reinforce our premise that generating a "good" plan before evaluating a query is beneficial and the execution time of the plan is directly proportional to the number of intermediate substructures generated. The consistent performance of the plan generator across different types of queries and databases establishes the validity of our proposed approach of cost-based plan generation for graph querying.

All experiments have been carried out on Dual Core AMD Opteron 2 GHz processor machine with 4 GB memory. To evaluate the performance of the plan generator, we used IMDB [2] and DBLP [1] data sets. DBLP data set contains the information of publications along with the information of their authors, conferences, and years. IMDB graph database contains the information of movies, actors, genres, year, company, etc. Since the focus is on good plan generation, we carefully extracted a section from both the graphs for query answering with at least one known query pattern. To see the performance of the system on graphs of various sizes, we extracted small graphs (12,000 vertices and 30,000 edges) to big graphs (350,000 vertices and 1100,000 edges.) This gives us better control in doing targeted queries with the assurance that they would be discovered in the graph.

For the above mentioned graph databases, we used queries having different characteristics such as queries with a comparison operator $(<, \leq, >, \geq, \neq, =)$, queries with a combination of multiple comparison operators, queries with logical operator (OR, AND) and queries with a combination of logical and comparison operators. In an AND query all query conditions need to be satisfied, while an OR query internally translates into an union of individual queries. Note that, our system is parameterized to generate minimum, maximum, or all cost plans. Note the our cost metrics is based on the number of intermediate substructure generated, but can be tweaked to match the expansion used in the algorithm for query evaluation. In order to see the effectiveness of our cost formulas among various plans, we picked minimum (**Min**), median (**Med**) and maximum (**Max**) cost plans for experimental analysis. We do not include the catalog creation cost and graph loading time (both being one time operations) as part of the query response time across all our experiments.

5.1   Performance of Plans

We used the following queries to verify our query plan.

Query 1: *"Find tv-series and its company name, where Kelsey, Wagner has worked as an animator and genre of the tv-series is animation and comedy"* is an example query to IMDB graph database containing AND logical operator. This query inspects how our system performs on queries that needs multiple satisfiability. Experiments in Fig. 4 shows us that our best plan performs significantly better than others. The max cost plan typically started exploring unknown nodes first, thereby generating a much larger set of intermediate results which increased the runtime.

Query 2: *"Find tv-series and its company by Soler, Rebecca where the genres are drama and family and the year is not equal to 1996"* is a query to IMDB graph database which contains a combination of both comparison ($\neq$) and logical operator (AND). This query verifies how our system handles combination of elementary query types. Results in Fig. 4 show the effectiveness of the plan found by our system. The inequality operation (because of selectivity) adds a little bit to the response time as compared to the equality condition.

Query 3: *"Find tv-series and its company where Kelsey, Wagner has worked as an animator OR Soler, Rebecca has worked as an actress"* is a query to IMDB graph database having an OR operator. This query helps understand the difference in runtime between queries with logical operators. Note that the OR query in Fig. 5 takes considerably more time than the AND query in Fig. 4.

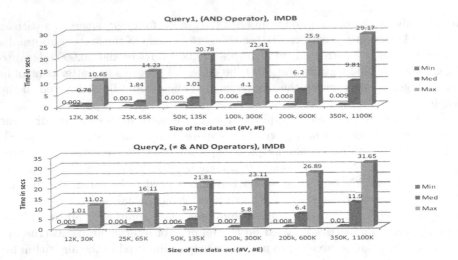

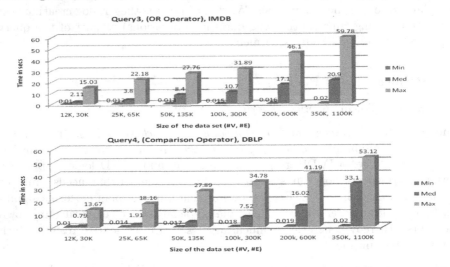

Fig. 4. Query response times for queries 1 and 2 with growing graph sizes

Fig. 5. Query response times for queries 3 and 4 with growing graph sizes

Since OR queries are divided into multiple sub plans based on the OR condition, execution of an OR query is internally translated as union of multiple plans thereby needing more time. We still see that our system generated query plan is the best among all query plans in terms of runtime.

Query 4: *"Find papers published by the author Eric Hanson prior to the year 2009"* is an example of query which contains a range ($<$) operator on the DBLP graph database. This query is important to understand how our system performs for range queries. Once again Fig. 5 shows that minimum cost plan executes in considerably less amount of time compared to other plans.

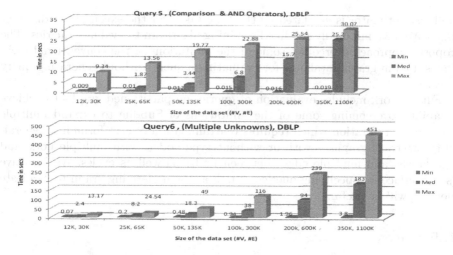

Fig. 6. Query response times for queries 5 and 6 with growing graph sizes

Query 5: *"Find papers where Yuri Breitbart AND Abraham Silberschatz have collaborated together after the year 1980"* shows an example a query with a combination of comparison and logical operators. In Fig. 6, the time varies for minimum cost plan from 9 ms to 19 ms on different data sets, while for the maximum cost plan, it varies approximately from 9 s to 30 s. The order of magnitude difference is indicative of the need for this approach for graph query processing.

Query 6: *"Find authors with their papers and conference information in year 2005"* is an example of a query which has multiple unknown nodes (authors, papers, conference). This query helps us to validate the quality of our generator in the presence of a high number of unknown values in the query. Figure 6 shows that our method still performs better than the median and maximum cost plans. However, the time taken to answer this query is significantly greater than other queries due to the presence of multiple unknowns in the query. Still the difference is in order of magnitude.

All the queries show that the best cost plan generated by our system always generates minimum query response time. And the median and max plans are orders of magnitude slower than the best plan. This validates our premise that different, but appropriate cost measures are needed for estimating graph query plans. The consistent performance of the plan generator validates the feasibility of the proposed approach for graph databases. Moreover, our plan generator is able to handle different query types.

6   Conclusions and Future Work

We have developed an initial framework that allows us to generate query plans for various types of queries containing one or more comparison and logical operations. Our choice of the number of intermediate substructures generated as a good cost estimator is validated by the time taken for the execution of different plans. We have minimally modified a mining algorithm to evaluate queries. The proposed approach overcomes some of the limitations of the conventional techniques used in graph databases for evaluating a query without generating query plans.

Future work includes evaluation of queries on partitioned graphs to achieve scalability, overcoming some of the limitations of Subdue to expand multiple edges in each iteration which requires re-examination of cost formulas, generalize to arbitrary graphs – with or without type nodes, with multiple node and edge labels, and multiple edges between nodes as well as cycles. An intuitive User interface for specifying queries as well as the underlying graph to facilitate querying will be useful.

References

1. http://www.informatik.uni-trier.de
2. http://www.imdb.com/
3. http://neo4j.com/
4. http://ailab.wsu.edu/subdue
5. Batra, S., Tyagi, C.: Comparative analysis of relational and graph databases. Int. J. Soft Comput. Eng. (IJSCE) **2**(2), 509–512 (2012)
6. Bollacker, K., Evans, C., Paritosh, P., Sturge, T., Taylor, J.: Freebase: a collaboratively created graph database for structuring human knowledge. In: SIGMOD 2008, pp. 1247–1250. ACM, New York (2008)
7. Cheng, J., Ke, Y., Ng, W., Lu, A.: Fg-index: towards verification-free query processing on graphdatabases. In: SIGMOD, pp. 857–872 (2007)
8. Garey, M.R., Johnson, D.S.: Computers and Intractability: A Guide to the Theory of NP-Completeness. W. H. Freeman & Co., New York (1979)
9. Giugno, R., Shasha, D.: Graphgrep: a fast and universal method for querying graphs. In: 16th International Conference on Pattern Recognition, ICPR 2002, Quebec, Canada, 11–15 August 2002, pp. 112–115 (2002)
10. Goyal, A.: QP-SUBDUE: Processing Queries Over Graph Databases. Master's thesis, The University of Texas at Arlington, December 2015
11. Holder, L.B., Cook, D.J., Djoko, S.: Substructure discovery in the SUBDUE system. In: Knowledge Discovery and Data Mining, pp. 169–180 (1994)
12. Holzschuher, F., Peinl, R.: Performance of graph query languages: comparison of cypher, gremlinand native access in neo4j. In: Proceedings of the Joint EDBT/ICDT 2013 Workshops, pp. 195–204. ACM (2013)
13. Jarke, M., Koch, J.: Query optimization in database systems. ACM Comput. Surv. (CsUR) **16**(2), 111–152 (1984)
14. Suri, S., Vassilvitskii, S.: Counting triangles and the curse of the last reducer. In: World Wide Web Conference Series, pp. 607–614 (2011)
15. Tian, Y., Patel, J.M.: TALE: a tool for approximate large graph matching. In: ICDE 2008, 7–12 April 2008, Cancún, México (2008)
16. Tong, H., Faloutsos, C., Gallagher, B., Eliassi-Rad, T.: Fast best-effort pattern matching in large attributed graphs. In: SIGKDD, pp. 737–746 (2007)

Ontology-Based Trajectory Data Warehouse Conceptual Model

Marwa Manaa[1(✉)] and Jalel Akaichi[2]

[1] BESTMOD, Université de Tunis, ISG, 2000 Le Bardo, Tunisia
manaamarwa@gmail.com
[2] College of Computer Science, King Khaled University, Abha, Saudi Arabia
j.akaichi@gmail.com

Abstract. The enormous evolution of positioning technologies and remote sensors is leading to big amounts of disparate mobility data. Collected mobility data generates the need of modelling of such behaviour and the understanding of them which gave the rise of different models achieved either by classical conceptual modelling or by those based on ontology. Modelling and analysing trajectory data are still challenging because of the heterogeneity of trajectory data models and the complexity of establishing choices about domain's consensual knowledge. To fulfil this objective, we propose a generic ontology that explains the semantics of these data and we define a trajectory data warehouse conceptual model based on the shared ontology in order to analyse trajectory data going from users' short transactions to complex queries involving decision makers. The shared ontology that we propose is an OWL-DL formalism that covers common structures encountered in trajectories. We illustrate our work with a real case study.

Keywords: Data warehouse · Ontology · OWL-DL formalism · Semantic modelling · Trajectory data

1 Introduction

Data driven scientific discovery approach has been an important paradigm for computing in many central areas including Internet of Things, social networks, remote sensors, etc. Under this paradigm, mobility data commonly named trajectory data is the core that reveals the details of instantaneous behaviours conducted by mobile entities. Basically, *trajectory data is a record of the evolution of the position (perceived as a point) of an object that is moving in space during a given time interval in order to achieve a given goal* [9]. Actually, working on this field is a fresh but active matter which is essentially due to the rise of applications, pervasive devices and positioning technologies offering mobility data. The management of collected mobility data is expected to extract useful knowledge about moving object and facilitates, then, the understanding of their behaviour from analytic and cognitive perspectives. Therefore, collected mobility data gave rise to different trajectory data models achieved either by enhancing classical

© Springer International Publishing Switzerland 2016
S. Madria and T. Hara (Eds.): DaWaK 2016, LNCS 9829, pp. 329–342, 2016.
DOI: 10.1007/978-3-319-43946-4_22

"conceptual models" used for designing database schema or by proposing new ones such as "ontology-based representations". Yet, disparate trajectory data, stored and manipulated in classical and semantic databases provides a limited support for analysing and understanding mobile objects behaviour and activities represented by heterogeneous trajectory data models.

To fulfil this need, the design model of trajectory data can be expressed through formal languages, for instance by using Description Logics formalism (DL) or one of its fragments [4]. Together with DL, one may also consider developing and using ontologies. Some studies argue that ontologies exceed conceptual models by making them consensual and enriching them by reasoning mechanisms [2]. In addition, Ontologies allow a shared understanding and may offer a common model for different structures and representations of trajectory data where designers can pick the appropriate knowledge to define trajectories in view of share, exchange or integration. Alongside, Trajectory Data Warehouse (TrDW) is considered as an efficient tool for analysing and extracting valuable information from heterogeneous trajectory data sources.

For this purpose, this paper sheds light on a Semantic Trajectory Data Warehouse (STrDW) using the best of both *Data Warehousing* and *Semantic Data Modelling* worlds. The STrDW will mainly (i) emphasize a generic shared trajectory ontology that explains the semantics of these data in an unambiguous way and (ii) defines the STrDW conceptual model. Our proposal permits to save too much designers efforts and time needed to acquire domain knowledge since the latter is extracted from the generic ontology. The STrDW will mainly highlights the trajectory to be seen as a first class semantic concept, providing an ontology-based multidimensional model. The generic shared ontology that we propose is an OWL-DL formalism that covers the most important existing conceptual and ontological trajectory data models. We focus on DL formalism because DL is able to capture the most popular data class-based modelling formalisms frequently used in databases, warehouses and information systems analysis [4].

The outline of this paper is structured as follows. In Sect. 2, we analyse the evolution of warehousing approaches towards the birth of trajectory data warehousing approaches. Section 3 outlines basic notions required to the understanding of our approach. Section 4 pinpoints an ontology driven approach that describes a STrDW conceptual model by using a generic trajectory ontology. Section 5 illustrates our work with a case study dealing with Edinburgh informatics forum. Section 6 concludes the paper and suggests some future work.

2   Related Research: Towards a STrDW

The predominant step for extracting knowledge from trajectory data is to provide a design model able to represent moving objects. In that the database community has stored and managed such type of data in Spatio-Temporal Databases (STDB) [18] and Moving Object Databases (MOD) [13,21] by the definition of spatio-temporal data types inter alia,*moving point* and *moving region data types*. However, current DBMS ability, even extended to support spatio-temporal data, is limited only on storing raw trajectory which omits any semantic information and/or analysis capabilities. To make an efficient exploitation of this data, there were attempts to enrich it with semantic annotations in order to support different views of knowledge. For that purpose, recently, ontology building and logics attracted researches aimed at supporting trajectory-based applications with semantic approaches [1,16,20,22]. The majority of these approaches deal with ontology as a storage repository and not as a *domain ontology*, where designers can pick concepts and properties to represent trajectory data.

There are many ways for efficiently analysing trajectory data. Warehousing and mining techniques are, among others, supporting the extraction of valuable information from disparate raw trajectory. Focusing on our research area, TrDW is the application of data warehousing techniques on trajectory data [7,19]. Before getting to the TrDW, research communities were interested in analysing spatio-temporal data in Spatio-Temporal Data Warehousing (STDW). There have been various proposals of multidimensional models for STDW [23] aiming at the integration of various data sources containing spatio-temporal data. Trajectory data is a particular case of spatio-temporal data characterizing objects mobility. Then, a TrDW is obviously a particular case of STDW where trajectory is the fact [3,7,17]. However, obtaining an implementation of the DW is a complex task that often forces designers to acquire wide knowledge of the domain, thus requiring a high level of expertise and becoming it a prone-to-fail task. In real-world projects, we have faced up with a set of situations i.e., additivity and conformed dimensions in which we believe that the use of some kind of knowledge resources will improve the design task of data warehouses.

In the light of these issues, ontologies seems to be a promising solution, since they are *common* **conceptualization** *of a universe of discourse representing shared knowledge in terms of classes and properties that is* **formal, consensual** *and referenceable* [14]. The first attempt to set a Semantic Spatio-temporal Data Warehouse is given by authors in [5] which annotate the datacube elements with domain ontologies as well as mathematical ontology. On top of this, substantial research has been conducted on methods and tools for designing the DW through ontologies. The team in [10,11] gathers domain ontologies and semantically annotated data resources. Authors in [12] presents the OLAP cube in the basis of an OLAP design ontology. The work of [6] defines a DW in the basis of a global ontology integrating local ontologies of ontology-based database sources participating in the integration process. [15] defines the DW multidimensional model from an ontology by identifying functional dependencies between concepts. In the following (Table 1), we summarize the evolution of reviewed warehousing approaches according to these criteria: type of warehouse (DW, STDW, TrDW) and the used technique for designing the DW (ontology, conceptual).

In this context, we hold a different point of view for unifying the modelling and the analysis of trajectory data. The innovation of our work consists of offering a generic trajectory ontology that describes heterogeneous mobility data sources. The shared trajectory ontology covers most important existing formalisms and representations of trajectory concept. This ontology serves as semantic layer for the STrDW allowing the analysis of heterogeneous trajectory data sources.

Table 1. Evolution of warehousing approaches

Criteria	Type of warehouse			Used technique	
	DW	STDW	TrDW	Ontology	Conceptual
[3]			✓		✓
[15]	✓			✓	
[11]	✓			✓	
[17]			✓		✓
[23]		✓			✓
[6]	✓			✓	
[19]			✓		✓
[7]			✓		✓
[16]			✓	✓	
[5]	✓			✓	
[12]	✓			✓	

(Left margin label: *Authors*)

3 Problem Definition

We focus in our study on different representations of trajectory data. In the following, we outline basic facts (representations) relevant to our work:

Definition 1 (Raw trajectory). *A sequence of spatio-temporal position recording the trace of a moving object i.e., $\{(x_0, y_0, t_0), ..., (x_n, y_n, t_n)\}$, where $x_i, y_i, t_i \in \Re$ for $i=0, ..., N$ and $t_0 < t_n$.*

Definition 2 (Structured trajectory). *A set of sub-trajectories according to predefined paths. A sub-trajectory includes strictly one Begin and one End. It includes also at least one Stop. Moves are used to connect stops to other elements (Stop, Begin, End) i.e., $\{(Sub\text{-}trajectory_1, ..., Sub\text{-}trajectory_n), Sub\text{-}trajectory = \{Begin, Move_1, ..., Stop_{n-1}, Move_n, End\}, Begin=\{x_0, y_0, t_0\}, Stop_{n-1}=\{x_{n-1}, y_{n-1}, t_{n-1}\}, End=\{x_n, y_n, t_n\}\}$ where $x_i, y_i, t_i \in \Re$ for $i = 0, ..., N$ and $t_0 < t_{n-1} < t_n$.*

Definition 3 (Trajectory with ROI). *A sequence of visited places (regions) and intervals. A region is a set of consecutive line segments i.e., $\{(ROI_1, ..., ROI_n), ROI_i=(Region_i, Interval_i)\}$ where $i \in \Re$ for $i= 1, ..., N$ and $Interval_1$ before $Interval_N$.*

Definition 4 (Semantic trajectory). *A structured trajectory where spatio-temporal positions are annotated. Begin, stop, move, and end become geographical concepts linked to points of interest rather than spatio-temporal data i.e., Semantic Trajectory=$\{(SemanticSub\text{-}trajectory_1,..., SemanticSub\text{-}trajectory_n), SemanticSub\text{-}trajectory = \{SemanticBegin, SemanticMove_1, ..., SemanticStop_{n-1}, SemanticMove_n, SemanticEnd\}, SemanticBegin= \{x_0, y_0, t_0, Point of Interest\}, SemanticStop_{n-1}= \{x_{n-1}, y_{n-1}, t_{n-1}, PointofInterest\}, SemanticEnd= \{x_n, y_n, t_n, PointofInterest\}\}$ where $x_i, y_i, t_i \in \Re$ for $i=0, ..., N$, $t_0 < t_{n-1} < t_n$, and PointofInterest is a geographical place.*

Definition 5 (Semantic ROI). *A trajectory with ROI annotated with semantic information i.e., trajectory with Semantic ROI={ (SemanticROI$_1$, ..., SemanticROI$_n$), SemanticROI$_i$=(Region$_i$, Interval$_i$, Pointof Interest)} where i $\in \Re$ for i=0, ...,N, Interval$_1$ before Interval$_N$, and PointofInterest is a geographical place.*

Definition 6 (Space-time path). *A semantic trajectory extended with mobile object activity i.e., space-time path={ (Space-time$_1$, Activity$_1$), ..., (Space-time$_n$, Activity$_n$), Space-time$_i$={x$_i$, y$_i$, t$_i$, PointofInterest}} where i $\in \Re$ for i=0, ..., N. PointofInterest is a geographical place, and activity is a contextual information about moving object activity.*

4   STrDW Approach

The outburst of ontologies in web applications and their use by different companies leads to the creation of important amount of web data referencing ontologies. These data are called Ontology-Based Moving Object Data (OBMOD). Some solutions proposed to manage OBMOD in main memory like Protégé are primarily used for designing ontologies and Jena TDB, Virtuoso for publishing triples cannot offer affordable performance for handling huge amount of trajectory datasets. To overcome this problem, data warehousing solutions have been proposed offering efficient storage and querying mechanisms for heterogeneous OBMOD. The generated data warehouses are called Semantic Trajectory Data Warehouses (STrDWs) which are data warehouses storing both ontology and trajectory data (Fig. 1). Our objective in this paper is to define an ontology-based design approach for modelling and analysing heterogeneous OBMOD. To fulfil this objective, we need to define: (i) a global trajectory shared ontology that explains the semantics of these data and (ii) the structure of the STrDW conceptual model. The following subsections will describe each step.

4.1   Trajectory Global Ontology

In this section, we define the global ontology. We adopt a modular approach to facilitate reusability and possible design extension. The proposed model holds three modules: Geometric Trajectory Ontology (GTO), Geographic Ontology (GO) and Application Domain Ontology (ADO).

Geometric Module. GTO holds resources to describe how moving objects movement can be understood and trajectories can be represented. It covers most important trajectory data works like raw trajectory which describes trajectory as (position, timestamps). Structured trajectory that organizes trajectory in sub-trajectories which include a begin, a set of stops, moves and End. Also, it provides a set of relations between concepts like hasBegin, hasEnd, hasStop, hasMove. Then, trajectory with ROI represents trajectory in the form of moving regions. In addition, it enriches structured trajectory and trajectory with

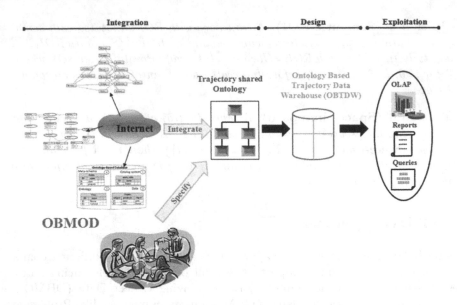

Fig. 1. Proposed framework

ROI with geographical features through the relation hasGeometry to give birth respectively to Semantic Trajectory and semantic ROI. The proposed model presents also trajectory in the form of space-time path which annotates semantic stops with corresponding activities. This set of resources allowed us to define a generic ontology-based geometric facet for trajectory data and supports linking trajectory concepts to application and geographical concepts (Fig. 2).

Application Domain Module. ADO contains resources relevant to a field such as traffic management, bird migration, transportation, etc. This module describes the mobile object i.e., animal or person and possible activities related to the displacement of the moving object like physical activities i.e., reading and virtual activities i.e., mailing. The module presents also points of interest relevant to the application domain i.e., university.

Geographic Module. GO contains concepts about the geographic environment in which mobile objects involved. Concepts are likely to include those describing the topography of the land (e.g. mountain, lake), networks (e.g. road network, railway network), building places (e.g. home, work, supermarket) and anything else that is of interest to the application. This module is closely related to the geometric trajectory module, as each trajectory concept that has a spatial implication is to be linked to a type of geography that is used by the application to specify the corresponding spatial measure. The module is also related with the application domain module as its concepts may also have a thematic description providing application information beyond geographic and geometric facets.

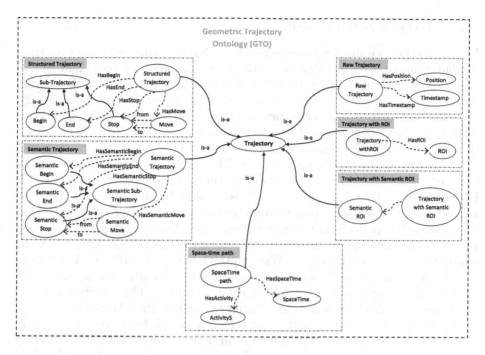

Fig. 2. Geometric Trajectory Ontology

For example, concepts about building places may include standard schemes defined in the geographic module in addition to other features specific to the application domain (Table 2).

Table 2. A description of geographic ontology

Concepts	Description
Building place	represents places where mobile object moves such work, home or supermarket.
Topography	represents natural topography such as mountain or lake.
StreetG	represents road network that follows mobile object while moving

Combining the GTO, GO and ADO together leads to the final overall Generic Semantic Trajectory Ontology. This final ontology maintains interoperability since it is a modular approach (domain oriented), ensures genericity as it covers most important trajectory data works and assures consensuality because it is based on commonly and shared conceptualizations by mobility data community.

4.2   STrDW Schema Design

In this section, we discuss the design of the STrDW supposed to be tailored around semantic concepts that allow the specification of thematic and spatio-temporal aspects of the moving object and its trajectory. The STrDW design starts from the annotation process using an ontology-based design methodology.

Aforementioned works in the state of the art proposed ontology-based methods for the design of semantic data warehouses. Most of these works hold a single domain ontology and threshold values must be set by the designer for the annotation process of the warehouse resources. This, clearly increases the complexity of design task for autonomous designers. In this work, the design of the STrDW is derived from the global ontology. This is done by importing all resources related to the chosen trajectory representation by the designer (i.e., raw trajectory, space-time path) from the generic ontology defined in the previous section to the STrDW conceptual model. A sub-ontology model is then extracted to be called Semantic Trajectory Data Warehouse Ontology (STrDWO). The following extracted model need then to be annotated by multidimensional roles such as: fact, dimension, measure and dimension attribute. The annotation phase identifies the multidimensional role of each resource in the STrDWO.

Broadly, what is most evident about semantic models built around trajectory data is that there are always spatio-temporal resources representing time varying geometry nature of trajectory data, added to the thematic part its application-specific aspect. For that, inhere we suppose that a generic ontology is generally composed of four types of resources: fact, thematic, temporal and spatial which are represented respectively within the following modules (sub-ontologies):

- GTO: The fact is the trajectory representation type selected by the designer i.e., structured trajectory;
- ADO &GO: Application domain concepts gathers at a time resources relevant to the mobile object, its activities during the travel and visited points of interest.
- Temporal Ontology: Temporal concepts and roles are based on the standard Time-owl ontology[1] developed by W3C.
- Spatial Ontology: Spatial concepts and roles are based on Geo ontology[2] developed likely by the W3C standard.

The question to be asked here is then:

"How to extract STrDW design model from these 4 parts and make resulting model take into consideration the special nature of disparate trajectory data?"

By analysing the generic ontology, spatial and temporal sub-ontologies, temporal and spatial concepts are identified: Instant, Interval for the temporal sub-ontology, and Point, Line, Region for the spatial sub-ontology. An Interval is described by a Start Date and an End Date instantiating the Instant concept. So, a first step for the annotation process related to the STrDW is to identify

---

[1] http://www.w3.org/TR/owl-time/.

[2] http://www.w3.org/2005/Incubator/geo/XGR-geo-ont-20071023/.

concepts and relations from the GTO, ADO and GO sub-ontologies, that could be assimilated to the aforementioned spatial and temporal concepts.

Indeed, the following Algorithm 1 is proposed to bring out resources. Since the thematic part is the subject of analysis, the STrDW's Fact might be extracted from the GTO concepts according to the type of trajectory chosen by the designer. A STrDW is, also, to be mainly composed of spatial and temporal dimensions, extracted from GTO concepts that are assimilated to concepts from the spatial and temporal sub-ontologies. The fact measures are time and space represented by spatial and temporal concepts from the GTO sub-ontology. To construct fact, spatial, temporal and thematic dimensions the following algorithm is proposed:

Algorithm 1. STrDW Schema Algorithm

    Input : GTO, ADO, GO, Spatial/Temporal Ontology
    Output: Fact, Spatial/Temporal/Thematic dimensions

1 **begin**
2   Fact= Identify the Fact from the GTO
3   Annotate the Trajectory Concept in the GTO as a Fact ;
4   Annotate each Role in the GTO as a Fact measure;
5   ThCR=Identify Thematic Concepts and Roles from the ADO and the GO;
6   Annotate each Concept in the ThCR as a Thematic Dimension candidate;
7   Annotate each Role in the ThCR as a Thematic Dimension Attribute candidate;
8   S/TCR=Identify Spatial/Temporal Concepts and Roles from the Spatial/Temporal Ontology related to GTO;
9   Annotate each Concept in the S/TCR as a Spatial/Temporal Dimension candidate;
10   Annotate each Role in the S/TCR as a Spatial/Temporal Dimension Attribute candidate;
11   Construct Hierarchies between Spatial/Temporal Dimension candidates by looking for (1,n) relationships between concepts identified as Spatial/Temporal Dimensions candidates;
12 **end**

---

5   Case Study: Edinburgh Informatics Forum

We illustrate the generic semantic trajectory modelling approach by using a case study related to *Edinburgh informatics forum*[3] [8]. In the remaining subsections, we present the application scenario and we drive the STrDW conceptual model.

---

[3] http://homepages.inf.ed.ac.uk/rbf/FORUMTRACKING/.

5.1   Application Scenario

This subsection is aimed to illustrate the application scenario. The hereinafter conducted researches are motivated by the scenario related to a set of pedestrians trajectories walking through the Informatics Forum, the main building of the School of Informatics at the University of Edinburgh. Data holds several months of observation which has resulted in about 1000 observed trajectories each working day.

The source trajectory datasets are time-stamped locations. Additional information related to pedestrian, and its activity during the trip, are provided too. The main components of the trajectory dataset are [8]:

- *Reference$_i$*: the trajectory's reference;
- Long and Lat: are respectively longitude and latitude, the spatial coordinates of the pedestrian's position;
- Start-date and End-date: are respectively the start and the end temporal coordinates of a pedestrian's trajectory.

Actually, the movement of pedestrians is still relatively unknown. In this work, our research team is interested in collecting and analysing data becoming from these pedestrians to understand their behaviour from cognitive and analytic perspectives. Clearly, it is hard to exploit raw trajectory data to that end. For that, a semantic layer was added to trajectory data and prominent semantic components were revealed. The STrDW design we developed is tailored around main concepts from the aforementioned ontological model, and that's what makes this former support trajectory semantic concepts.

5.2   The Design Model

The STrDW model is derived from the already existing semantic layer including thematic, spatial and temporal ontologies. The designer identifies resources and their coordinates according to the mentioned trajectory representation type. For example, we consider in this case the trajectory representation type "Space-time path". A *Space-time path* is defined as follows:

```
Space-time path equivalentTo SemanticStop ∪ Activity
Semantic Stop isa Stop.hasGeometry PointofInterest
Stop isa Point ∩ Interval
```

In addition, the designer instantiates the ADO according to the case study. In our case, the mobile object (pedestrian), activities (phone call, drink coffee, walking, eating) and visited points of interest (stairs, night exit, coffee, elevator, labs, front door) as illustrated in (Fig. 3).

The ADO is linked to GTO (i) and GO (ii) respectively by using the following statements:

(i) Pedestrian <u>hasTrajectory</u> Trajectory
    Activity <u>equivalentTo</u> ActivityS
    SemanticROI <u>IsLocatedIn</u> PointOfInterest
    SemanticSub-Trajectory <u>IsLocatedIn</u> PointOfInterest
    SpaceTimepath <u>IsLocatedIn</u> PointOfInterest
(ii) StreetA <u>equivalentTo</u> StreerG
    PointOfInterest <u>equivalentTo</u> BuildingPlace

The projection of resources allows then the extraction of sub-ontology STrDWO from the global ontology. This step is of paramount importance because it will permit, later, the definition of the STrDW conceptual model based on ontological concepts that express as much as possible effective user's requirements (trajectory representation type). In addition, user's requirements are also used for the annotation of the STrDWO by multidimensional concepts such as fact, dimension, measures and dimension attributes, to result on the STrDW conceptual model. A first possible design model for the application scenario is given in (Fig. 4).

5.3 Analysis

Here is a statement that incorporates a user requirement example:

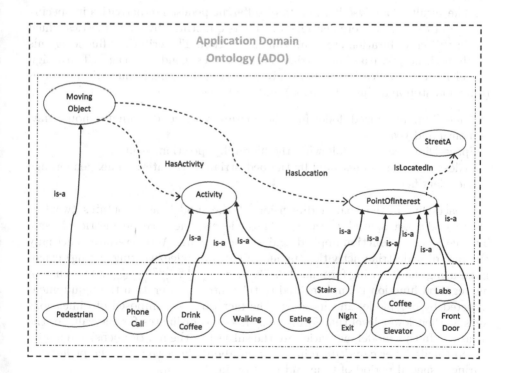

Fig. 3. Application domain ontology

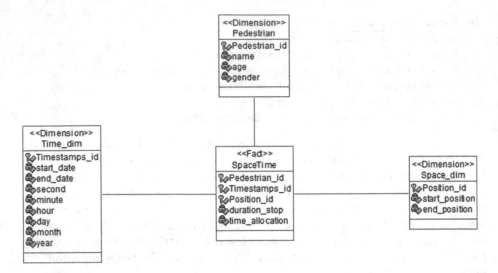

Fig. 4. Proposed model of the STrDW

Q:*"Analyse pedestrian activities in a given time interval in a specific point of interest"*

The result to analyse is the rate of different pedestrian activities in specific place and time. The aforementioned result is quantified by some metrics which are in this case duration_stop and time_allocation. The criteria influencing this result are time, space and pedestrian characteristics (gender and age). The design model for the application scenario given in (Fig. 4) appeals numeric measures (duration_stop and time_allocation) and 3 dimensions:

– Time-Dim: organized following the hierarchy: second, minute, hour, day, month and year;
– Space-Dim: organized following the hierarchy: position, stop;
– Pedestrian-Dim: represented by the pedestrian' attributes: name, gender, age and identifier.

Space-Dim is the spatial dimension of the model, and contains two levels related to a spatial hierarchy. Those levels reference geometric objects. The aggregation function applied against the measure Activity-Rate is actually the rate of pedestrian activities (phone call/drink coffee/walking/eating) calculated using the following formula $Activity\text{-}rate = \frac{Walking - Sum}{All - Activities - Sum}$. This custom aggregation function is implemented to take into consideration the requirement inflicted by our model and its aims. The fact table is composed of dimensions keys at their lower level that form the symbolic coordinates for the value of the measure. In this model, activities are the subject of the multidimensional analysis, so the designer can deduce information about the activity of the pedestrian during a special period of time and location in the forum.

6   Conclusion and Future Work

Throughout this work, we have been motivated by the need to support applications dealing with heterogeneous trajectory data sources. To meet this need, we first presented a trajectory shared ontology which served as semantic layer, where designers can pick resources to represent their trajectories. Then, we offered a STrDW ontology-based approach for modelling and analysing heterogeneous OBMODs allowing interoperability, reusability, and maintenance between applications supporting trajectory data.

Research on this topic is crucial for expanding the usefulness of multidimensional models to non-traditional applications. The STrDW contains huge amounts of mobility data, so optimization issues are of paramount importance either for data storage and retrieval issues.

Acknowledgements. The authors would like to thank the creator of the dataset Barbara Majecka as part of her MSc projects [8].

References

1. Baglioni, M., de Macêdo, J.A.F., Renso, C., Wachowicz, M.: An ontology-based approach for the semantic modelling and reasoning on trajectories. In: ER Workshops, pp. 344–353 (2008)
2. Bellatreche, L., Dung, N.X., Pierra, G., Hondjack, D.: Contribution of ontology-based data modeling to automatic integration of electronic catalogues within engineering databases. Comput. Ind. **57**(8), 711–724 (2006)
3. Braz, F.J.: Trajectory data warehouses: Proposal of design and application to exploit data. In: GeoInfo, pp. 61–72 (2007)
4. Calvanese, D., Lenzerini, M., Nardi, D.: Description logics for conceptual data modeling. In: Logics for Databases and Information Systems, pp. 229–263 (1998)
5. Diamantini, C., Potena, D.: Semantic enrichment of strategic datacubes. In: ACM 11th International Workshop on Data Warehousing and OLAP, DOLAP, Napa Valley, California, USA, pp. 81–88 (2008)
6. Khouri, S., Boukhari, I., Bellatreche, L., Sardet, E., Jean, S., Baron, M.: Ontology-based structured web data warehouses for sustainable interoperability: requirement modeling, design methodology and tool. Comput. Ind. **63**(8), 799–812 (2012)
7. Leonardi, L., Orlando, S., Raffaetà, A., Roncato, A., Silvestri, C., Andrienko, G., Andrienko, N.: A general framework for trajectory data warehousing and visual OLAP. GeoInformatica **18**(2), 273–312 (2014)
8. Majecka, B.: Statistical models of pedestrian behaviour in the Forum. Ph.D. thesis, University of Edinburgh (2009)
9. Manaa, M., Akaichi, J.: Unifying mobility data warehouse models using UMLprofile. In: Kozielski, S., Mrozek, D., Kasprowski, P., Małysiak-Mrozek, B., Kostrzewa, D. (eds.) 2014 10th InternationalConference on Beyond Databases, Architectures, and Structures, 82–91. CCIS, vol. 424, pp. 82–91. Springer, Heidelberg (2014)
10. Martinez, J.M.P., Berlanga, R., Aramburu, M.J., Pedersen, T.B.: Integrating data warehouses with web data: A survey. IEEE Trans. Knowl. Data Eng. **20**(7), 940–955 (2008)

11. Nebot, V., Berlanga, R.: Building data warehouses with semantic web data. Decis. Support Syst. **52**(4), 853–868 (2012)
12. Niinimäki, M., Niemi, T.: An ETL process for OLAP using RDF/OWL ontologies. In: Spaccapietra, S., Zimányi, E., Song, I.-Y. (eds.) Journal on Data Semantics XIII. LNCS, vol. 5530, pp. 97–119. Springer, Heidelberg (2009)
13. Pelekis, N., Theodoridis, Y., Vosinakis, S., Panayiotopoulos, T.: Hermes - A framework for location-based data management. In: Ioannidis, Y., Scholl, M.H., Schmidt, J.W., Matthes, F., Hatzopoulos, M., Böhm, K., Kemper, A., Grust, T., Böhm, C. (eds.) EDBT 2006. LNCS, vol. 3896, pp. 1130–1134. Springer, Heidelberg (2006)
14. Pierra, G.: Context representation in domain ontologies and its use for semantic integration of data. J. Data Semant. **10**, 174–211 (2008)
15. Romero, O., Abelló, A.: A framework for multidimensional design of data warehouses from ontologies. Data Knowl. Eng. **69**(11), 1138–1157 (2010)
16. Sakouhi, T., Akaichi, J., Malki, J., Bouju, A., Wannous, R.: Inference on semantic trajectory data warehouse using an ontological approach. In: Andreasen, T., Christiansen, H., Cubero, J.-C., Raś, Z.W. (eds.) ISMIS 2014. LNCS, vol. 8502, pp. 466–475. Springer, Heidelberg (2014)
17. Campora, S., Fernandes, J.A., de Macedo, L.S.: St-toolkit: A framework for trajectory data warehousing. In: AGILE, pp. 18–22 (2011)
18. Tryfona, N., Price, R., Jensen, C.S.: Conceptual models for spatio-temporal applications. In: Spatio-Temporal Databases: The CHOROCHRONOS Approach, pp. 79–116 (2003)
19. Wagner, R., de Macêdo, J.A.F., Raffaetà, A., Renso, C., Roncato, A., Trasarti, R.: Mob-warehouse: A semantic approach for mobility analysis with a trajectory data warehouse. In: Advances in Conceptual Modeling - ER 2013 Workshops, Hong Kong, China, 11–13 November 2013, pp. 127–136 (2013)
20. Wannous, R., Malki, J., Bouju, A., Vincent, C.: Modelling mobile object activities based on trajectory ontology rules considering spatial relationship rules. In: Amine, A., Mohamed, O.A., Bellatreche, L. (eds.) Modeling Approaches and Algorithms. SCI, vol. 488, pp. 249–258. Springer, Heidelberg (2013)
21. Xu, J., Güting, R.H.: A generic data model for moving objects. GeoInformatica **17**(1), 125–172 (2013)
22. Yan, Z., Chakraborty, D.: Semantics in Mobile Sensing. Synthesis Lectures on the Semantic Web: Theory and Technology. Morgan & Claypool Publishers, San Rafel (2014)
23. Zimányi, E.: Spatio-temporal data warehouses and mobility data: Current status and research issues. In: 19th International Symposium on Temporal Representation and Reasoning, TIME 2012, Leicester, United Kingdom, September 12–14, 2012, pp. 6–9 (2012)

Data Intelligence and Technology

Discovery, Enrichment and Disambiguation of Acronyms

Jayendra Barua[✉] and Dhaval Patel

Indian Insitute of Technology, Roorkee, India
jayendrabarua@gmail.com

Abstract. Acronym disambiguation is the process of linking an acronym in a given text to its intended expansion in the text. Acronyms are frequently used in short-texts such as news headlines and tweets. The direct application of state-of-art named entity disambiguation approaches on short text results in poor performance since, entities are not associated with their acronyms in the Knowledge Bases. Also, many acronyms in short-text represent out of Knowledge Base entities. Existing acronym dictionaries such as Acronymfinder also cannot be used for disambiguation as contextual information requires for disambiguation is absent in them. In this paper, we propose a system for effective disambiguation acronyms in short-text. In particular, we built an Acronym dictionary that is automatically updated with new acronyms by continuous monitoring of news media. Each acronym in our Acronym dictionary is enriched with additional meta information comprised of category, location and context words extracted from news articles. We use our enriched Acronym dictionary for disambiguation of acronyms in short-texts. Experimental results shows that our system is efficient in discovery and disambiguation of acronyms.

Keywords: Acronym · Abbrevation · Acronym disambiguation

1 Introduction

In this age of social media, short texts such as tweets and news headlines frequently contain short terms like hashtags and acronyms. Acronyms are playing an important role to convey the information in compact manner. By observing tweets and news streams, we found that on an average 1 tweet out of 10 contains an acronym and about 20 % of news headlines contain acronyms. Generally, acronyms are abbreviations and constructed by using the initial letters of words in a phrase e.g. BJP stands for "Bharatiya Janata Party". Acronyms are widely used in short texts such as news headlines and tweets to cope with word/character limit restrictions. For e.g. in the following headline with three acronyms.

PDP, BJP face April 9 deadline to form govt in JK

© Springer International Publishing Switzerland 2016
S. Madria and T. Hara (Eds.): DaWaK 2016, LNCS 9829, pp. 345–360, 2016.
DOI: 10.1007/978-3-319-43946-4_23

'PDP', 'BJP' and 'J&K' are acronyms which stands for 'Peoples Democratic Party', 'Bharatiya Janata Party' and 'Jammu & Kashmir' respectively.

Some acronyms are popular whereas many acronyms are unpopular and newly formed. Thus, when users encounter such unpopular or novel acronyms in tweets or headlines, they may not be able to interpret those acronyms. Moreover, many acronyms have more than one full form, for e.g. acronym "PDP" in the headline "_PDP, BJP face April 9 deadline to form govt in JK_" may refer to "Policy Decision Point", "Plasma Display Panel", "People's Democratic Party", "Parallel Distributed Processing" and many other expansions. Such ambiguity in acronyms creates serious understanding difficulties for readers and automated natural language understanding systems [1,2].

Given an acronym present in a given text, Acronym disambiguation is the problem of linking acronym with intended expansion in the text. Disambiguation of acronyms in short-text can be seen as problem of word-sense disambiguation where acronyms need to be linked to its correct sense (expansion) in the Knowledge base. Existing works [3–5] focuses on disambiguating acronyms in a collection of documents, where expansion of an acronym is assumed to be present in some documents in the collection. However, these methods are not applicable in short texts, because expansion of acronyms are rarely found in news headlines and tweets itself. Some other researches [6,7] have been proposed to address the acronym disambiguation problem in Medical domain which uses a medical domain Knowledge Base UMLS (Unified Medical Language System). All the aforementioned researches aim to disambiguate acronyms in large text corpus, where sufficient context is available to disambiguate an acronym. To the best of our knowledge we are not aware of any work that aims to disambiguate acronyms in short texts.

Since acronyms in news are mostly entities such as organization and locations, applying existing entity disambiguation systems such as AIDA [8], TAGME [9] and Wikifier [10] for acronym yields poor results. This is due to the fact that many entities in Knowledge Bases are not associated with their corresponding acronyms. For e.g. entities "Prime Minister" and "High Court" are not associated with acronyms "PM" and "HC" respectively in YAGO and Wikipedia Knowledge Bases. Moreover, Knowledge bases are incomplete as well as does not update regularly with newly generated acronyms, for e.g. OROP (one rank one pension) is not updated in YAGO Knowledge Base.

1.1 Proposed Work

In this paper, we present a system for acronym disambiguation in short-texts. We observed that new acronyms-expansion pairs consistently occurs in news articles, whereas short text such as news headlines and tweets contain only acronym words. We used this observation to build our acronym disambiguation system. Our system (Fig. 1) is based on Data-centric architecture which builds an Acronym dictionary by continuously discovering acronym-expansion pairs from the stream of long text (news articles). The newly discovered acronyms in

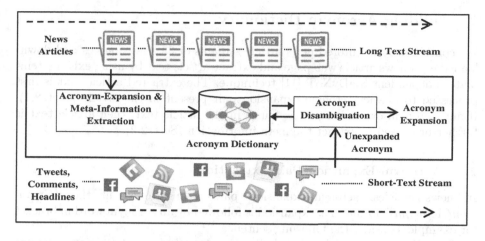

Fig. 1. Overall Approach

Acronym dictionary are enriched with additional meta information, such as category, location and contextual words which are also extracted from the news articles stream. The enriched Acronym dictionary is used to disambiguate acronym words in short text streams (tweets and news articles). The acronyms are disambiguated on the basis of matching contextual information present in short-text with meta information of expansions present in Acronym dictionary.

1.2 Technical Contribution

We summarize the technical contributions of our work as follows:

1. We propose a system for disambiguation of acronyms in short-text.
2. We build Acronym dictionary, which is updated continuously with new acronym-expansion pairs appearing in news articles.
3. We provide a way to enrich acronyms with meta-information such as category, location and context words. The category of acronyms is extracted from the URLs of news articles in which acronyms-expansion pair are discovered. Location is extracted from URL and article text, and contextual words are extracted from headline of the news article.
4. We propose a method to disambiguate the acronyms occurs in the short texts such as news headlines and tweets using our Acronym dictionary.
5. Extensive experiments are performed for determining the accuracy of discovered acronym-expansion pairs in Acronym dictionary and disambiguation accuracy of acronyms in short texts. The accuracy of acronym-expansion pairs in our dictionary is 85 %. Whereas disambiguation accuracy in headline and tweets is 97 % and 79 % respectively.

2 Building Acronym Dictionary

We continuously monitored news article streams of 33 news sources and downloaded each news article web page and extract article text by using existing template independent StaDyNoT [11] technique. The extracted article text is first processed to extract acronym-expansion pair present in the article (Sect. 2.1). If any acronym-expansion pair is found in the article text, then article text is further processed to extract the meta-information (Sect. 2.2).

2.1 Acronym-Expansion Pair Extraction

In news articles, acronym-expansion pair is mentioned in the *"Expansion(Acronym)"* format, i.e. expansion is followed by an acronym word in bracket for example, consider the following sentence.

Sony Pictures Network (SPN) India has launched its television campaign (TVC) for the Vivo Indian Premier League (IPL) on March 8, across its network.

The use of inverse format i.e. *"Acronym(Expansion)"* is very rare in news articles and therefore we neglect them.

We first divided the given article text T into sentences, and detect the presence of acronym inside the brackets, i.e. *"(Acronym)"* in each sentence. Next, we filter out the bracketed words from a sentence and apply rule based *AcroCheck* algorithm (Algorithm 1) to check whether it is an acronym or not. Acrocheck determine the word as Acronym on the basis of fraction of uppercase letters in that word. We experimentally determine 0.65 as best threshold value for *caps_fraction*.

Algorithm 1. *AcronymCheck*

1: **Input:** *word*
2: **Output:** {*true, false*} //*true* indicates *word* is acronym, *false* otherwise
3: *char_count= Count_Chars(word)*
4: *caps_count = Count_Uppercase(word)*
5: *caps_fraction = caps_count/char_count*
6: **If**(*char_count = 2* and *caps_count = 2*) **then** return *true*
7: **else If**(*char_count = 3* and *caps_count>1*) then return *true*
8: **else If**(*char_count>3* and *caps_fraction>0.65*) **then** return *true*
9: **else** return *false*

Expansion Extraction. If an acronym A in detected in a sentence S, then we extract the expansion of A from the sentence S. We know that the expansion just precedes its bracketed acronym in the sentence as acronym-expansion pair follows *"Expansion(Acronym)"* format. If the number of letters in A is n, then we extract a substring S' from S containing n+2 words and ending on word before bracketed acronym in S as shown in the following example.

$S=$*After crossing a lot of hurdles of late, now the Delhi and District Cricket Association (DDCA) is facing yet another issue as their staff has gone on an indefinite strike.*

$$A = DDCA \ (n = 4), \qquad S' = the \ Delhi \ and \ District \ Cricket \ Association$$

We have extracted n+2 words because the expansion may contain one or two additional stop-words which are not used in forming acronym as shown in aforementioned example. Next, we extract the correct expansion for acronym A from S'. Sometimes bracketed acronyms are not preceded by their expansions as shown in the following sentences.

– *A total of 13 suspected Islamic State (ISIS) militants were detained*
– *Mr. Praveen Dixit (IPS) is appointed as Maharashtra director general of Police*

Therefore, while extracting proper expansion for A from S', we also need to check that bracketed acronym is preceded by its expansion or not. Given acronym A having n characters, and String S' first, we create all possible candidate expansions of A by making substrings of S' using following rules:

1. Expansion must start with a word having first letter as of A and it should not be a stopword.
2. Expansion must include the last word in S'
3. Minimum length of expansion $= n - 2$ (2 in case $n < 4$) and maximum length of expansion $= n + 2$

for e.g. Acronym = 'DDCA' and $S' = $ "the Delhi and District Cricket Association", the candidate expansions are:

1. Delhi and District Cricket Association
2. District Cricket Association

If there are more than one candidate expansions, then we generate acronyms from each candidate expansion and find its Levenshtein distance with original acronym A. Notice that, stop-words present in candidate expansion may or may not be the part of its corresponding acronym. Therefore, in case stop-words are present in expansion, then we generate two acronyms for a candidate expansion (1)acronym including stop-words and (2)acronym excluding stop-words. For example, generated acronyms from candidate expansions '*Delhi and District Cricket Association*' including and excluding stop words are DADCA and DDCA respectively. Next, we remove the acronyms having Levenshtein distance greater than minimum threshold[1]. The candidate expansion corresponding to generated acronym having least Levenshtein distance with A is identified as correct expansion for A. For example, the Levenshtein distance of generated acronyms DADCA, DDCA, and DCA is 1,0 and 1 respectively. Hence the correct expansion for DDCA is Delhi and District Cricket Association.

---

[1] Threshold is 0, 1 and 2 for acronyms with two letters, three letters and greater than three letters respectively.

Different news articles may contain different variants for same expansions corresponding to an acronym. for e.g. for acronym 'IOCL' have following variants for the same expansion.

−Indian Oil Corporation Ltd   − Indian Oil Corp Ltd
−Indian Oil Corporation Limited − Indian Oil Corporation

In order to correctly define acronym-expansion pair correctly in Acronym dictionary, we need to cluster the expansion variants, such that each cluster represent different expansion of acronym.

Expansion Clustering. Given an acronym A having a set of expansion variants $E = \{e_1, e_2, e_3...e_k\}$. Each expansion $e_i \in E$ have an occurrence count f_i in news articles. We have used Hierarchical clustering algorithm to cluster the expansions of an acronym. Each cluster has an expansion variant e_i having highest occurrence count f_i and is the centroid and representative expansion for that cluster.

Distance Metric for Clustering. First, we have used following Levenshtein edit distance based similarity metric with a minimum threshold of 0.80 similarity.

$$Leven\_Sim(s_1, s_2) = 1 - \frac{Levenshtein Distance(s_1, s_2)}{Max(l_1, l_2)} \tag{1}$$

where, l_1 and l_2 are length of input strings s_1 and s_2.

We observed that $Leven\_Sim$ works properly in expansion variants having minor spelling mistakes. But it is inefficient in cases where lengthy expansions having different meaning are dissimilar only in initiating word of expansion. For e.g. expansions "*Srinagar Municipal Corporation*", "*Silliguri Municipal Corporation*" and "*Surat Municipal Corporation*" for acronym 'SMC' are being placed in the same cluster by using $Leven\_Sim$ as it satisfies the threshold constraint. Another problem is that sometimes an expansion may contain short-words which increase the dissimilarity between expansions representing the same meaning for e.g. expansions "*Indian Oil Corp Ltd*" and "*Indian Oil Corporation Ltd*" expansions represent same meaning but are placed in different clusters by $Leven\_Sim$ as it does not satisfies threshold constraint. To deal with these type of expansion variants, we use *first word similarity (isFWSim())* constraint and *Contain()* constraint with $Leven\_Sim$ as follows:

$$isFWSim(s_1, s_2) = \begin{cases} 1 & \text{if } Leven\_Sim(sf_1, sf_2) \geq 0.80 \\ 0 & else \end{cases} \tag{2}$$

sf_1 and sf_2 are first words of strings s_1 and s_2 respectively.

$$isContain(s_1, s_2) = \begin{cases} 1 & \text{if all words of } s_1 \text{ are present in} \\ & s_2 \text{ or vice versa} \\ 0 & else \end{cases} \tag{3}$$

$$isSimilar(s_1, s_2) = Contain(s_1, s_2) \quad \lor \quad (isFWS(s_1, s_2)$$
$$\land \quad (Leven\_Sim(s_1, s_2) \geq 0.80)) \tag{4}$$

isFWSim() handles the first word dissimilarity problem while *Contain()* deals with short-word problem in expansion. We have used $isSimilar(s_1, s_2)$ function in hierarchical clustering algorithm to check whether two expansions are similar to each other or not.

2.2   Meta Information Extraction

An acronym-expansion pair may occur in multiple news articles. The meta information such as location, category and context words present in these news articles may related to discovered acronym-expansion pairs in the news article. We have used different components of news articles such as URL, headline and article text to extract meta information related to the acronym. Location and category information is extracted from URL and article text, whereas Context words are extracted from headline of news article.

Extraction of Location and Categories. We have observed that URLs of news articles contains category and location information. A typical news article URL is comprised of three components (1)Domain name, (2)Path and (3)Query as shown below.

$$\underbrace{http://www.thehindu.com}_{\textbf{Domain Name}} / \underbrace{sport/cricket/world-cup/article8404802.ece}_{\textbf{Path}}$$
$$\underbrace{? homepage = true}_{\textbf{Query}}$$

The location and categories are found in the 'Path' component URL. For example Table 1 lists some URLs where bold and underlined words in 'Path' of URL are either location or categories. We present a general approach that can be applied to URLs of any news source to identify the category and location words.

Our approach uses a set of URLs of a news source for extracting location and category information. The input set of URLs should encover all types of news article URLs generated by that source. The property of category and location words in URL is that they occur at same position in different URLs of a news source. Our approach uses this property to identify the positions of location and category information in 'Path' component of the URL. Our approach is described as follows.

Given a set of news article URL $U = \{u_1, u_2, u_3...u_k\}$ of a news source N, we create a set of paths $P = \{p_1, p_2, p_3...p_k\}$ by extracting 'Path' component from each URL. Each path $p_i \in P$ have a length l_i, which depicts the number of elements (tokenized by '/') present in path p_i for e.g. length of path "*sport/cricket/world-cup/article8404802.ece*" is 4. Next, we divide the set P into

Table 1. Category and location information in URLs of News articles

http://indianexpress.com/article/**sports**/**ashes**/2448831/ ashes-2015-i-need-to-turn-my-form-around-says-michael-clarke/
http://timesofindia.indiatimes.com/**business**/**india-business**/Vijay-Mallya-offers-to-repay-Rs-4000-crore-to-banks-by-September/articleshow/51612258.cms
http://www.hindustantimes.com/**punjab**/state-higher-education-council-to-set-up-office-in-ut/story-tR8mncdE0b1fvF7dDewSvK.html
http://www.thehindu.com/news/national/**karnataka**/ indians-abducted-in-libya-two-families-swing-between-hope-and-uncertainty/article7486366.ece

n subsets, where each subset contains the paths of the same length. Path subsets are denoted as $p_1^i, p_2^j, p_3^k...p_n^l$ where i,j,k,l denotes the length of path in the subset. Each subset is processed in following manner.

Given a subset p_i^l having m paths, each of length l, we count the number of unique elements at each position of the path. Suppose c_j is the number of unique elements at position j, then position j contain category or location information if $1 < c_j < m$. for e.g. Table 2 shows 15 paths of length 4 where number of unique elements at positions 1,2,3 and 4 are 1,6,11,15 respectively. Therefore, only positions 2 and 3 contains the category or location information. Categories/location present in Paths shown in Table 2 at position 2 and 3 are as follows.

Position 2: {*sports, cities, lifestyle, world, entertainment, business*}.

Position 3: {*cricket, football, tennis, delhi, mumbai, books, asia, americas, bollywood, business-others, economy*}

To distinguish between category and location information we use location dictionary[2] which contains more than 2.6 million locations (city and country) names from all over the world. We check the extracted words from article URLs in the location dictionary. If the word is found in the location dictionary, we extract that word as 'Location', else we extract it as 'Category'. The extracted categories and locations from Table 2 are as follows:

Locations: *delhi, mumbai asia, americas*

Categories: *sports, cities, lifestyle, world, entertainment, business, cricket, football, tennis, books, bollywood, business-others economy*

The association of a Location or category term with an acronym-expansion pair is the occurrence frequency of that term in URLs of the news articles that contain acronym-expansion pair. Table 3 lists top 6 locations and categories associated with acronym-expansion pairs present in our Acronym Dictionary.

Extracting Location from Article Text. While reporting about a new incident, it is a common convention in news articles to initiate the article with name of city or location, where incident has been taken place. For e.g. following initial lines of a news articles initiates with "Lahore/Islamabad".

---

[2] https://www.geodatasource.com/world-cities-database/free.

Table 2. A set of 15 'Path' of length 4 extracted from URLs of Indianexpress.com

article/sports/cricket/ashes-2015-england-beat-australia-
article/sports/football/rickie-lambert-leaves-liverpool-to-join-west-brom
article/sports/cricket/ashes-2015-i-need-to-turn-my-form...
article/sports/tennis/rust-off-rafael-nadal-embraces-dirt
article/cities/delhi/first-year-students-join-protest-against-cbcs
article/cities/delhi/french-womans-arrest-justified-under-extra....
article/cities/mumbai/best-lanes-to-make-commute-faster
article/lifestyle/books/those-that-time-forgot
article/lifestyle/books/the-theatre-of-cruelty
article/world/asia/reports-of-fearsome-haqqani-network-founders-....
article/world/americas/were-expecting-a-baby-girl-announce-mark-zuck...
article/entertainment/bollywood/drishyam-actor-ajay-devgn-mee...
article/entertainment/bollywood/reactions-to-salman-khans-tweets...
article/business/business-others/more-funds-for-psu-banksfinmin-....
article/business/economy/ahead-of-policy-meet-on-aug-4-rajan-me

LAHORE/ISLAMABAD: A suicide bomber killed at least 65 people, mostly women and children, at a park in Lahore on Sunday in an attack.....

We extract first three words from article text and search them into location dictionary. If any word is found in the location dictionary, we extract that word as 'Location'.

Extracting Context Words. We have used headline of news article to extract the context words of an acronym. The context words are the dictionary words (excluding stop-words) that are present in the headline. For e.g. Given a headline

Missing Infosys employee confirmed dead in Brussels terror attacks

The context words are: *Missing, employee, confirmed, dead, terror, attack*. The association of a context-word with an acronym-expansion pair is the is the occurrence frequency of that context-word in headlines of the news articles that contain acronym-expansion pair. Table 3 lists top 12 context words along with their occurrence frequency) with acronym-expansion pairs present in our Acronym Dictionary.

3 Acronym Disambiguation

The expansion of an acronym can be found in long text such as news articles, however short texts such as news headlines, tweets and comments contains acronyms without expansion. An acronym may have multiple expansions and

Table 3. Some Acronym-Expansion Pair with their Meta-Information in our Acronym Dictionary

Acronym-Expansion Pair	Locations	Categories	Context Words	
RBI (Reserve Bank of India)(2821)	Delhi(171)	business(478)	rate(328)	policy(139)
	Mumbai(161)	markets(414)	bank(261)	inflation(128)
	Anand(55)	industry(346)	cut(223)	Govt(102)
	India(46)	economy(322)	Nifty(203)	growth(101)
	Jammu(36)	indian-markets(269)	Live(158)	payments(99)
	Chennai(35)	finance(193)	Rupee(148)	Fed(97)
NIA(National Investigation Agency)(821)	India(441)	india-others(50)	attack(140)	probe(43)
	Delhi(220)	nation(18)	terrorist(97)	terrorists(42)
	Pathankot(219)	other-states(15)	terror(73)	accused(37)
	Jammu(198)	city(10)	blast(61)	arrested(30)
	Udhampur(151)	politics(6)	Pakistani(45)	bail(28)
	Mumbai(144)	economy(5)	court(44)	Police(27)
NDA(National Democratic Alliance)(550)	Delhi(127)	politics(169)	polls(49)	Parliament(18)
	India(53)	economy-policy(25)	govt(40)	session(18)
	Patna(35)	Industry(18)	Congress(35)	Seat(18)
	Jammu(28)	economy(10)	Bill(27)	Special(17)
	Mumbai(27)	elections(9)	Land(22)	Assembly(16)
	Hyderabad(20)	lead(9)	elections(22)	Opposition(14)
				poll(13)
NDA(National Defence Academy)(35)	India(22)		cadets(5)	burns(3)
	Jammu(13)	india-others(1)	recruitment(4)	dies(3)
	Jind(11)	education(1)	quota(4)	fighting(3)
	Delhi(8)	life-style(1)	Army(3)	Artillery(3)
	Pune(6)	books(1)	inquiry(3)	prowess(3)
	Srinagar(5)		officers(3)	martyr(3)

therefore, it is a non-trivial task to automatically identify the correct expansion of acronym present in short texts. We identify the expansion of acronyms present in short text by using two measures (1)Using Popularity of expansion in news articles. (2)Matching the contextual information around that acronym with the meta information of the candidate expansions present in our Acronym dictionary.

Given a short text $T = \{t_1, t_2, t_3...t_n\}$, where each $t \in T$ is a word in the short text. Suppose a $t \in T$ is an acronym A, whose expansion is to be identified. First, we create a set $C_{text} \subset T$, which represent the surrounding context of A. $C_{text} = \{t'_1, t'_2, t'_3...t'_k\}$, where a $t' \in C_{text}$ is a term $t \in T$ such that $t \neq A$ and t is not a stop word.

Next, we retrieve candidate expansions of A from Acronym dictionary. Suppose $E_{cand} = \{e_1, e_2, e_3...e_m\}$ be the set of candidate expansions for Acronym A, where each $e \in E_{cand}$ is associated with following two informations.

1. Popularity (P_e): Popularity P_e of expansion e is frequency of occurrence of expansion e with acronym A in the form of acronym-expansion pair in news articles.
2. Context information (C_e): C_e for a candidate expansion e is a set of terms, which is built up by using location and context words of expansion e in Acronym dictionary. i.e. $C_e = Location_e \cup ContextWords_e$.

Popularity based Expansion Identification. A candidate expansion $e \in E_{cand}$ having maximum popularity P_e is the correct expansion for Acronym A.

Context based Expansion Identification. We find the contextual overlapping of each candidate expansion $e \in E_{cand}$ with surrounding context C_{text} of A using following equation.

$$ContextOverlap_e = C_e \cap C_{text} \tag{5}$$

The candidate expansion e having maximum value for $|ContextOverlap|$ is the correct expansion of acronym A.

4   Experiments

In this section, we present the details of experiments conducted to evaluate our system. We have divided experiments into four subsections. The first section describes the acronym-expansion extraction statistics. In second section, we evaluate the accuracy of acronym-expansion pairs present in our Acronym dictionary through a user study. In the third section, we compare our Acronym dictionary with existing Acronymfinder.com dictionary. In last section, we describe the acronym disambiguation experiment performed on news headline and twitter dataset.

4.1   Acronym-Expansion Pair Extraction Statistics

Our system is continuously extracting acronym-expansion pair from 1st Aug 2015 to 15th March 2016 from news articles of 34 news sources. We have downloaded and processed about 680 K news articles, and extracted more than 200 K acronym-expansion pairs. Out of 680 K news articles, 132K (20 % of total news articles) contains acronym-expansion pairs. After expansion clustering, we obtain about 18600 unique acronym-expansion pairs and 12800 unique acronym words. Out of 12800 acronym words, 2930 acronym words, i.e. 22 % of total unique acronym words have more than one expansion. On an average our system daily discovers 23 new acronym words as shown in Fig. 2.

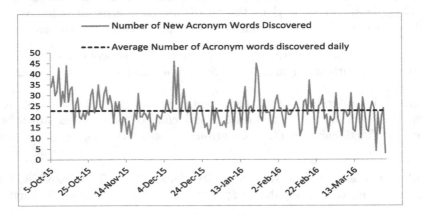

Fig. 2. News Acronym Words discovered per day

4.2 Extraction Accuracy of Acronym-Expansion Pair

To evaluate the extraction accuracy of acronym-expansion pairs in our Acronym dictionary, we perform a user study with 105 users. We randomly present 5 acronym-expansion pairs to each user from our Acronym dictionary of 18600 acronym-expansion pair. Each user is asked to check whether expansions for given acronyms is correct or incorrect by searching on Google search engine. The acronym-expansion pairs for users are generated by random weighted sampling, where weight of an acronym-expansion pair is inversely proportional to their popularity. Here, popularity of acronym expansion pair is their occurrence frequency in news articles. Therefore, unpopular acronym-expansion pairs had a higher probability to get selected in a random sample of 5 acronyms. The unpopular acronyms are presented to users as they have less probability to be found in search engine results and thus have high probability to be incorrect. A sample of 525 acronym-expansion pairs of Acronym dictionary were checked by users, out of which 75 (14.2 %) acronym-expansion pairs were found incorrect and 450 (85.8 %) acronym-expansion pairs were found correct by users. Thus about 85 % of acronym-expansion pairs are estimated to correct.

4.3 Comparision with Acronymfinder.com

Acronymfinder.com is the largest known Acronym dictionary which contains more than 1000 K human edited acronym-expansion pairs. Acronymfinder do not provide any API to retrieve the acronym-expansion pairs. However, we have manually retrieved all the acronym words from the index webpages[3] of Acronymfinder. Acronymfinder.com contains 379309 unique acronym words. Total number of unique acronym words in our Acronym dictionary are 12868. Unique acronym words in our Acronym dictionary that are not present in Acronymfinder.com is 1710 (13 % of unique acronym words in Acronym dictionary). These 1710 acronym words constitute 1781 acronym-expansion pairs (9 % of acronym-expansion pairs in Acronym dictionary). Thus, 13 % acronym words and 9 % acronym-expansion pairs of our Acronym dictionary are not present in Acronymfinder.com.

4.4   Acronym Disambiguation in Short Text

We have evaluated acronym disambiguation on two different short text dataset.

1. Headline Dataset: This dataset contains 2000 news headlines each contain an acronym. The expansion of each acronym is present in their corresponding news article text. We have extracted expansion of acronyms in headline from article text and use it as a baseline for evaluating our acronym disambiguation approach.
2. Tweet Dataset: This dataset contains 100 tweets of popular Indian personalities[4], where each tweet contains an Acronym. We have manually identified the expansion of acronym in each tweet and use it as a baseline to evaluate our acronym disambiguation approach

Recall, that our acronym disambiguation method identifies the expansion from Acronym dictionary by using popularity based method and context based method. We apply acronym disambiguation methods on both the datasets and evaluate them against their corresponding baseline methods. Results are shown in Table 4. We can see that popularity based method performs better than context based method. This indicates that acronyms without expansion in short text such as twitter and headlines, represents the popular expansion for that acronym.

Table 4. Acronym Disambiguation results on Headline and Tweet dataset

Disambiguation Method	F1 -Score	
	Headline Dataset	Tweet Dataset
Popularity based	92.55 %	76.04 %
Context based	90.41 %	72.92 %
Popularity+Context	97.88 %	79.17 %

---

[3] http://www.acronymfinder.com/Index--.html.
[4] Narendra Modi, Smriti Irani, MS Dhoni, Arun Jaitely, Rahul Gandhi, Arvind Kejriwal and Ravishankar Prasad.

We found that context based method is not able to map some acronyms to any expansion in the Acronym dictionary due to no contextual overlapping. To overcome this we combine context based method with popularity *(Context+Popularity)*. In *Context+Popularity* method, we first disambiguate acronyms with context based method, if no expansion is found, then we use Popularity to identify the expansion for acronym. The result of *Context+Popularity* on both the datasets is shown in Table 4. We can see that *Context+Popularity* performs better than both Popularity and Context based method.

5   Related Work

We describe the previous research works on acronym mining in two Sections (1) Acronym discovery and (2) Acronym disambiguation.

Acronym Discovery. Acronymfinder.com is the world's largest collection of 379,309 acronyms with more than 1000 K human edited expansions. Although, acronyms in Acronymfinder.com are categorized in one of the 6 pre-defined categories, it is not sufficient for disambiguating an Acronym. It is updated voluntarily by some contributors manually. Existing researches [3,12–16] discovers acronym-expansion pairs in different types of text documents by rule based or supervised methods. Taneva et al. in [12] uses user query click log of search engines, Feng et al. [3] uses web pages and Ehrmann et al. [13] uses news articles for discovering acronym-expansion pairs. However, none of these works focused on retrieving meta-information required for disambiguating the acronyms.

Acronym Disambiguation. Feng et al. [3] proposed a graph based technique to disambiguate ambiguous acronyms, which makes use of hyperlinks of webpages to perform the acronym disambiguation. Li et al. [4] proposed an acronym disambiguation approach using word embedding is provided in which use TF-IDF based word embedding for proper disambiguation of acronyms. Zahariev [17] and proposed a machine learning system, which uses other terms occurring with acronyms in the same documents and distances of those terms with acronyms to disambiguate the expansion of acronyms. Zhang et al. in [5] introduced a supervised learning algorithm to identify the correct expansion for acronym words in text. Choi et al. [18] present a word co-occurrence based technique to disambiguate the acronyms in Wikipedia text.

All the aforementioned acronym disambiguation techniques perform inter document acronym disambiguation however, we are not aware of any technique that builds acronym repository specifically for disambiguation of acronyms. Also, we did not find any technique that disambiguates the acronyms in short texts such as tweet or news headlines.

6   Conclusion

In this paper, we proposed a system for discovery, enrichment and disambiguation of acronyms. Our system builds an Acronym dictionary by continuous monitoring of new article stream. Each acronym in our Acronym Dictionary is enriched with meta information such as category, location and context words. The meta-info enriched Acronym dictionary is used for disambiguation of acronym words that occurs in short texts. We have also conducted a series of experiments to evaluate our system. Our experimental results show that our system efficiently discover and disambiguate the acronym in news media.

References

1. Silva, G., Montgomery, C.A.: Knowledge representation for automated understanding of natural language discourse. Comput. Humanit. **11**(4), 223–243 (1977)
2. Lavi, O., Auerbach, G., Persky, E.: Dynamic natural language understanding. US Patent 7,840,400, 23 Nov 2010
3. Feng, S., Xiong, Y., Yao, C., Zheng, L., Liu, W.: Acronym extraction and disambiguation in large-scale organizational web pages. In: Proceedings of the 18th ACM Conference on Information and Knowledge Management, CIKM 2009, ACM (2009)
4. Li, C., Ji, L., Yan, J.: Acronym disambiguation using word embedding. In: Twenty-Ninth AAAI Conference on Artificial Intelligence (2015)
5. Zhang, W., Sim, Y.C., Su, J., Tan, C.L.: Entity linking with effective acronym expansion, instance selection and topic modeling. In: Proceedings of Internationl Joint Conference on Artifical Intelligence, IJCAI 2011, AAAI Press (2011)
6. HaCohen-Kerner, Y., Kass, A., Peretz, A.: Combined one sense disambiguation of abbreviations. In: Proceedings of the 46th Annual Meeting of the Association for Computational Linguistics on Human Language Technologies, ACL (2008)
7. McInnes, B.T., Pedersen, T., Liu, Y., Pakhomov, S.V., Melton, G.B.: Using second-order vectors in a knowledge-based method for acronym disambiguation. In: Proceedings of the Fifteenth Conference on Computational Natural Language Learning, CoNLL 2011, Association for Computational Linguistics (2011)
8. Nguyen, D.B., Hoffart, J., Theobald, M., Weikum, G.: Aida-light: high-throughput named-entity disambiguation. In: Linked Data on the Web at WWW (2014)
9. Ferragina, P., Scaiella, U.: Fast and accurate annotation of short texts with wikipedia pages. Proceedings of arXiv preprint (2010)
10. Ratinov, L., Roth, D., Downey, D., Anderson, M.: Local and global algorithms for disambiguation to wikipedia. In: Proceedings of ACL (2011)
11. Barua, J., Patel, D., Agrawal, A.K.: Removing noise content from online news articles. In: Proceedings of the 20th International Conference on Management of Data, COMAD 2014, Computer Society of India (2014)
12. Taneva, B., Cheng, T., Chakrabarti, K., He, Y.: Mining acronym expansions and their meanings using query click log. In: Proceedings of the 22nd International Conference on World Wide Web, WWW 2013, ACM (2013)
13. Ehrmann, M., Della Rocca, L., Steinberger, R., Tanev, H.: Acronym recognition and processing in 22 languages. arXiv preprint arXiv:1309.6185 (2013)
14. Dannélls, D.: Automatic acronym recognition. In: Proceedings of the Eleventh Conference of the European Chapter of the Association for Computational Linguistics. EACL 2006, Association for Computational Linguistics (2006)
15. Sánchez, D., Isern, D.: Automatic extraction of acronym definitions from the web. Appl. Intell. **34**(2), 311–327 (2011)

16. Nadeau, D., Turney, P.D.: A supervised learning approach to acronym identification. In: Kégl, B., Lee, H.-H. (eds.) Canadian AI 2005. LNCS (LNAI), vol. 3501, pp. 319–329. Springer, Heidelberg (2005)
17. Zahariev, M.: Automatic sense disambiguation for acronyms. In: Proceedings of the 27th Annual International ACM SIGIR Conference on Research and Development in Information Retrieval. SIGIR 2004, ACM (2004)
18. Choi, D., Kim, P.: Identifying the most appropriate expansion of acronyms used in wikipedia text. Softw. Pract. Experience **45**(8), 1073–1086 (2015)

A Value-Added Approach to Design BI Applications

Nabila Berkani[1(✉)], Ladjel Bellatreche[2], and Boualem Benatallah[3]

[1] École nationale Supérieure d'Informatique (ESI), Algiers, Algeria
n_berkani@esi.dz
[2] LIAS/ISAE-ENSMA–Poitiers University, Poitiers, France
bellatreche@ensma.fr
[3] University of New South Wales, Sydney, Australia
boualem@cse.unsw.edu.au

Abstract. Big Data Era has largely contributed in accelerating the development of large, high quality and valuable Knowledge Bases ($\mathcal{KB}$) by academicians (e.g., *Cyc*, *DBpedia*, *Freebase*, and *YAGO*) and industrials (e.g., Knowledge Graph). On the other hand, serious studies have identified the crucial role of $\mathcal{KB}$ for analytical tasks, by offering analysts more entities (people, places, products, etc.). The availability of a huge, high quality and valuable $\mathcal{KB}$ may contribute on designing *value added approaches* for business intelligence applications. In this paper, we first propose a novel approach for semantic $\mathcal{DW}$ design that considers $\mathcal{KB}$ in the life cycle. Secondly, based on graph formalization adapted to $\mathcal{KB}$, we produce conceptual multidimensional design and a semantic ETL process that orchestrates the graph data flows from data sources to the $\mathcal{DW}$ storage. Finally, all steps of our approach are illustrated using the YAGO $\mathcal{KB}$ and deployed in Oracle RDF Semantic Graph 12c.

1 Introduction

The deluge of data available in various heterogeneous, autonomous, distributed and evolving data/Web sources has to be organized and materialized for analysis purposes through the construction of *valuable* data warehouse ($\mathcal{DW}$) systems. The design of $\mathcal{DW}$ passes through two main generations: *traditional* design and *semantic* design. Contrary to the first generation, the semantic design incorporates ontologies in the all phases of the life cycle of $\mathcal{DW}$ design (refer the survey paper for details [1]): user requirements formalization and explicitation of their semantics [7,14], the definition of multidimensional concepts (facts and dimensions) [18], the data integration [4] and the automation of ETL process [3,19]. Nobody can deny the large contribution of ontologies in designing $\mathcal{DW}$ systems requiring one domain ontology, with reduced size of data and few restrictions. However, decision makers are always looking for more valuable knowledge to improve the competence of their companies to make relevant decisions. The value of intangible assets to the organization. Recently, in ACM SIGMOD blog[1],

---

[1] http://wp.sigmod.org/?p=1519.

© Springer International Publishing Switzerland 2016
S. Madria and T. Hara (Eds.): DaWaK 2016, LNCS 9829, pp. 361–375, 2016.
DOI: 10.1007/978-3-319-43946-4_24

an interview of several prominent members of the data management community around a theme *"The elephant in the room: getting value from Big Data"* arguy that the **V - value** is not talked about as much as it should be. In the context of $\mathcal{DW}$, the value is only obtained from data sources. The spectacular development of valuable $\mathcal{KB}$ can be an asset for decision makers to enhance the value of sources by the value that $\mathcal{KB}$.

Actually, the increasing of knowledge-sharing communities such as *Wikipedia* and the improvement of information extraction from the Web have enabled the automatic construction of very large $\mathcal{KB}$s. *DBpedia* [11], *Freebase.com*, *Know-ItAll* [8,9], and *YAGO* [10,12,21] are examples of these $\mathcal{KB}$s. They contain millions of entities and hundreds of millions of facts which increase their richness. Most of these $\mathcal{KB}$ represent facts in the form of *subject-property-object* triples according to the RDF data model. Prominent examples of how $\mathcal{KB}$s can be capitalized and harnessed are confirmed by commercial interests, strongly growing with projects like the *Google Knowledge Graph*, the *EntityCube* and *Renlifang* project at Microsoft Research [16], and the IBM Watson question answering system.

Such $\mathcal{KB}$s contain different application domains and are linked to Web data. It offers a deeper a valuable knowledge interpretation and reasoning capabilities. For that, we claim that the presence of $\mathcal{KB}$ can be an asset for $\mathcal{DW}$ building. Particular emphasis on the role of $\mathcal{KB}$s in the process of constructing $\mathcal{DW}$ concerns two main aspects: (i) augmenting the value of extracted facts and dimensions of the target $\mathcal{DW}$, and (ii) leveraging existing ETL algorithms to capture this value, by adapting them to deal with the graph data structure of the used $\mathcal{KB}$. In this paper, we propose a third generation of $\mathcal{DW}$ design, characterized by the use of huge $\mathcal{KB}$ in all steps of designing.

Note that $\mathcal{DW}$ contains concepts and instances heavily cleaned by using advanced algorithms (statistical learning, inferences and crowdsourcing) [5]. As a consequence they offer a great opportunity to the $\mathcal{DW}$ community to consider them during the construction of their applications. In this paper, we propose a new methodology of the third generation of the construction of $\mathcal{DW}$ based on $\mathcal{KB}$. Through this paper, we concentrate on four main phases of the life cycle of $\mathcal{DW}$ design which are: (i) user requirements, (ii) conceptual phase, (iii) logical phase and (iv) deployment phase. To the best of our knowledge, this work is the sole that heavily uses $\mathcal{KB}$s during $\mathcal{DW}$ design.

The paper is organized as follows: Sect. 2 presents related work. Section 3 presents the main concepts related to our approach. Section 4 presents in details our approach. Section 5 presents a case study. In Sect. 6, intensive experiments are conducted to validate our proposal. Section 7 concludes the paper.

2   Related Work

In this section, we review the most important works related to the second generation of $\mathcal{DW}$ design that considers ontologies.

In [20], authors built an ontology for automatizing ETL process. They used a reasoner to identify the operations needed to load data. [19] proposes a template-based natural language mechanism in order to describe data sources and ETL operations into a narrative textual report more suitable for the user. Both works assume that schema of data sources and $\mathcal{DW}$ are previously known and they deal only with ETL phase. [17] defined a multidimensional design schema starting from an OWL ontology describing data sources. They identified a central concept under analysis (fact concept) and concepts connected to it through *one-to-many* relationships (dimensions and hierarchies). The output is a star or snowflake schema suitable to be instantiated in a traditional multidimensional database. [15] proposed a formal-based method using OWL/DL to transfer combination of data into fact tables and to build dimensions. The analyst defines the multidimensional model from existing domain ontology.

Romero et al. described in [18] a method that combines multidimensional conceptual design and ETL process. They start by analyzing the data sources mapped to a domain ontology. Then, they define steps that lead to identifying facts, dimensions, and dimension hierarchies. User confirmation for suggested concepts is needed before to produce a multidimensional schema and a flow of conceptual ETL operations. However logical and physical design steps are not considered in this work. [14] considers semantic data provided by the semantic Web and annotated by OWL ontologies, from which a $\mathcal{DW}$ model is defined and populated. However, the ETL process in this work is dependent of a specific instance format (triples). [3] proposes an ontology-based design approach, where ontologies are confronted to each phase of the life-cycle. The ETL process is defined at the conceptual level allowing a deployment *à la carte*.

The main limitation of the discussed approaches is the fact that they use a single ontology with a reduced size and few restrictions. However, actual real applications reference different domain ontologies with considerable large size structured around $\mathcal{DW}$.

3   Background

This section presents the essential concepts and notions related to $\mathcal{KB}s$ and RDF/RDFS.

A $\mathcal{KB}$ typically contains a set of concepts $\mathcal{C} : \{C_1, C_2, \ldots, C_n\}$, $\mathcal{I}$: instances $\{I_1, I_2, \ldots, I_m\}$, and relations $\mathcal{R} : \{R_1, R_2, \ldots, R_k\}$ [13]. It allows describing a domain and providing reasoning capabilities. The Description Logic (DL) formalism (one of its fragments) is used to capture the meaning of the most popular data class-based modeling formalisms presently used in databases and information system analysis [4]. A $\mathcal{KB}$ in DL is composed of two components: the *TBOX* (Terminological Box) stating the intentional knowledge and the *ABOX* (Assertion Box) stating the extensional knowledge or the instances (e.g., *Physicist(Einstein) denotes that Einstein is an instance of the concept Physicist*). Terminological axioms have the form of inclusions: $C \sqsubseteq D$ or equalities: $C \equiv D$ (Eg. Physicist $\sqsubseteq$ Person).

The RDF[2] is a graph-based model accepted as the W3C standard for Semantic Web applications. An RDF graph is a set of triples of the form $< s, o, p >$. A triple states that its subject s has the property p, and the value of that property is the object o. They are used to describe resources, and property assertions. The RDF standard provides a set of built-in classes and properties, with predefined namespaces. For example *rdf:type* specifies the classes to which a resource belongs. An RDF graph encodes a graph structure in which every triple describes a directed *edge* labeled with p from the *node* labeled with s to the *node* labeled with o.

RDF Schema (RDFS)[3] enriches the descriptions in RDF graphs, by declaring semantic constraints between the classes and properties used in the graph. Such constraints can define *subClass, subProperty relations, domain* and *range* of properties. The RDFS constraints are interpreted under the open-world assumption (OWA) [2].

Formally, a $\mathcal{KB}$ is defined as 5-tuples: $< \mathcal{C}, \mathcal{R}, \mathcal{I}, C_i, R_i >$, the $\mathcal{C}, \mathcal{R}$ and $\mathcal{I}$ are above defined. $C_i : C \rightarrow Rep(I)$ corresponds to the concept instantiation. *Rep* associates instances to classes using DL constructors. For example, based on RDF formalism, Student = Rep(Student#1) where Rep = rdf:resource). R_i: $\mathcal{R} \rightarrow Rep(\mathcal{I}^2)$ role instantiation, where $\forall x \in \mathcal{R} : R_i(x) \subseteq C_i(domain(x)) \times C_i(range(x))$.

4   Our Proposal: Graph $\mathcal{DW}$ Modeling

Before detailing our approach, let us formalize it. Given an existing $\mathcal{KB}$ referenced by a set of semantic databases ($\mathcal{SDB}$) alimenting the target $\mathcal{DW}$. Figure 1 illustrates the steps of our approach that we describe in next sections.

User Requirement Definition. Note that user requirements represent a crucial role in designing $\mathcal{DW}$ applications [3]. In this study, we consider a functional requirement model defined by [7]. It is composed by three main elements: **(a)** actors issuing the requirements, **(b)** requirements, and **(c)** relationships between requirements. Each requirement corresponds to **(i)** a set of tasks $\mathcal{T}$, where each one is represented by: $< Subject, Action, Object >$, where *Subject*: is an actor who uses the requirement to achieve a result. *Action* is an action that a system performs to yield to an observable result. *Object* is the concept concerned by the requirement action. **(ii)** the results offered by the system to be designed, **(iii)** a set of criteria which quantifies a result, and **(iv)** the formalism used to model the requirements (e.g. UML use case, goal oriented formalism). Finally, relationships may exist between requirements such as *Equal, Contain, Refine, Require, Conflicts with, partially Refine.*

Example 1. Let us take an example of a functional requirement: *the system computes the average of students integrating university.* This requirement is then represented by the above model as follows: $((subject : students), (action : integrate), (object : university), (criterion : \emptyset), (result : average of students))$.

---

[2] https://www.w3.org/RDF/.
[3] http://www.w3.org/TR/rdf-schema/.

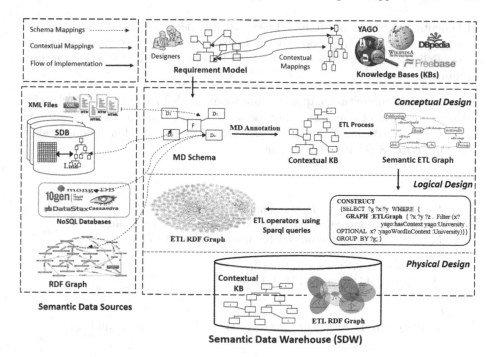

Fig. 1. Design method proposed

Note that $\mathcal{KB}$ may be very large in terms of concepts and facts. It has to be personalized in order to respond to the application to be designed. This personalization may be performed by projecting requirements on the $\mathcal{KB}$. This will be done using the *context* operator, recently introduced in Yago2 $\mathcal{KB}$ [21].

Definition 1. *The context is defined in [6] as any information that can be used to characterize the situation of entities (i.e., whether a person, place or object) that are considered relevant to the interaction between a user and an application.*

In our case, the context is a set of mappings between the $\mathcal{KB}$ and requirements. More concretely, these mappings concern elements of each requirement *(subject, action, object, Result and Criteria)* and elements of $\mathcal{KB}$ *(concepts and roles)*. Three mapping are distinguished: **(i)** *class-mappings* between $\mathcal{KB}$ classes and requirement entities, **(ii)** *attribute-mappings* between $\mathcal{KB}$ roles (data-type) and requirement attributes and **(iii)** *relation-mappings* between $\mathcal{KB}$ roles (object-property) and requirement relations. The mapping can be unidirectional or bidirectional. Expressions can be: equivalence ($\equiv$), generic/specific ($\subseteq$ or $\supseteq$) or operators for numeric constants ($=, <, >, \leq, \geq$). Three scenarios can be defined to define these mappings:

- Elements of $\mathcal{KB}$ correspond exactly to user requirements (equivalent contextual mapping);
- Elements of $\mathcal{KB}$ cover more than user requirements (complete contextual mappings);

– Elements of $\mathcal{KB}$ does not fulfill all users requirements and need to be enriched (sound contextual mappings). For that, we propose to use existing reasoners (e.g. Pellet, Fact[4]) in order to infer more elements extending $\mathcal{KB}$ to cover user requirements.

In our work, contextual mapping expressions are defined as axioms and added to $\mathcal{KB}$ in the form of semantic annotations. This will make possible to discover them at any time during $\mathcal{DW}$ construction and maintenance. The following algorithm annotates $\mathcal{KB}$ with requirements:

> **begin**
> > **Inputs:** $\mathcal{KB}$, a set of UR, MappingList, Context
> > **Output:** $\mathcal{KB}$ annotated with *user requirements (UR)* mappings
> > M:= GetMapping(quintuplet(UR), Context, MappingList);
> > **for** *each $M_i \in M$* **do**
> > > E := ExtractElement $(\mathcal{KB}, M_i)$;
> > > $\mathcal{KB}$:= AddAnnotation $(\mathcal{KB}, E, M_i)$;
> >
> > **end**
>
> **end**

Algorithm 1. Algorithm for definition of contextual mappings

Conceptual Design: Multidimensional Schema Definition. The definition of Multidimensional (MD) schema is done based on contextual $\mathcal{KB}$ and data sources schemes. As a result, concepts and properties are identified and then annotated with MD concepts (*Measure, Fact* and *Dimension*). This is done as follows: numerical attributes represented by a data-property are considered as candidate measures and their related classes as candidate facts. They are validated based on *measure-type* attribute of *Result* requirement entity. We annotate classes and properties identified respectively with *Fact* and *measure* annotations. After that, all classes related to the fact entity with an object property are candidate dimensions. This is done by checking existing *many-to-one* relationships between fact and classes (all axioms having min and max cardinality restrictions). Dimensions are validated if they are domains or ranges of requirement attributes. They are annotated with *dimension*. Finally, dimension levels are identified based on *one-to-many* relationships between dimensions, by checking max-cardinality of related axioms.

On Graph ETL Process. Generally, ETL process is usually represented as a directed acyclic graph (DAG) containing a set of nodes (that represent schema attributes or instances) and edges (describing data flow among the nodes using RDFS taxonomy). The data flow is a set of operations needed for transformation of input data. The operations are applied at node level and generates a new nodes forming an ETL graph. Based on this observation, we formally define an ETL process as: $< InputSet, OutputSet, Operator, Function, ETLGraph >$, where

---

[4] http://owl.cs.manchester.ac.uk/tools/fact/.

Input-Set represents a finite set of input nodes (concepts and instances) corresponding to the warehouse schema and data sources, *Output-Set* is a finite set of intermediate or target nodes (concepts and instances), *Operator* is a set of operators commonly encountered during ETL process, *Function* is a function over a subset of Input-Set nodes applied in order to generate data satisfying restrictions defined by ETL operators. *ETL-Graph* is a set of output nodes (concepts and instances) added to final warehouse schema and linked by edges (roles).

In [20], authors have defined 10 generic conceptual ETL operators. We have overloaded them to consider the characteristics of a semantic graph G representing $\mathcal{KB}$ whose nodes (N), edges (E) and labels (L) represent respectively classes, instances and data properties, object properties and **DL** constructors.

$\vdash$ *Retrieve*(G, n_j, L_j): retrieves a node n_j having an edge labeled by L_j of G.

$\vdash$ *Extract*(G, n_j, CS): extracts, from G, the node n_j satisfying constraint CS.

$\vdash$ *Convert*(G, G_T, n_i, n_j): converts the format of the node $n_i \in G$ to the format of the target node $n_j \in G_T$. The conversion operation is applied at instance level.

$\vdash$ *Filter*(G, n_i, CS): is applied on the nodes n_i and allows only the part satisfying constraint CS. Filter can also be applied on instance level based on defined axioms.

$\vdash$ *Merge*(G, n_j, I_j): adds instances I_j as nodes n_j in same graph G.

$\vdash$ *Union*(G, G_T, n_i, n_j, E_j): links nodes that belongs to different graphs. It adds edge E_j that link the node $n_i \in G$ to the node $n_j \in G_T$ in the target graph G_T.

$\vdash$ *Join*(G, G_T, n_i, n_j, E_j): joins instances whose corresponding nodes are $n_i \in G$ and $n_j \in G_T$. They are linked by an *object property* defined by an edge E_j.

$\vdash$ *Store*(G_T, n_j, I_j): associates instances I_j to the nodes $n_j \in G_T$ added to the target graph G_T.

$\vdash$ *DD*(G_T, CS): starts by sorting the graph G_T based on constraint CS and detects nodes associated to duplicated instances.

$\vdash$ *Aggregate*(G_T, n_j, Op): aggregates instances represented by the nodes n_j.

Some primitives have been added to manage the ETL operations required to build the ETL graph such as: *AddNode, UpdateNode, RenameNode, DeleteNode, SortGraph*.

$\vdash$ *AddNode*(G_T, n_j, E_j): adds an edge E_j and a node n_j to G_T.

$\vdash$ *UpdateNode*(G_T, n_j, E_j): updates node n_j and edge E_j in G_T.

$\vdash$ *RenameNode*(G_T, n_j, E_j): renames node n_j and edge E_j in G_T.

$\vdash$ *DeleteNode*(G_T, n_j, E_j): deletes node n_j and edge E_j from G_T.

$\vdash$ *SortGraph*(G_T, n_j, CS): sorts nodes of G_T based on some criteria CS to improve search performance.

Our goal is to facilitate, manage and optimize the conceptual design of the ETL process during the initial design, deployment phase and during the continuous evolution of the data warehouse. For that, we enrich the conceptual ETL operators with *split, context* and *Link* operators elevating the cleanup and deployment of ETL process at the semantic level.

– $Split(G, G_i, G_j, CS)$: splits G into two sub-graphs G_i and G_j based on some criteria CS.

– $Link(G_T, n_i, n_j, CS)$: links two nodes n_i and n_j using the rule CS.

– $Context(G, G_T, CS)$: extracts from G a sub-graph G_T that satisfies the context CS defined by some restrictions. CS is defined as an axiom applied on G.

Our ETL algorithm takes as inputs the $\mathcal{KB}$ representing a global schema and a set of semantic data sources (SDB) whose schemes are a fragment of $\mathcal{KB}$. For lack of space, the details of this algorithm are given at google drive following this link: https://drive.google.com/file/d/0B4K41KMxS6mZUjVqb0l2YUpITW8/view?u sp=sharing.

Logical and Physical Design. The logical design phase requires the translation of $\mathcal{DW}$ schema into logical model (example relational model) and translation of ETL process into logical workflow. The translation of ETL process into logical workflow consists in translating conceptual ETL operators into SPARQL[5] queries. Each query result is represented through an RDF graph. The ETL process is defined by the link of all those graphs. The following example translates the *Filter* operator into an RDF graph:

```
Select ?instance ?P where {
GRAPH   :?G {?Instance rdf:type name-space:Class . name-space:P ?P .
FILTER (?P op value_condition)}}
```

5   A Case Study

In order to demonstrate the feasibility of our approach, we consider a case study, where an education ministry wants to create a ministry central $\mathcal{DW}$ to perform some analysis studies (identified by a set of user requirements *UR*) to take relevant decisions. These data sources need thus to be integrated following an ETL process. Let assume the existing of four Oracle $\mathcal{SDB}$ (S_1, S_2, S_3 and S_4) and populated locally using YAGO $\mathcal{KB}$. Oracle delivers *RDF Semantic Graph features* as part of Oracle Spatial and Graph. With native support for RDF and OWL standards for representing semantic data, with *SPARQL* for query language. Oracle has defined two subclasses of DLs: *OWLSIF* and a richer fragment *OWLPrime*. We use *OWLPrime* fragment which offers the following constructors: *rdfs:domain, rdfs:range, rdfs: subClassOf, rdfs:subPropertyOf, owl:equivalentClass, owl: equivalentProperty, owl: sameAs, owl:inverseOf, owl: TransitiveProperty, owl:SymmetricProperty, owl: FunctionalProperty, owl: InverseFunctionalProperty*. Note that *OWLPrime* limits the expressive power of the DL formalism in order to ensure decidable query answering.

---

[5] https://www.w3.org/TR/rdf-sparql-query/.

Currently, YAGO has knowledge of more than 10 million entities (like persons, organizations, cities, etc.) and contains more than 120 million facts about these entities. It has achieved a precision of 95 %. Compatible with RDF, it allows generating data in different formats: *Turtle, N-Triple, Literals*, etc. *YAGO* contains knowledge about the real world including entities (e.g., university, people, cities, countries, etc.) and facts about these entities (city of each university, program of each level in the university, which city is located in a country, etc.).

Multidimensional Schema Definition. The typical requirements could be: UR_1: number of students by university and year, UR_2: average of publications per program and UR_3: number of professors per university and per city.

Using the pivot model of [7] (developed in our laboratory), we translate all those requirements in the format explained in Sect. 4. We obtained the following quintuplets for each requirement:

- UR_1: ((subject: *student*), (action: *master-degree and Time.level*), (object: *university and city*), (criterion: ∅), (result: *number-student.sum(student.name)*);
- UR_2: ((subject: *publication*), (action: *assigned*), (object: *program*), (criterion: ∅), (result: *publication.avg(publication.id)*));
- UR_3: ((subject,*professor*), (action,*assigned and lives*), (object,*university and city*), (criterion, ∅), (result, *publication.sum(publication.id)*)).

Then, we elaborate the contextual mappings between user requirements and YAGO $\mathcal{KB}$. It is worth repeating that designer should use the terms of YAGO $\mathcal{KB}$ to define the contextual mappings. This is to avoid a representation conflict between the concepts. After that, Algorithm 1 is applied to define the contextual mappings as annotations. We obtained a contextual YAGO $\mathcal{KB}$ annotated with user requirements. We apply algorithm presented in our technical report. We consider in our example the case that $\mathcal{SDB}$ schema corresponds to the scenario in which: $\mathcal{SDB} \subset \mathcal{KB}$.

ETL Process and Deployment. Storage deployment models can follow different representations according to specific requirements. A $\mathcal{DW}$ can be deployed using several storage layouts: horizontal, vertical, hybrid models, NoSQL, etc. In our case, we choose to deploy the semantic $\mathcal{DW}$ into vertical storage in Oracle to represent graphs by means *Oracle RDF Semantic Graph*. On this basis, we translated the $\mathcal{SDB}$ schema into vertical relational model. We generated an N-Triple file and load it into a staging table using Oracle's *SQL*Loader* utility. After that, we applied the ETL Algorithm to populate the target schema. Note that generic ETL operators defined are expressed on the conceptual level. Therefore, each operator has to be translated according the logical level of the target DBMS using Sparql. Here an example of aggregation ETL operator translation to Sparql:

```
PREFIX yago: http://yago-knowledge.org/resource/yago.owl#
AGGREGATE: Aggregates incoming record-set.
Select (Count(?Instance) AS ?count) Where {
GRAPH  :?G {?Instance rdf:type yago:Class}}
Group By ?Instance.
```

The translation of Context operator in our case study to Sparql query generates the ETL Graph from YAGO $\mathcal{KB}$ using *yagoWordInContext* object property as predicate, university value as object, and subject being a class of resources. It will provide an additional information required for RDF data analysis.

```
PREFIX  rdf : <http ://www.w3.org/1999/02/22-rdf-syntax-ns#>
PREFIX  yago: http://yago-knowledge.org/resource/yago.owl#
CONSTRUCT {SELECT  ?g ?x ?y
WHERE {GRAPH:ETLGraph { ?x ?y ?z .
Filter (?x rdf:type  owl:DatatypeProperty . x? yago:hasContext yago:university .
OPTIONAL  x? yago:yagoWordInContext yago:university)}} GROUP  BY ?g ?c ;}
```

6   Experimentation

In this section, we present the performance of our approach through a set of experiments considering large $\mathcal{KB}$. Four criteria are used to evaluate our proposal: (i) the complexity of the proposed ETL algorithm, (ii) the data load performance during integration process, (iii) inference performance and (iv) query response time of requirements. Finally, a comparison is done between our approach and a state-of-art work that covers the whole life cycle [3].

6.1   Environment of Our Experiments

Implementation Settings. Our experiments is based on YAGO $\mathcal{KB}$ (version 3.0.2). The architecture of the YAGO system is based on themes. Each theme is a set of facts. A fact is the equivalent of an RDF triple: $< s, p, o >$. YAGO has defined the context relation between individuals [21]. We used those relations to extract the set of themes related to our context study. The resulting contextual YAGO $\mathcal{KB}$ contains around $5, 9 \times 10^6$ triples.

Data-Sets. We have generated from contextual YAGO $\mathcal{KB}$, 5 data-sets of instances containing respectively 3, 6, 9, 12 and 15 universities (same number of universities than our previous work [3] in order to make a comparison). Each university contains different dependencies with the classes defined in the multidimensional schema such as: Students, Courses, cities, publications, etc. Number of instances (N-Triple format) used in both approaches is shown in Table 1. However, we have added some contradictory cases to test their influence on the semantic data integration.

Deployment of $\mathcal{SDB}$ and Semantic $\mathcal{DW}$. We have created five Oracle $\mathcal{SDB}$ using generated data-sets. We have also deployed the schema of semantic $\mathcal{DW}$ using Oracle $\mathcal{SDB}$. It offers different format for data loading such as: RDF/XML, N-TRIPLES, N-QUADS, TriG and Turtle. We choose N-Triple format (.nt) to load instances using Oracle SQL*Loader.

Oracle Database Tuning. $\mathcal{SDW}$ schema was optimized using Btree indexing triples and sparql query hints. In the other hand, some PL/SQL APIs are invoked after each load of significant amount of data (integration of each data source). The API SEM_PERF.GATHER_STATS allows collecting stats for

Table 1. Dataset characteristics.

Concepts	Ontological Instances	$\mathcal{KB}$ Instances
3 Universities	21 057	15988
6 Universities	42 115	96 962
9 Universities	63 173	123 004
12 Universities	84 231	$3,9 \times 10^5$
15 Universities	105 289	$2,5 \times 10^6$

the data sources models and SEM_APIS.ANALYZE_MODEL for $\mathcal{DW}$ model in the semantic network graph. The memory SGA and PGA are also increased to 2 GB.

Inference Engine. Oracle has incorporated a reasoner engine defined based on *TrOWL* and *Pellet* reasoners. Oracle provides full support for native inference in the database for RDFS, RDFS++, OWLPRIME, OWL2RL, etc. It uses forward chaining to do the inference. It compiles entailment rules directly to SQL and uses Oracle's native cost-based SQL optimizer to choose an efficient execution plan for each rule. The following is an example of user defined rules applied, they are saved as records in tables.

$Rule_1$: co-author rule:

$$authorOf(?A1, ?P) \wedge authorOf(?A2, ?P) \rightarrow CoAuthor(?A1, ?A2).$$

Hardware. Our evaluations were performed on a laptop computer (HP Elite-Book 840 G2) with an Intel(R) CoreTM i5-5200U CPU 2.20 GHZ and 8 GB of RAM and a 500 GB hard disk. We use Windows10 64bits. We use Oracle Database 12c release 1 that offers RDF Semantic Graph features of Oracle Spatial and Graph. Cytoscape[6] is used for visualization.

6.2 Obtained Results

As we said before, we evaluate our proposal based the above four criteria.

Criterion 1: ETL Algorithm Complexity. The algorithm is implemented based on graph theory, where nodes represents concepts and instances, edges for roles and labels for definitions. First of all, we examine the number of iterations of our algorithm to generate ETL graph (semantic $\mathcal{DW}$ populated). In this case, we are interesting on time complexity. The algorithm is based on concepts searches (Tbox for intentional mappings) and not instances. The time complexity is $O(n)$, where n is the number of involved nodes (which means $\mathcal{KB}$ concepts). It depends on the resolving of constraints defined on data sources, which is at least $O(n)$, where n is the number of involved schemes. Figure 2 shows the number of iterations by $\mathcal{KB}$ classes involved in the multidimensional schema. It indicates a polynomial time. This finding shows the feasibility and efficiency of our approach.

---

[6] http://www.cytoscape.org/.

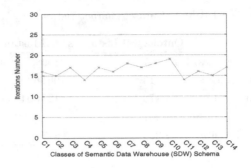

Fig. 2. Complexity of the proposed ETL algorithm

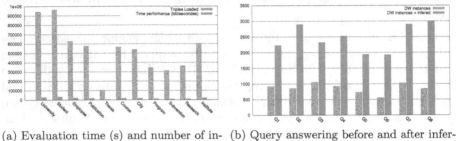

(a) Evaluation time (s) and number of instances loaded in SDW per concept

(b) Query answering before and after inference

Fig. 3. Evaluation time (s) and number of instances loaded in SDW per concept, Query answering before and after inference

Criterion 2: Performance load of ETL Process. We run the ETL Algorithm to populate the target schema of semantic $\mathcal{DW}$ by $\mathcal{SDB}$ data and contextual $\mathcal{KB}$ instances for new dimensions. We measure the time spent to integrate instances into each multidimensional concept (facts and dimensions). Note that the time spent to load all instances is equal 3, 2 min. Figure 3a illustrates the results, where for each concept, number of instances is shown in thousands and the time performance in milliseconds. The result remains reasonable w.r.t. the size of the stored instances.

Criterion 3: Inference Performance. We evaluate inference performance using OWLPrime fragment and user defined rules. First, we define two different models, where each one stores sub-graph of semantic $\mathcal{DW}$. The first model stores instances integrated from $\mathcal{SDB}$ sources, the second one stores instances extracted from both semantic $\mathcal{SDB}$ sources and contextual $\mathcal{KB}$. Then, we used reasoner mechanism to infer instances from both models to show number of instances inferred using $\mathcal{SDB}$ sources unified with $\mathcal{KB}$ instances. Table 2 shows results obtained. It clearly demonstrates that number of triples inferred is important when it comes to use contextual $\mathcal{KB}$ instances. It includes a new dimensions and thus allows more graph analysis.

Table 2. Inference performance: Time and number of Triples.

Criteria	SDW	SDW + $\mathcal{KB}$
Integrated instances	315 865	$5,4 \times 10^6$
Inferred instances	6k	34K
Time inference (minutes)	5,4	11,2

Criterion 4: Response time of Requirements. We consider a set of queries involving multidimensional concepts of semantic $\mathcal{DW}$. We executed the queries from three different perspectives: first taking in account only $\mathcal{SDB}$ instances integrated, then considering also contextual $\mathcal{KB}$ instances added (about new dimensions), and last including all instances of $\mathcal{SDB}$, $\mathcal{KB}$ and inferred one using OWLPrime fragment and user defined rules. Figure 3b shows that query result size is most strongly important using $\mathcal{KB}$ and inferred instances.

6.3 Our Approach vs. an Ontology-Based Approach

Our experiments demonstrate the feasibility of our full $\mathcal{KB}$ warehousing approach, which exploits standard RDF functionalities offered by Oracle such as: triple storage, graph definition and reasoning. We showed scalable performance when loading and integrating data sources. We proved that semantic data warehouse are enriched with concepts and instances from $\mathcal{KB}$ and inference mechanism. This will give more possibility for query answering and data analysis.

Table 3. Comparison between Ontological and $\mathcal{KB}$ approaches.

Criteria	Ontology based	$\mathcal{KB}$ Approach
Number of Dimensions	3	9
Input size	322 887	$5,9 \times 10^6$
Selectivity	293	1053
Complexity	polynomial	polynomial
Response time (minutes)	1,1	3,2

Table 3 demonstrates a comparison between the two approaches on the basis of some criteria identified during the experiment. These results clearly indicate that our approach outperforms the ontological approaches by offering value-added in terms of the final number of dimensions and the size of the target warehouses.

7 Conclusion

In this paper, we show the interest of considering $\mathcal{KB}$ in the construction of $\mathcal{DW}$ applications. This is a continuity of our previous works performed in designing semantic $\mathcal{DW}$. The spectacular development of $\mathcal{KB}$ with very large number of concepts and instances with high quality contributes in augmenting the quality and richness of the final $\mathcal{DW}$. We first leverage the traditional life cycle of the $\mathcal{DW}$ design, by incorporating the graph concepts related to knowledge base. Four phases have been revisited: user requirements that play a crucial role in contextualizing the $\mathcal{KB}$, (b) conceptual phase, (c) ETL, where the ten existing operators were specified on graphs, and (d) logical phase and (e) deployment phase. The obtained results are encouraging and really show the great interest of $\mathcal{KB}$ on the design. The most spectacular results are the augmentation of the number of dimensions and instances of the target $\mathcal{DW}$ compared to the semantic solutions of the second generation. Currently, we are working on optimization issues of our proposal, by concentrating on physical design and the storage layouts.

References

1. Abelló, A., Romero, O., Pedersen, T.B., Llavori, R.B., Nebot, V., Cabo, M.J.A., Simitsis, A.: Using semantic web technologies for exploratory OLAP: a survey. IEEE TKDE **27**(2), 571–588 (2015)
2. Abiteboul, S., Hull, R., Vianu, V.: Foundations of Databases. Addison-Wesley, Boston (1995)
3. Berkani, N., Bellatreche, L., Khouri, S.: Towards a conceptualization of ETL and physical storage of semantic data warehouses as a service. Cluster Comput. **16**(4), 915–931 (2013)
4. Calvanese, D., Lenzerini, M., Nardi, D.: Description logics for conceptual data modeling. In: Chomicki, J., Saake, G. (eds.) Logics for Databases and Information Systems, pp. 229–263. Springer, Heidelberg (1998)
5. Chu, X., Morcos, J., Ilyas, I.F., Ouzzani, M., Papotti, P., Tang, N., Ye, Y.: Katara: a data cleaning system powered by knowledge bases and crowdsourcing. In: ACM SIGMOD, pp. 1247–1261 (2015)
6. Dey, A.K., Abowd, G.D., Salber, D.: A conceptual framework and a toolkit for supporting the rapid prototyping of context-aware applications. Hum.-Comput. Interact. **16**(2–4), 97–166 (2001)
7. Djilani, Z., Khouri, S.: Understanding user requirements iceberg: semantic based approach. In: Bellatreche, L., Manolopoulos, Y., Zielinski, B., Liu, R. (eds.) MEDI 2015. LNCS, vol. 9344, pp. 297–310. Springer, Heidelberg (2015). doi:10.1007/978-3-319-23781-7_24
8. Etzioni, O., Cafarella, M.J., Downey, D., Popescu, A., Shaked, T., Soderland, S., Weld, D.S., Yates, A.: Unsupervised named-entity extraction from the web: an experimental study. Artif. Intell. **165**(1), 91–134 (2005)
9. Fader, A., Soderland, S., Etzioni, O.: Identifying relations for open information extraction. In: EMNLP, pp. 1535–1545 (2011)
10. Hoffart, J., Suchanek, F.M., Berberich, K., Lewis-Kelham, E., de Melo, G., Weikum, G.: YAGO2: exploring and querying world knowledge in time, space, context, and many languages. In: WWW, pp. 229–232 (2011)
11. Lehmann, J., Isele, R., Jakob, M., Jentzsch, A., Kontokostas, D., Mendes, P.N., Hellmann, S., Morsey, M., van Kleef, P., Auer, S., Bizer, C.: Dbpedia - a large-scale, multilingual knowledge base extracted from wikipedia. Semant. Web **6**(2), 167–195 (2015)

12. Mahdisoltani, F., Biega, J., Suchanek, F.M.: YAGO3: a knowledge base from multilingual wikipedias. In: CIDR (2015)
13. Nath, R.P.D., Seddiqui, M.H., Aono, M.: An efficient and scalable approach for ontology instance matching. JCP **9**(8), 1755–1768 (2014)
14. Nebot, V., Llavori, R.B.: Building data warehouses with semantic web data. Decis. Support Syst. **52**(4), 853–868 (2012)
15. Nebot, V., Llavori, R.B., Pérez-Martínez, J.M., Aramburu, M.J., Pedersen, T.B.: Multidimensional integrated ontologies: a framework for designing semantic data warehouses. J. Data Semant. **13**, 1–36 (2009)
16. Nie, Z., Ma, Y., Shi, S., Wen, J., Ma, W.: Web object retrieval. In: WWW, pp. 81–90 (2007)
17. Romero, O., Abelló, A.: Automating multidimensional design from ontologies. In: ACM DOLAP, pp. 1–8 (2007)
18. Romero, O., Simitsis, A., Abelló, A.: *GEM*: requirement-driven generation of ETL and multidimensional conceptual designs. In: Cuzzocrea, A., Dayal, U. (eds.) DaWaK 2011. LNCS, vol. 6862, pp. 80–95. Springer, Heidelberg (2011)
19. Simitsis, A., Skoutas, D., Castellanos, M.: Natural language reporting for ETL processes. In: ACM DOLAP, pp. 65–72 (2008)
20. Skoutas, D., Simitsis, A.: Designing ETL processes using semantic web technologies. In: ACM DOLAP, pp. 67–74 (2006)
21. Suchanek, F.M., Kasneci, G., Weikum, G.: Yago: a core of semantic knowledge. In: WWW, pp. 697–706 (2007)

Towards Semantification of Big Data Technology

Mohamed Nadjib Mami[1,2(✉)], Simon Scerri[1,2], Sören Auer[1,2],
and Maria-Esther Vidal[1,2]

[1] University of Bonn, Bonn, Germany
{mami,scerri,auer,vidal}@cs.uni-bonn.de
[2] Fraunhofer IAIS, Sankt Augustin, Germany

Abstract. Much attention has been devoted to support the volume and velocity dimensions of Big Data. As a result, a plethora of technology components supporting various data structures (e.g., key-value, graph, relational), modalities (e.g., stream, log, real-time) and computing paradigms (e.g., in-memory, cluster/cloud) are meanwhile available. However, systematic support for managing the *variety* of data, the third dimension in the classical Big Data definition, is still missing. In this article, we present SeBiDA, an approach for managing *hybrid* Big Data. SeBiDA supports the *Semantification of Big Data* using the RDF data model, i.e., non-semantic Big Data is semantically enriched by using RDF vocabularies. We empirically evaluate the performance of SeBiDA for two dimensions of Big Data, i.e., volume and variety; the Berlin Benchmark is used in the study. The results suggest that even in large datasets, query processing time is not affected by data variety.

1 Introduction

Before 'Big Data' became a phenomenon and a tremendous marketplace backed by ever-increasing research efforts, Gartner suggested a novel data management model termed *3-D Data Management* [3]. At the basis of this model are three major challenges to the potential of data management systems: increasing volume, accelerated data flows, and diversified data types. These three dimensions have since become known as Big Data's 'three V's': *volume*, *velocity*, and *variety*. Other dimensions have been added to the V-family to cover a broader range of emerging challenges such as *veracity* and *value*.

In the last few years, a number of efforts have sought to design *generic* re-usable architectures that tackle the aforementioned Big Data challenges. However, they all miss the *explicit support* for semantic data integration, querying, and exposure. The *Big Data Semantification* introduced in this paper is an umbrella term that covers the previous operations in a large scale. It also includes the ability to semantically enrich non-semantic input data using Resource Description Framework (RDF) ontologies. In this paper, we propose a realization of this vision by presenting a blueprint of a generic semantified Big Data architecture. All while preserving semantic information, so semantic data can be exported in its natural form, i.e., RDF. Although both an effective management of volume and velocity has gained a considerable attention both from

S. Madria and T. Hara (Eds.): DaWaK 2016, LNCS 9829, pp. 376–390, 2016.
DOI: 10.1007/978-3-319-43946-4_25

academia and industry, the variety dimension has not been adequately tackled; even though it has been reported to be the top big challenge by many industrial players and stakeholders[1].

In this paper we target the lack of variety by suggesting unique data model and storage, and querying interface for both semantic and non-semantic data. SeBiDA provides a particular support for Big Data semantification by enabling the semantic lifting of non-semantic datasets. Experimental results show that (1) SeBiDA is not impacted by the variety dimension, even in presence of an increasing large volume of data, and (2) outperforms a state-of-the-art central-ized triple in several aspects.

Our contributions can be summarized as follows:

- The definition of a blueprint for a semantified Big Data architecture that enables the ingestion, querying and exposure of heterogeneous data with vary-ing levels of semantics (hybrid data), while ensuring the preservation of seman-tics (Sect. 3).
- SeBiDA: A proof-of-concept implementation of the architecture using Big Data components such as, Apache Spark &Parquet and MongoDB (Sect. 4).
- Evaluation of the benefits of using the Big Data technology for the storage and processing of hybrid data (Sect. 5).

The rest of this paper is structured as follows: Sect. 2 presents a motivation example and the requirements of a Semantified Big Data Architecture. Section 3 presents a blueprint for a generic Semantic Big Data Architecture. SeBiDa imple-mentation is described in Sect. 4, while the experimental study is reported in Sect. 5. Section 6 summarizes related approaches. In Sect. 7, we conclude and present an outlook to our future work.

2   Motivating Example and Requirements

Suppose there are three datasets (Fig. 1) that are large in size (i.e., volume) and different in type (i.e., variety): (1) MOBILITY: an RDF graph containing transport information about buses, (2) REGIONS: a JSON encoded data about one country's regions semantically described using ontologies terms in JSON-LD format, and (3) STOP: a structured (GTFS-compliant[2]) data describing Stops in CSV format. The problem to be solved is to provide unified data model to *store* and *query* these datasets, independently of their dissimilar types.

A Semantified Big Data Architecture (SBDA) allows for efficiently ingesting and processing the previous heterogeneous data in large scale. Previously, there has been a focus on achieving an efficient big data loading and querying for RDF data or other structured data *separately*. The support of variety that we claim in this paper is achieved through proving (1) a unified data model and storage

---

[1] http://newvantage.com/wp-content/uploads/2014/12/Big-Data-Survey-2014-Sum mary-Report-110314.pdf,   https://www.capgemini-consulting.com/resource-file-access/resource/pdf/cracking_the_data_conundrum-big_data_pov_13-1-15_v2.pdf.

[2] https://developers.google.com/transit/gtfs/.

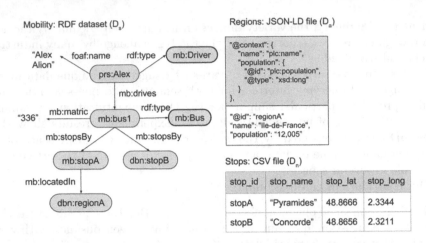

Fig. 1. Motivating Example. MOBILITY: semantic RDF graph for buses; REGIONS: semantic JSON-LD data about country's regions; and (3) STOP: non-semantic data about stops, presented using the CSV format

that is adapted and optimized for RDF data and structured and semi-structured non-RDF data, and (2) a unified query interface over the whole stored data.

SBDA meets the next requirements to provide the above-mentioned features:

R1: Ingest Semantic and Non-semantic Data. SBDAs must be able to process arbitrary types of data. However, we should distinguish between semantic and non-semantic data. In this paper, semantic data is all data, which is either originally represented according to the RDF data model or has an associated mapping, which allows to convert the data to RDF. Non-semantic data is then all data that is represented in other formalisms, e.g., CSV, JSON, XML, without associated mappings.

The semantic lifting of non-semantic data can be achieved through the integration of mapping techniques e.g., R2RM[3], CSVW[4] annotation models or JSON-LD contexts[5]. This integration can lead to either a representation of the non-semantic data in RDF, or its annotation with semantic mappings so as to enable full conversion at a later stage. In our example MOBILITY and REGIONS are semantic data. The former is originally in RDF model, while the latter is not in RDF model but has mappings associated. STOPS in the other hand is a non-semantic dataset on which semantic lifting can be applied.

R2: Preserve Semantics, and Metadata in Big Data Processing Chains. Once data is preprocessed, semantically enriched and ingested, it is paramount to preserve the semantic enrichment as much as possible. RDF-based data representations and mappings have the advantage (e.g., compared to XML) of

---

[3] http://www.w3.org/TR/r2rml/.

[4] http://www.w3.org/2013/csvw/wiki/Main_Page.

[5] http://www.w3.org/TR/json-ld/.

using fine-grained formalisms (e.g., RDF triples or R2RML triple maps) that persist even when the data itself is significantly altered or aggregated. Semantics preservation can be reduced as follows: *(1) Preserve IRIs and literals.* The most atomic components of RDF-based data representation are IRIs and literals[6]. Best practices and techniques to enable storage and indexing of IRIs (e.g., by separately storing and indexing namespaces and local names) as well as literals (along with their XSD or custom datatypes and language tags) in an SBDA, need to be defined. In the dataset MOBILITY, the Literals "Alex Alion" and *"12,005"xsd:long*, and the IRI 'http://xmlns.com/foaf/0.1/name' (shortened foaf:name in the figure), should be stored in an optimal way in the big data storage. *(2) Preserve triple structure.* Atomic IRI and literal components are organized in triples. Various existing techniques can be applied to preserve RDF triple structures in SBDA components (e.g., HBase [2,6,14]). In the dataset MOBILITY, the triple (prs:Alex mb:drives mb:Bus1) should be preserved by adopting a storage scheme that keeps the connection between the subject *prs:Alex*, the property *mb:drives* and the object *mb:Bus1*. Preserve mappings. Although conversion of the original data into RDF is ideal, it must not be a requirement as it is not always feasible (due to limitations in the storage, or to time critical use-cases). However, it is beneficial to at least annotate the original data with mappings, so that a transformation of the (full or partial) data can be performed on demand. R2RML, JSON-LD contexts, and CSV annotation models are examples of such mappings, which are usually composed of fine-grained rules that define how a certain column, property or cell can be transformed to RDF. The (partial) preservation of such data structures throughout processing pipelines means that the resulting views can also be directly transformed to RDF. In REGIONS dataset, the semantic annotations defined by the JSON object *@context* should be persisted associated to the actual data it describes: *RegionA*.

R3: Scalable and Efficient Query Processing. Data management techniques like data caching, query optimization, and query processing have to be exploited to ensure scalable and efficient performance during query processing.

3   A Blueprint for a Semantified Big Data Architecture

In this section, we provide a formalisation for an SBDA blueprint.

Definition 1 (Heterogeneous Input Superset). *We define a heterogeneous input superset HIS, as the union of the following three types of datasets:*

- $D_n = \{d_{n_1}, \ldots, d_{n_m}\}$ *is a set of non-semantic, structured or semi-structured, datasets in any format (e.g., relational database, CSV files, Excel sheets, JSON files).*
- $D_a = \{d_{a_1} \ldots, d_{a_q}\}$ *is a set of semantically annotated datasets, consisting of pairs of non-semantic datasets with corresponding semantic mappings (e.g., JSON-LD context, metadata accompanying CSV).*

---

[6] We disregard blank nodes, which can be avoided or replaced by IRIs [4].

- $D_s = \{d_{s_1}, \ldots, d_{s_p}\}$ is a set of semantic datasets consisting of RDF triples.

In our running example, STOPS, REGIONS, and MOBILITY correspond to D_n, D_a, and D_s, respectively.

Definition 2 (Dataset Schemata). *Given* HIS=$D_n \cup D_a \cup D_s$, *the dataset schemata of* D_n, D_a, *and* D_s *are defined as follows:*

- $S_n = \{s_{n_1}, \ldots, s_{n_m}\}$ *is a set of* **non-semantic schemata structuring** D_n, *where each* s_{n_i} *is defined as follows:*

  $$s_{n_i} = \{(T, A_T) \mid T \text{ is an entity type and } A_T \text{ is the set of all the attributes of } T\}$$

- $S_s = \{s_{s_1}, \ldots, s_{s_q}\}$ *is a set of the* **semantic schemata behind** D_s *where each* s_{s_i} *is defined as follows:*[7]

  $$s_{s_i} = \{(C, P_C) \mid T \text{ is an RDF class and } P_C \text{ is the set of all the properties of } C\}$$

- $S_a = \{s_{s_1}, \ldots, s_{s_p}\}$ *is a set of the* **semantic schemata annotating** D_a *where each* s_{s_i} *is defined the same way as elements of* S_s.

In the running example, the semantic schema of the dataset[8] MOBILITY is: $s_{s_1} = \{(mb : Bus, \{mb : matric, mb : stopsBy\}), (mb : Driver, \{foaf : name, mb : drives\})\}$

Definition 3 (Semantic Mapping). *A semantic mapping is a relation linking two semantically-equivalent schema elements. There are two types of semantic mappings:*

- $m_c = (e, c)$ *is a relation mapping an entity type e from* S_n *onto a class c.*
- $m_p = (a, p)$ *is a relation mapping an attribute a from* S_n *onto a property p.*

SBDA facilitates the lifting of non-semantic data to semantically annotated data by mapping non-semantic schemata to RDF vocabularies. The following are possible mappings($stop\_name, rdfs : label$), ($stop\_lat, geo : lat$), ($stop\_long, geo : long$).

Definition 4 (Semantic Lifting Function). *Given a set of mappings M and a non-semantic dataset* d_n, *a semantic lifting function SL returns a semantically-annotated dataset* d_a *with semantic annotations of entities and attributes in* d_n.

In the motivating example, dataset STOPS can be semantically lifted using the following set of mappings: { *(stop_name, rdfs:label),(stop_lat, geo:lat), (stop_long, geo:long)*}, thus a semantically annotated dataset is generated.

---

[7] A set of properties P_C of an RDF class C where: $\forall$ p $\in P_C$ (p rdfs:domain C).

[8] mb and foaf are prefixes for *mobility* and *friend of friend* vocabularies, respectively.

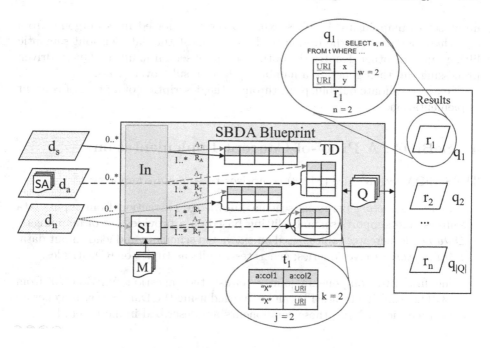

Fig. 2. A Semantified Big Data Architecture Blueprint

Definition 5 (Ingestion Function). *Given an element $d \in HIS$, an ingestion function $In(d)$ returns a set of triples of the form (R_T, A_T, f), where:*

- *T an entity type or class for data on d,*
- *A_T is a set of attributes $A_1, \ldots, A_n$ of T,*
- *$R_T \subseteq type(A_1) \times type(A_2) \times \cdots \times type(A_n) \subseteq d$, where $type(A_i) = T_i$ indicates that T_i is the data type of the attribute A_i in d, and*
- *$f : R_T \times A_T \rightarrow \bigcup_{A_i \in A_T} type(A_i)$ such that $f(t, A_i) = t_i$ indicates that $t_i \in$ tuple t in R_T is the value of the attribute A_i.*

The result of applying the ingestion function In over all $d \in HIS$ is the final dataset that we refer to as the Transformed Dataset TD.

$$TD = \bigcup_{d_i \in HIS} In(d_i)$$

The above definitions are illustrated in Fig. 2. The SBDA blueprint handles a representation (TD) of the relations resulting from the ingestion of multiple heterogeneous datasets in HIS (d_s, d_a, d_n). The ingestion (In) generates relations or tables (denoted R_T) corresponding to the data, supported by a schema for interpretation (denoted T and A_T). The ingestion of semantic (d_s) and semantically-annotated (d_a) data is direct (denoted resp. by the solid and dashed lines), a non-semantic dataset (d_n) can be optionally semantically lifted (SL) given an input set of mappings (M). This explains the two dotted lines outgoing from d_n, where one denotes the option to directly ingest the data without semantic lifting, and the other denotes the option to apply semantic lifting. Query-driven processing can then generate a number ($|Q|$) of results over (TD).

Next, we validate our blueprint through the description of a proof of concept implementation.

4   SeBiDA: A Proof-of-Concept Implementation

The SeBiDA architecture (Fig. 3) comprises three main components:

- *Schema Extractor*: performs schema knowledge extraction from input data sources, and supports semantic lifting based on a provided set of mappings.
- *Data Loader*: creates tables based on extracted schemata and loads input data.
- *Data Server*: receives queries; generates results as tuples or RDF triples.

The first two components jointly realise the ingestion function In from Sect. 3. The resulting tables TD can be queried using the Data Server to generate the required views. Next, these components are described in more detail.

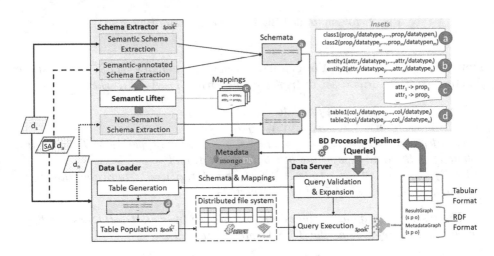

Fig. 3. The SeBiDA Architecture

4.1   Schema Extractor

We extract the structure of both semantic and non-semantic data to transform it into a tabular format that can be easily handled (stored and queried) by existing Big Data technologies.

(A) From each semantic or semantically-annotated input dataset (d_s or d_a), we extract *classes* and *properties* describing the data (cf. Sect. 3). This is achieved by first reformatting RDF data into the following representation:

$$(class, (subject, (property, object)^+)^+)$$

which reads: "each class has one or more instances where each instance can be described using one or more (property, object) pairs", and then we retain the *classes* and *properties*. The XSD datatypes, if present, are leveraged to type the properties, otherwise[9] string is used. The reformatting operation is performed using *Apache Spark*[10], a popular Big Data processing engine.

(B) From each non-semantic input dataset (d_n), we extract *entities* and *attributes* (cf. Sect. 3). As examples, in a relational database, table and column names can be returned using particular SQL queries; the entity and its attributes can be extracted from a CSV file's name and header, respectively. Similarly whenever possible, attribute datatypes are extracted, otherwise casted to string. Schemata that do not natively have a tabular format e.g., the case of XML, JSON, are also flattened into entity-attributes pairs.

As depicted in Fig. 3, the results are stored in an instance of *MongoDB*[11]. MongoDB is an efficient document-based database that can be distributed among a cluster. As schema can automatically be extracted (case of D_n), it is essential to store the schema information separately and expose it. This enables a sort of discovery, as one can navigate through the schema, visualize it, and formulate queries accordingly.

4.2   Semantic Lifter

In this step, SeBiDA targets the lifting of non-semantic elements: entities/attributes to existing semantic representations: classes/properties), by leveraging the LOV catalog API[12]. The lifting operation is *supervised* by the user and is *optional* i.e. the user can choose to ingest non-semantic data in its original format. The equivalent classes/properties are first fetched automatically from the LOV catalogue based on syntactical similarities with the original entities/attributes. The user next validates the suggested mappings or adjust them, either manually or by keyword-based searching the LOV catalogue. If semantic counterparts are undefined, internal IRIs are created by attaching a base IRI. Example 1 shows a result of this process, where four attributes for the GTFS entity 'Stop'[13] have been mapped to existing vocabularies, and the fifth converted to an internal IRI. Semantic mappings across the cluster are stored in the same MongoDB instance, together with the extracted schemata.

---

[9] When the object is occasionally not typed or is a URL.
[10] https://spark.apache.org/.
[11] https://www.mongodb.org.
[12] http://lov.okfn.org/dataset/lov/terms.
[13] http://developers.google.com/transit/gtfs/reference.

Example 1 (Property mapping and typing).

Source	Target	Datatype
stop_name	http://xmlns.com/foaf/0.1/name	string
stop_lat	http://www.w3.org/2003/01/geo/wgs84_pos#lat	double
stop_lon	http://www.w3.org/2003/01/geo/wgs84_pos#long	double
parent_station	http://example.com/sebida/20151215T1708/parent_station	string

4.3 Data Loader

This component loads data from the source HIS into the final dataset TD by generating and populating tables as described below. These procedures are also realised by employing Apache Spark. Historically, storing RDF triples in tabular layouts (e.g., Jena Property Table[14]) has been avoided due to the resulting large amount of *null* values in wide tables. This concern has largely been reduced following the emergence of NOSQL databases e.g., HBase, and columnar storage formats on top of HDFS (Hadoop Distributed File System) e.g. *Apache Parquet* and ORC files. HDFS is the *de facto* storage for Big Data applications, thus, is supported by the vast majority of Big Data processing engines.

We use Apache Parquet[15], a column-oriented tabular format, as the storage technology. One advantage of this kind of storage is the schema projection, whereby only projected columns are read and returned following a query. Further, columns are stored consecutively on disk; thus, Parquet tables are very compression-friendly (e.g., via Snappy and LZO) as values in each column are guaranteed to be of the same type. Parquet also supports composed and nested columns, i.e., saving multiple values in one column and storing hierarchical data in one table cell. This is very important since RDF properties are frequently used to refer to multiple objects for the same instance. State-of-the-art encoding algorithms are also supported, e.g., bit-packing, run-length, and dictionary encoding. In particular, the latter can be useful to store long string IRIs.

Table Generation. A corresponding *table template* is created for each derived *class* or *entity* as follows: **(A)** Following the representation described in Subsect. 4.1, a table with the same label is created for each *class* (e.g. a table 'Bus' from RDF class _:bus). A default column 'ID' (of type string) is created to store the triple's subject. For each *property* describing the class, an additional column is created typed according to the *property* extracted datatype. **(B)** For each *entity*, a table is similarly created as above, taking the entity label as table name and creating a typed column for each attribute.

---

[14] http://www.hpl.hp.com/techreports/2006/HPL-2006-140.html.
[15] https://parquet.apache.org.

Table Population. In this step, each table is populated as follows:
(1) For each RDF extracted class (cf. Subsect. 4.1) iterate throug its instance, a new row is inserted for each *instance* into the corresponding table: The instance IRI is stored in the 'ID' column, whereas the corresponding *objects* are saved under the column representing the property. For example, the following semantic descriptions (formatted for clarity):

```
(dbo:bus, [
(dbn:bus1, [(foaf:name,"OP354"), (dc:type,"mini")]),
(dbn:bus2, [(foaf:name,"OP355"), (dc:type,"larg")])
])
```

is flatted to a table "dbo:bus" in this manner:

ID	foaf:name	dc:type
dbn:bus1	"OP354"	"mini"
dbn:bus2	"OP355"	"larg"

(2) The population of the tables in case of non-semantic data varies depending on its type. For example, we iterate through each CSV line and save its values into the corresponding column of the corresponding table. XPath can be used to iteratively select the needed nodes in an XML file.

4.4   Data Server

Data loaded into tables is persistent, so one can access the data and perform analytical operations by way of ad-hoc queries.

Quering Interface. The current implementation utilizes SQL queries, as the internal TD representation corresponds to a tabular structure, and because the query technology used, i.e., Spark, provides only SQL-like query interface.

Multi-format Results. As shown in Fig. 3 query results can be returned as a set of tables, or as an RDF graph. RDFisation is achieved as follows: Create a triple from each projected column, casting the column name and value to the triple predicate and object, respectively. If the result includes the ID column, cast value as the triple subject. Otherwise, set the subject to $base\_IRI/i$, where the base IRI is defined by the user, and i is an incremental integer. To avoid the cumbersome need to access online schema descriptions when ingesting large datasets, property value ranges are elicited from the XSD datatypes defined for attached objects. If not available, the property is considered of type string. If the object is an IRI, this is also stored as a string.

5   Experimental Study

The goal of the study is to evaluate if SeBiDA[16] meets requirements **R1**, **R2**, and **R3** (cf., Sect. 2). We evaluate if data management techniques, e.g., caching data, allow SeBiDA to speed up both data loading and query execution time, whenever semantic and non-semantic data is combined.

Datasets: Datasets have been created using the Berlin benchmark generator[17]. Table 1 describes them in terms of number of triples and file size. We choose XML as non-semantic data input to demonstrate the ability of SeBiDA to ingest and query semi-structured data (requirement **R1**). Parquet and Spark are used to store XML and query nested data in tables.

Metrics: We measure the loading time of the datasets, the size of the datasets after the loading, as well as the query execution time over the loaded datasets.

Implementation: We ran our experiments on a small-size cluster of three machines each having DELL PowerEdge R815, 2x AMD Opteron 6376 (16 Core) CPU, 256 GB RAM, and 3 TB SATA RAID-5 disk. We cleared the cache before running each query. To run on warm cache, we executed the same query five times by dropping the cache just before running the first iteration of the query; thus, data temporally stored in cache during the execution of iteration i could be used in iteration $i + 1$.

Discussion. As can be observed in Table 2, loading time takes almost 3 h for the largest dataset. This is because one step of the algorithm involves sending part of the data back from the workers to the master, which incurs important network transfer (collect function in Spark). This can, however, be overcome by saving data to a distributed database e.g., Cassandra, which we intend to implement in the next version of the system. However, we achieved a huge gain in terms of disk space, which is expected from the adopted data model that avoids the repetition of data (in case of RDF data) and the adopted file format, i.e. Parquet, which performs high compression rates (cf. Subsect. 4.3).

Table 1. Description of the Berlin Benchmark RDF Datasets

RDF Dataset	Size (n triples)	Type	Scaling Factor
Dataset$_1$	48.9 GB (200M)	RDF	569,600
Dataset$_2$	98.0 GB (400M)	RDF	1,139,200
Dataset$_3$	8.0 GB (100M)	XML	284,800

---

[16] https://github.com/EIS-Bonn/SeBiDA.

[17] Using a command line: `./generate -fc -pc [scaling factor] -s [file format] -fn [file name]`, where file format is `nt` for RDF data and `xml` for XML data. More details in: http://wifo5-03.informatik.uni-mannheim.de/bizer/ berlinsparqlbenchmark/spec/BenchmarkRules/#datagenerator+.

Table 2. Benchmark of RDF Data Loading For each dataset the loading time as well as the obtained data size together with the compression ratio.

RDF Dataset	Loading Time	News size	Ratio
Dataset$_1$	1.1 h	389 MB	1:0.015
Dataset$_2$	2.9 h	524 MB	1:0.018
Dataset$_3$	0.5 h	188 MB	1:0.023

Table 3. Benchmark Query Execution Times (secs.) in Cold and Warm Caches. Significant differences are highlighted in **bold**.

Dataset$_1$ - Only Semantic Data (RDF)

	Q01	Q2	Q3	Q4	Q5	Q6	Q7	Q8	Q9	Q10	Q11	Q12	Geom. Mean
Cold Cache	3.00	2.20	1.00	4.00	3.00	0.78	**11.3**	6.00	16.00	7.00	11.07	11.00	4.45
Warm Cache	1.00	1.10	1.00	2.00	3.00	0.58	**6.10**	5.00	14.00	6.00	10.04	9.30	3.14

Dataset$_1$ ∪ Dataset$_3$–RDF XML Data

	Q1	Q2	Q3	Q4	Q5	Q6	Q7	Q8	Q9	Q10	Q11	Q12	Geom. Mean
Cold Cache	3.00	2.94	2.00	5.00	3.00	0.90	**11.10**	7.00	**25.20**	8.00	11.00	11.5	5.28
Warm Cache	2.00	1.10	1.00	5.00	3.00	1.78	**8.10**	6.00	**20.94**	7.00	11.00	9.10	4.03

Table 4. Benchmark Query Execution Times (secs.) in Cold and Warm Caches-Significant differences are highlighted in **bold**

Dataset$_2$ - Only Semantic Data (RDF)

	Q01	Q2	Q3	Q4	Q5	Q6	Q7	Q8	Q9	Q10	Q11	Q12	Geom. Mean
Cold Cache	5.00	3.20	3.00	8.00	3.00	1.10	20.00	7.00	18.00	7.00	13.00	11.40	6.21
Warm Cache	4.00	3.10	2.00	7.00	3.00	1.10	18.10	6.00	17.00	6.00	12.04	11.2	5.55

Dataset$_2$ ∪ Dataset$_3$–RDF XML Data

	Q01	Q2	Q3	Q4	Q5	Q6	Q7	Q8	Q9	Q10	Q11	Q12	Geom. Mean
Cold Cache	**11.00**	3.20	**7.20**	**17.00**	3.00	1.10	23.10	**16.00**	20.72	**10.00**	14.10	13.20	**8.75**
Warm Cache	**4.00**	3.20	**2.00**	**8.00**	3.00	1.10	21.20	**8.00**	18.59	**7.00**	12.10	11.10	**5.96**

Tables 3 and 4 report on the results of executing the Berlin Benchmark 12 queries against Dataset$_1$, Dataset$_2$, in two ways: first, the dataset alone and, second the dataset combined with Dataset$_3$ (using UNION in the query). Queries are run in cold cache and warm cache. We notice that caching can *improve* query performance significantly in case of hybrid large data (entries highlighted in bold). Among the 12 queries, the most expensive queries are Q7 and Q9. Q7 scans a large number of tables: five tables, while Q9 produces a large number of intermediate results. These results suggest that SiBiDA is able to scale to large hybrid data, without deteriorating query performance. Further, these results provide evidences of the benefits of loading query intermediate results in cache.

Discussion. Table 5 shows that RDF-3X loaded Dataset$_1$ within 93 min, compared to 66 min in SeBiDA, while it timedout loading Dataset$_2$. We set a timeout of 12 h, RDF-3X took more than 24 h; before we terminate it manually. Table 6 shows no definitive dominant, but suggests that SeBiDA in all queries does not exceed a threshold of 20 s, while RDF-3X does in four queries and even passes to the order of minutes. We do not report on query time of Dataset$_2$ using RDF-3X because the prohibitive time of loading it.

6 Related Work

There have been many works related to the combination of Big Data and Semantic technologies. They can be classified into two categories: MapReduce-only-based and non-MapReduce-based, where in the first category only Hadoop framework is used, for storage (HDFS) and for processing (MapReduce); and in the second, other storage and/or processing solutions are used. The major limitation of MapReduce framework is the overhead caused by materializing data between Map and Reduce, and between two subsequent jobs. Thus, works in the first category, e.g., [1, 8, 12], try to minimize the number of join operations, or maximize the number of joins executed in the Map phase, or additionally, use indexing techniques for triple lookup.

In order to cope with this, works in the second category suggest to store RDF triples in NoSQL databases (e.g., HBase and Accumulo) instead, where a variety of physical representations, join patterns, and partitioning schemes is suggested.

Centralized vs. Distributed Triple Stores. We can look at SeBiDA as a distributed triple store as it can load and query RDF triples separately. We thus try to compare against the performance of one of the fastest centralized triple stores: RDF-3X[18]. Comparative results can be found in Tables 5 and 6.

Table 5. Loading Time of RDF-3X.

RDF Dataset	Loading time	New size	Ratio
Dataset$_1$	93 min	21GB	1:2.4
Dataset$_2$	*Timed out*	-	-

For processing, either MapReduce is used on top (e.g., [9,11]), or the internal operations of the NoSQL store are utilized but with conjunction with triple stores (e.g., [6,10]). Basically, these works suggest to store triples in three-columns tables, called *triple tables*, using column-oriented stores. This latter offers enhanced compression performances and efficient distributed indexes. Nevertheless, using the tied three-columns table still entails a significant overhead because of the inter-join operations required to answer most of the queries.

---

[18] https://github.com/gh-rdf3x/gh-rdf3x.

Table 6. SeBiDA vs. RDF-3X Query Execution Times (secs.) in Cold Cache only on Dataset$_1$. Significant differences are highlighted in **bold**.

Q01	Q2	Q3	Q4	Q5	Q6	Q7	Q8	Q9	Q10	Q11	Q12
3.00	2.20	1.00	4.00	3.00	0.78	11.3	6.00	**16.00**	7.00	**11.07**	**11.00**
0.01	1.10	**29.213**	0.145	**1175.98**	2.68	**77.80**	**610.81**	0.23	0.419	0.13	1.58

There are few approaches that do not fall into one of the two categories. In [7], the authors focus on providing a real-time RDF querying, combining both live and historical data. Instead of plain RDF, RDF data is stored under the binary space-efficient format RDF/HDT. Although called Big Semantic Data and compared against the so-called Lambda Architecture, nothing is said about the scalability of the approach when the storage and querying of the data exceed the single-machine capacities. In [13], RDF data is loaded into property tables using Parquet. Impala, a distributed SQL query engine, is used to query those tables. A query compiler from SPARQL to SQL is devised. Our approach is similar as we store data in property tables using Parquet. However, we do not store all RDF data in only one table but rather create a table for each detected RDF class.

For more comprehensive survey we refer to [5]. In all the presented works, storage and querying were optimized for storing RDF data only. We, in the other hand, not only aimed to optimize for storing and querying RDF, but also to make the same underlying storage and query engine available for non-RDF data; structured and semi-structured. Therefore, our work is the first to propose a blueprint for an end-to-end semantified big data architecture, and realize it with a framework that supports the semantic data i.e. integration, storage, and exposure, along non-semantic data.

7   Conclusion and Future Work

The current version of semantic data loading does not consider an attached schema. It rather extracts this latter thoroughly from the data instances, which can scale proportionally with data size and, thus, put a burden in the data integration process.

Currently, in case of instances of multiple classes, the selected class is the last one in lexical order. We would consider the schema in the future—even if incomplete—to select the most specific class instead. Additionally, as semantic data is currently stored isolated from other data, we could use a more natural language, such as SPARQL, to query only RDF data. Thus, we envision conceiving a SPARQL-to-SQL converter for this sake. Such converters exist already, but due to the particularity of our storage model, which imposes that instances of multiple types to be stored in only one table while adding references to the other types (other tables), a revised version is required. This effort is supported by and contributes to the H2020 BigDataEurope Project (GA 644564).

References

1. Du, J.H., Wang, H.F., Ni, Y., Yu, Y.: HadoopRDF: a scalable semantic data analytical engine. ICIC 2012. LNCS, vol. 7390, pp. 633–641. Springer, Heidelberg (2012)
2. Franke, C., Morin, S., Chebotko, A., Abraham, J., Brazier, P.: Distributed semantic web data management in HBase and MySQL cluster. In: Cloud Computing (CLOUD), pp. 105–112. IEEE (2011)
3. Gartner, D.L.: 3-D data management: controlling data volume, velocity and variety. 6 February 2001
4. Hogan, A.: Skolemising blank nodes while preserving isomorphism. In: 24th International Conference on World Wide Web, 2015. WWW (2015)
5. Kaoudi, Z., Manolescu, I.: RDF in the clouds: a survey. VLDB J. **24**(1), 67–91 (2015)
6. Khadilkar, V., Kantarcioglu, M., Thuraisingham, B., Castagna, P.: Jena-HBase: a distributed, scalable and efficient RDF triple store. In: 11th International Semantic Web Conference Posters & Demos, ISWC-PD (2012)
7. Martínez-Prieto, M.A., Cuesta, C.E., Arias, M., Fernández, J.D.: The solid architecture for real-time management of big semantic data. Future Gener. Comput. Syst. **47**, 62–79 (2015)
8. Nie, Z., Du, F., Chen, Y., Du, X., Xu, L.: Efficient SPARQL query processing in mapreduce through data partitioning and indexing. In: Sheng, Q.Z., Wang, G., Jensen, C.S., Xu, G. (eds.) APWeb 2012. LNCS, vol. 7235, pp. 628–635. Springer, Heidelberg (2012)
9. Papailiou, N., Konstantinou, I., Tsoumakos, D., Karras, P., Koziris, N.: H2RDF+: High-performance distributed joins over large-scale RDF graphs. In: BigData Conference. IEEE (2013)
10. Punnoose, R., Crainiceanu, A., Rapp, D.: Rya: a scalable RDF triple store for the clouds. In: Proceedings of the 1st International Workshop on Cloud Intelligence, pp. 4. ACM (2012)
11. Schätzle, A., Przyjaciel-Zablocki, M., Dorner, C., Hornung, T., Lausen, G.: Cascading map-side joins over HBase for scalable join processing. In: SSWS+ HPCSW, pp. 59 (2012)
12. Schätzle, A., Przyjaciel-Zablocki, M., Hornung, T., Lausen, G.: PigSPARQL: a SPARQL query processing baseline for big data. In: International Semantic Web Conference (Posters & Demos), pp. 241–244 (2013)
13. Schätzle, A., Przyjaciel-Zablocki, M., Neu, A., Lausen, G.: Sempala: interactive SPARQL query processing on hadoop. In: Mika, P., Tudorache, T., Bernstein, A., Welty, C., Knoblock, C., Vrandečić, D., Groth, P., Noy, N., Janowicz, K., Goble, C. (eds.) ISWC 2014, Part I. LNCS, vol. 8796, pp. 164–179. Springer, Heidelberg (2014)
14. Sun, J., Jin, Q.: Scalable RDF store based on HBase and mapreduce. In: 3rd International Conference on Advanced Computer Theory and Engineering. IEEE (2010)

Author Index

Printed in the United States
By Bookmasters